31

Topics in Heterocyclic Chemistry

For further volumes:
http://www.springer.com/series/7081

Aims and Scope

The series Topics in Heterocyclic Chemistry presents critical reviews on present and future trends in the research of heterocyclic compounds. Overall the scope is to cover topics dealing with all areas within heterocyclic chemistry, both experimental and theoretical, of interest to the general heterocyclic chemistry community.

The series consists of topic related volumes edited by renowned editors with contributions of experts in the field.

Michael Schnürch • Marko D. Mihovilovic
Editors

Metalation of Azines and Diazines

With contributions by
F. Chevallier • C. Fruit • P.C. Gros • E. Horkel •
A. Kolarovič • C. Metallinos • F. Mongin • N. Plé •
M. Schlosser • K. Stromski

Springer

Editors
Michael Schnürch
Marko D. Mihovilovic
Institute of Applied Synthetic Chemistry
Vienna University of Technology
Vienna
Austria

ISSN 1861-9282 ISSN 1861-9290 (electronic)
ISBN 978-3-642-35021-4 ISBN 978-3-642-35022-1 (eBook)
DOI 10.1007/978-3-642-35022-1
Springer Heidelberg New York Dordrecht London

Library of Congress Control Number: 2013944365

Printed on acid-free paper

Springer is part of Springer Science+Business Media (www.springer.com)

Preface

The importance of heterocyclic chemistry is unquestioned in the chemical community. Heterocycles appear in compounds such as blockbuster drugs, organic materials, agrochemicals, or natural products. Amongst the vast variety of heterocycles it is generally acknowledged that nitrogen containing heterocycles are privileged scaffolds. Hence, their chemistry is a field of intense research in various aspects such as formation, application, and decoration. In the context of the latter subject, metalation reactions play an important role since they allow the introduction of various electrophiles and control of the regiochemistry depending on the base applied or directing groups present. Specifically, this volume is dedicated to the metalation of azines and diazines covering mainly research from the past 5–10 years.

We decided to cover only reactions using stoichiometric amounts of metal reagents since a separate volume on catalytic transformations is currently under preparation. Regarding the metal reagents, naturally lithiations and magnesiations play a prominent role, but it was our intention to cover also the new developments such as the use of ate-bases. Regarding the arrangement of chapters, we decided to make separate chapters according to the type of heterocycle. In case of pyridine and pyrimidine, we additionally divided the material according to the metal bases used, otherwise the chapters would have been too elaborate. So the content of the chapters is as follows: the first three chapters report on ring systems containing one nitrogen atom. Chapter 1 covers the reactions of pyridines, benzopyridines, and azapyridines with organolithiums and organomagnesiums. In this chapter also fundamental reaction pathways are outlined. In Chapter 2 the benzofused compounds quinolones and isoquinolines are treated, again their lithiation and magnesiation reactions. Chapter 3 moves on to other metal reagents discussing the metalation reactions of pyridines, quinolines, and isoquinolines with ate-bases and their alkali metal salt-modified congeners.

In the next chapters heterocycles containing two nitrogen atoms are treated. It starts with Chapter 4 where lithiations and Grignard reactions of pyrimidine and quinazoline are treated. Subsequently, Chapter 5 is dedicated to the reactions of the very same ring systems with other metal reagents such as borylation, stannylation, zincation, or silylation. Chapter 6 covers all kinds of metalation reactions carried out on pyrazine and quinoxaline. The same topic is presented in Chapter 7 but on pyridazine, cinnoline, and phtalazine.

Of course many people have been involved in preparation and I am very grateful to all of them. Most importantly I would like to thank all the authors for doing an excellent job and in my belief making this volume an outstanding one in the series. Additionally, I would like to thank Tanja Jäger and Elizabeth Hawkins from Springer who dealt with many organizational issues which facilitated my work significantly. Finally, I would like to thank my family for their support and patience in the time composing this volume.

Vienna, Austria Michael Schnürch

Contents

Top Heterocycl Chem (2013) 31: 1–20
DOI: 10.1007/7081_2012_93

Published online: 8 November 2012

Other Stoichiometric Metalation Reactions on Pyrimidine and Quinazoline

Philippe C. Gros

Abstract Pyrimidines and quinazolines are important diazines with interesting electronic and biologic properties and promising applications that require the development of new methodologies to synthesize derivatives in efficient and selective ways. This chapter covers the preparation of useful metalated diazines including B-, Sn-, Si-, Zn-, Cd-, and Cu-based derivatives.

Keywords Pyrimidine · Quinazoline · Sensitive heterocycles · Deprotonation · Halogen-metal exchange · Transmetalation

Contents

P.C. Gros (✉)
SOR-SRSMC, Université de Lorraine, CNRS, Faculté des Sciences, Boulevard des Aiguillettes, 54506 Vandoeuvre-Lès-Nancy, France
e-mail: philippe.gros@univ-lorraine.fr

Abbreviations

Ac	Acetyl
Bu	Butyl
DME	Dimethoxyethane
DMF	Dimethylformamide
dppf	1,1′- Bis(diphenylphosphino)ferrocene
equiv	Equivalent(s)
h	Hour(s)
i-Pr	Isopropyl
LDA	Lithium diisopropylamide
LTMP	Lithium 2,2,6,6-tetramethylpiperidide
MeO	Methoxy
MeOH	Methanol
min	Minute(s)
rt	Room temperature
s	Second(s)
s-Bu	*sec*-Butyl
t-Bu	*tert*-Butyl
THF	Tetrahydrofuran
TMEDA	*N*,*N*,*N*′,*N*′-Tetramethyl-1,2-ethylenediamine
TMP	2,2,6,6-Tetramethylpiperidyl
Ts	Tosyl

1 Introduction

Metalated heterocycles are highly important intermediates since they allow numerous functionalizations via electrophilic trapping or metal-catalyzed coupling processes. While highly developed in the pyridine series, work on pyrimidine and especially quinazoline is less abundant essentially due to the high electrophilicity of π-deficient heteroaromatic rings precluding the use of usual alkyllithiums as metalating agents. Also the stability of the formed organometallics remains problematic. However, some success has been obtained in the preparation of useful organometallic intermediates that are described in this chapter.

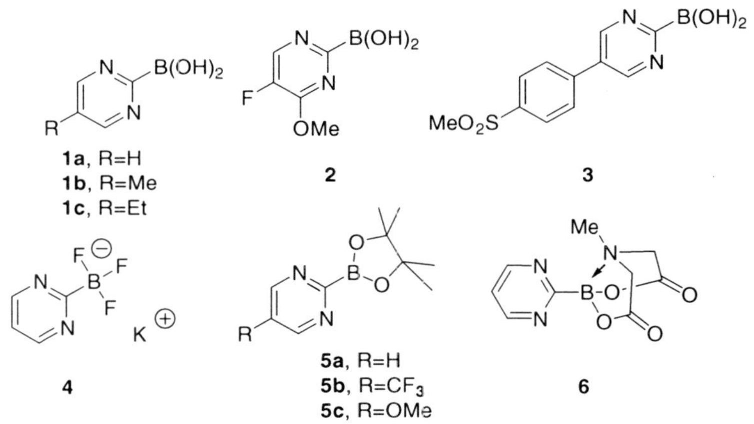

Fig. 1 2-Pyrimidinyl boron derivatives

2 Pyrimidine

2.1 Borylation

2.1.1 Borylation at C-2

The Suzuki–Miyaura coupling is now a well-known and extensively used cross-coupling methodology. Unfortunately, its application in nitrogen-containing heteroaromatic chemistry is precluded by the marked trend for protodeborylation of heteroarylboron compounds when boron is present at the alpha position to nitrogen [1]. In contrast with the large amount of work devoted to pyridine derivatives, the chemistry of pyrimidylboronic acids has been neglected for a long time. Pyrimidine displays two available nitrogen alpha positions at C-2 and C-4. Despite these sources of instability, efforts have been made to introduce and stabilize boron-based moieties at these positions and perform subsequent cross-coupling reactions. A survey of literature on C-2 boron compounds recorded only a few publications hits but essentially patents without detailed experimental procedures or in suppliers' chemicals catalogues (Fig. 1).

2-pyrimidinyl boronic acids **1a–c** have been patented [2–6] as well as fluoro **2** [7] and aryl **3** [8]. One example of the trifluoroborate derivative **4** is also mentioned in a recent patent [9]. The use of N-methyliminodiacetic acid (MIDA) known as an efficient protective group in the pyridine series [10] has been reported for the preparation of derivative **6** in one patent [11] as well.

The pinacolate protected derivatives (**5a–c**) have also been reported. The preparation involved a palladium-catalyzed borylation from 2-bromopyrimidine and dipinacolborane (Scheme 1) [12, 13].

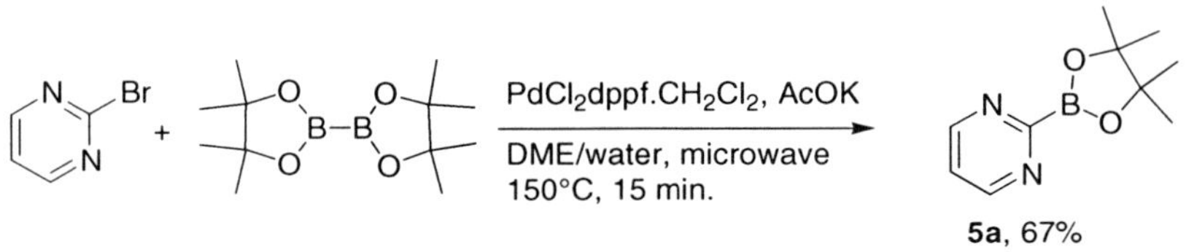

Scheme 1 Synthesis of 2-pyrimidinyl pinacol borane **5a**

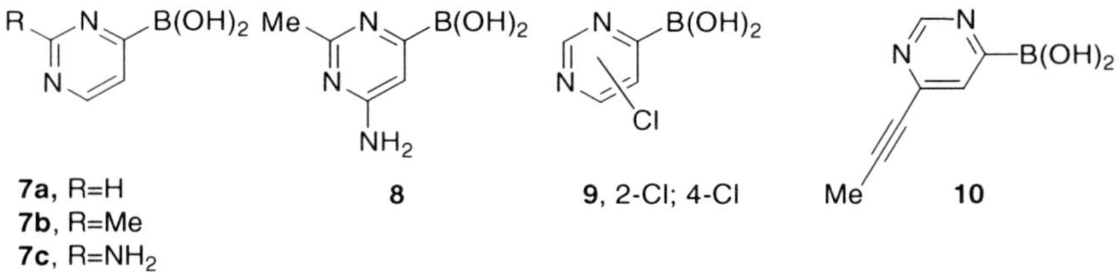

Fig. 2 4-Pyrimidinyl boron derivatives

2.1.2 Borylation at C-4

As for the C-2 borylated compounds, the C-4 borylated ones can be found in patents and publications without any detail on their preparation (Fig. 2). Besides the unsubstituted boronic acid **7a** [14], several functionalities have been introduced on the pyrimidine ring such as methyl (**7b**) [15], amine (**7c–8**) [16, 17], chlorine (**9**) [18], and alkynyl groups (**10**) [19].

2.1.3 Borylation at C-5

The C-5 borylated compounds where boron is far from the nitrogens offered an easier access and better stability and, therefore, have been well characterized. The main methodology to access pyrimid-5-yl boronic acid involved a bromine–lithium exchange of 5-bromopyridine using BuLi and reaction with triisopropyl borate (Scheme 2). This reaction was studied in detail on 3-bromopyridine by Li and Nelson at Merck [20] and was found to be highly sensitive to the order of addition of reagents. Poor yields were obtained following a sequential procedure, i.e., bromine–lithium exchange preceding addition of borate. The trick was to perform an in situ quenching of the lithiated intermediate by the borate thus preventing side reactions and a final crystallization in acetonitrile as a stable boroxin **11**. The boron intermediate can be used as such in Suzuki-type couplings, hydrolyzed to the boronic acid, or protected as pinacol derivative. Boronic acid **12a** can be obtained in 76% yield while the sequential method only gave 28%. Notably, the synthesis can be realized in kg scale.

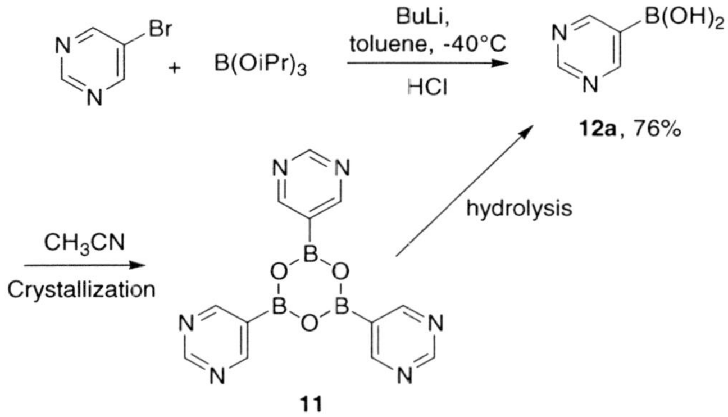

Scheme 2 Synthesis of 5-pyrimidinyl boronic acid from boroxin **11**

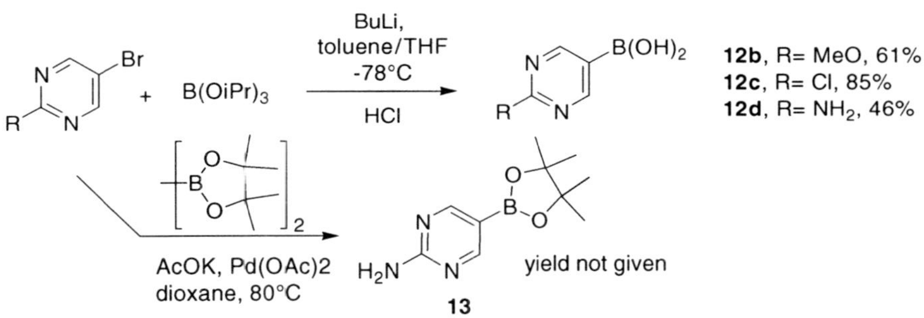

Scheme 3 Synthesis of substituted 5-pyrimidinyl boronic acids

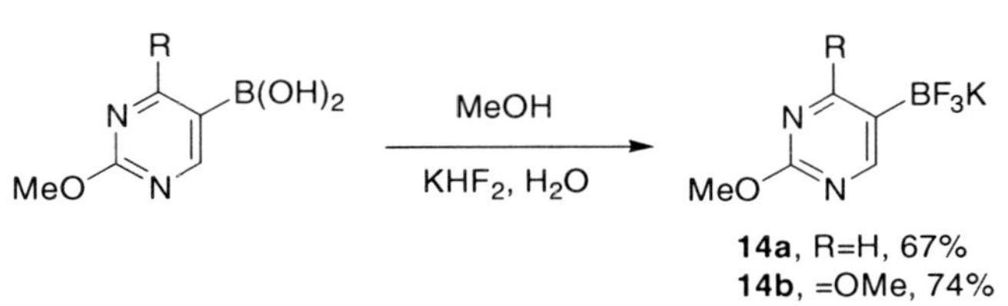

Scheme 4 Synthesis of 5-pyrimidinyl trifluoroborates

Methoxy (**12b**) [21, 22], chloro (**12c**) [22], and amino (**12d**) [22] pyrimidyl boronic acids were next prepared using this methodology. Interestingly, the reaction can be performed in the presence of the free amine (Scheme 3). The pinacol derivative **13** was also prepared via a palladium catalyzed process [22].

Molander reported the preparation of the trifluoroborates **14a–b** from the boronic acids (Scheme 4) [23]. These compounds are much more stable and did not undergo any deborylation. The trifluoroborate can be used as such in

Table 1 Nucleophilic stannylation of halogenopyrimidines

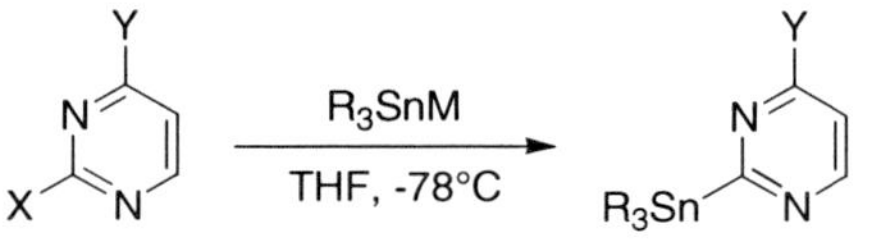

15a-d

X	Y	R	M	Yield (%)
Cl	H	Bu	Li	**15a**, 84
Cl	H	Me	Li	**15b**, 46
Cl	H	Ph	Li	**15c**, 35
Cl	Cl	Me	Li or Na	Degradation
Br	Br	Me	Na	**15d**, 78
Br	Br	Me	Li	Degradation

cross-couplings [24, 25]. Additional chemical modifications can be made on the heterocycle without affecting the C–B bond and if necessary the boronic acid can be regenerated on silica gel at the end of the sequence [22].

2.2 *Stannylation*

2.2.1 Nucleophilic Stannylation

A special feature of pyrimidine is a marked electrophilicity due to π-deficiency of the aromatic ring. This has been exploited to introduce tin via nucleophilic substitution of halogenated pyrimidines. In pioneering works, Udheim and coworkers [26] have studied the stannylation at C-2 or C-4 electrophilic positions of substituted pyrimidines using various tin sources (Table 1). R_3SnM was generated from R_3SnCl and the appropriate metal (Li or Na).

2-Tributyl stannyl pyrimidine **15a** was obtained in good yield while the trimethyl (**15b**) and triphenyl (**15c**) analogues were formed in lower yields. Interestingly, the nucleophilic substitution regioselectively occurred at C-2 in 2,4-dibromopyridine leading to stannane **15d** in good yield using trimethylstannylsodium while trimethylstannyllithium only led to degradations.

Substitution of halogen at C-4 was also studied on sulfur-stabilized 4-chloropyrimidine (Scheme 5) but the regioselectivity was poor leading to methylthio substitution (product **16**), besides expected **17a**, due to the enhancement of the electrophilicity at C-2 by both azomethine bonds. Selectivity was achieved by switching from chlorine to iodine as better leaving group giving exclusively **17a** in 52% yield.

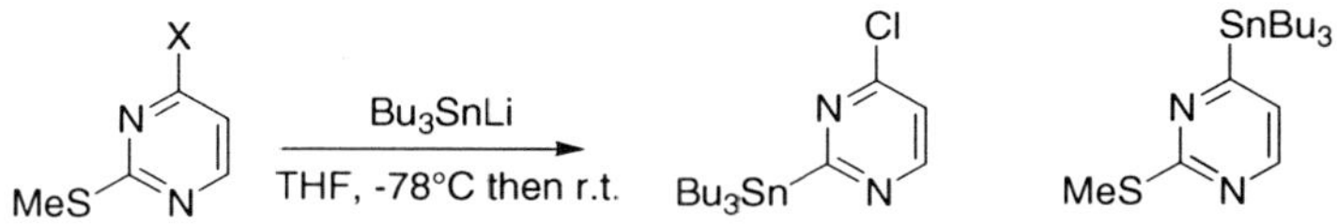

X=Cl	**16**, 35%	**17a**, 20%
X=I	-	**17a**, 52%

Scheme 5 Selectivity in stannylation of methylthio halogenopyrimidines

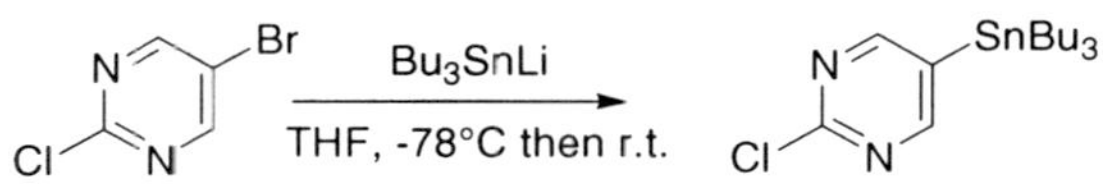

18, 53%

Scheme 6 Stannylation at C-5 of 2-chloro-5-bromopyrimidine

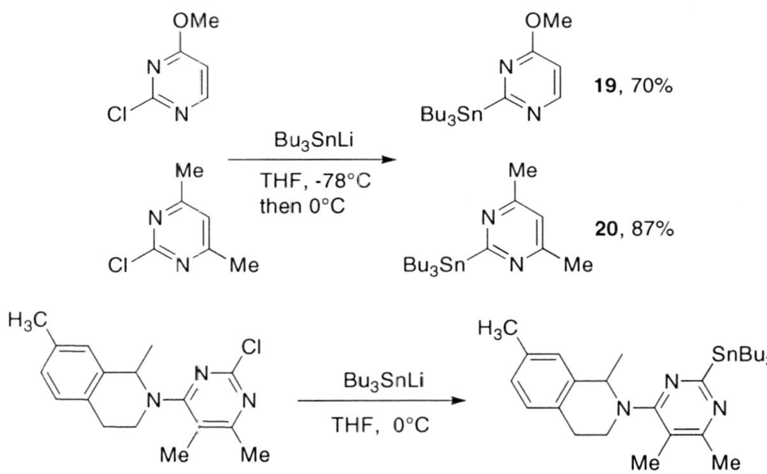

19, 70%

20, 87%

21, 45%

Scheme 7 Stannylation at C-2 of substituted pyrimidines

Finally, tin was also introduced at C-5 using the same methodology (Scheme 6). While 2,5-dichloropyrimidine did not give any product, 2-chloro-5-bromo-pyrimidine gave exclusively stannylation product **18**.

The reaction was next exploited for the preparation of variously substituted stannyl pyrimidines bearing methoxy (**19**) [27] or methyl groups (**20**) [28] or used for functionalization of more sophisticated compounds such as **21** (Scheme 7) [29]. In these examples, Bu_3SnLi was generated from Bu_3SnH and LDA (lithium diisopropylamide) as lithiating agent.

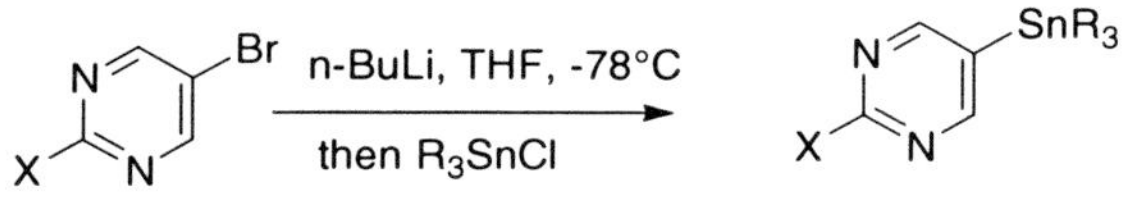

22a, X=SMe, R=Me, 82%
22b, X=SMe, R=Bu, 70%
22c, X=OTBDMS, R=Me, 81%

Scheme 8 Stannylation at 2- substituted 5-bromo-pyrimidines

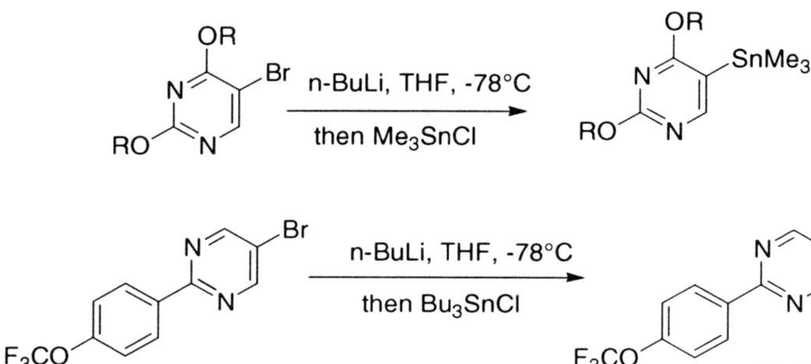

23a, R=Me, 23%
23b, R=PhCH$_2$, 29%

24, 30%

Scheme 9 Stannylation of alkoxy and aryl pyrimidines

2.2.2 Stannylation via Deprotonation or Halogen–Metal Exchange

Pyrimidines are sensitive substrates and their direct lithiation is problematic due to their high electrophilicity. The strategies involved bromine–lithium exchange using *n*BuLi or ortholithiation first reported by Quéguiner and coworkers using LiTMP (lithium 2,2,6,6-tetramethylpiperidide) as a non-nucleophilic base [30]. The tin compounds were finally obtained by reaction of lithiopyrimidines with trialkyltin chloride.

Bromine–lithium exchange/stannylation sequences have been performed with success on a range of substituted 5-bromopyrimidines (Scheme 8). Alkylthio moieties [31, 32], and silyl protected alcohols [31], were reacted efficiently with trimethyl or tributyltin chlorides leading to expected stannanes **22a–c** in excellent yields.

Dialkoxy moieties [33] or aryl groups [34] also gave the expected products **23a–b** and **24** but in moderate yield (Scheme 9).

The C-4 position was also stannylated using this methodology giving **25** in good yield (Scheme 10) [26].

Attempts to obtain 2-trialkylstannylpyrimidine via halogen–metal exchange led to only 10% yield of the expected tin compound **26** starting from 2-iodopyrimidine [26]. Thus, the nucleophilic route is definitely the best route to access 2-stannylpyrimidine. The lithio derivative can be, however, generated from the tin compound and trapped cleanly by ketones leading to alcohols **27** in good yields (Scheme 11) [26].

The deprotonative approach has also been reported via ortholithation (Scheme 12). The best base reported to date for this reaction is LiTMP [30].

Scheme 10 Stannylation at C-4 from 2-methylthio-4-iodo-pyrimidine

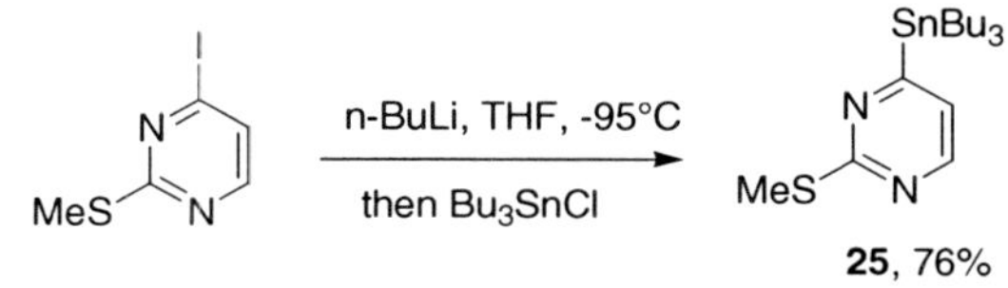

Scheme 11 Iodine- and tin-lithium exchange

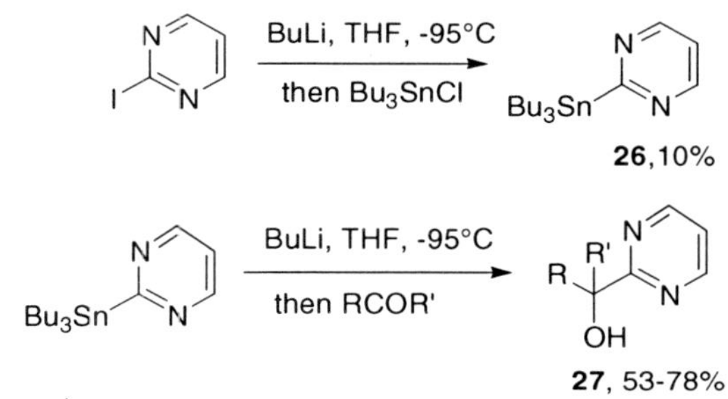

Scheme 12 Stannylation via chlorine and methoxy directed lithiation

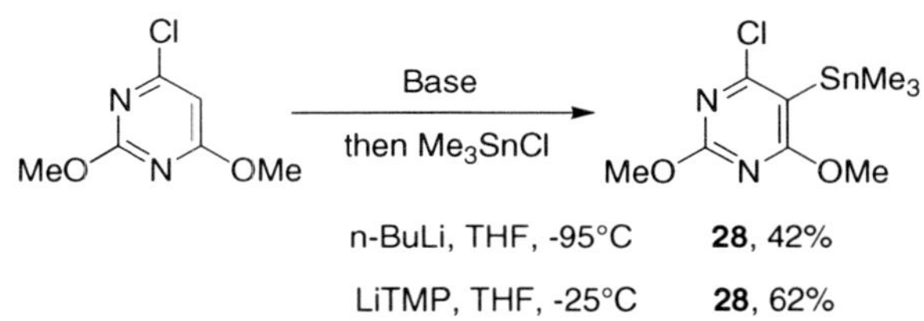

Quéguiner and coworkers have reported the preparation of tin compound **28** in 62% yield [35]. *n*BuLi has also been used as base leading to the expected tin compounds, however, in lower yield [33].

The methoxy group has been used as an efficient ortho directing group for introduction of tin in product **29** (Scheme 13) [36].

2.3 Silylation

Silicon is usually introduced in heteroaromatic compounds via reaction of the electrophilic trialkylsilyl chlorides with lithiated intermediates. While it was essentially used as a protecting group for selective metalations, it has gained new interest for Hiyama cross-coupling reactions in aromatic [37–39] and more recently heteroaromatic series [40–42].

Radinov has studied in detail the silylation of chloropyrimidines via metalation with LDA (Scheme 14) [43]. While **30a** was obtained in poor yield, the presence of three chlorine atoms on the ring led to silane **30c** in good yield probably due to a better stabilization of the lithiated intermediate.

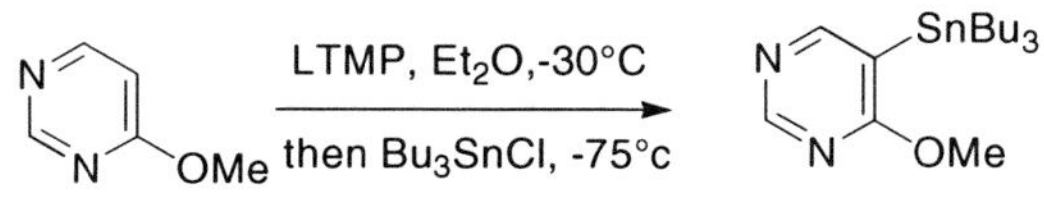

29, 30%

Scheme 13 Stannylation via methoxy directed lithiation

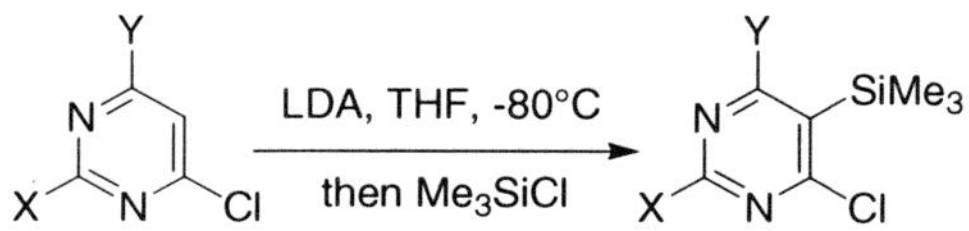

30a, X=Cl, Y=H, 6%
30b, X=H, Y=Cl, 44%
30c, X=Y=Cl, 67%

Scheme 14 Silylation via ortholithiation

Scheme 15 Silylation via nitrogen directed lithiation

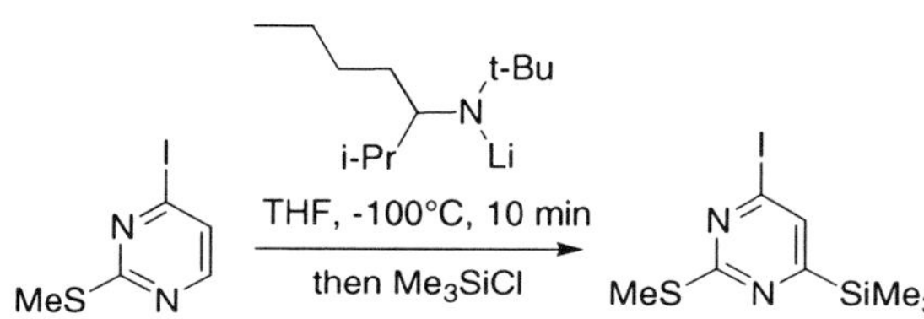

31, 40%

Scheme 16 Silylation via nitrogen directed magnesation

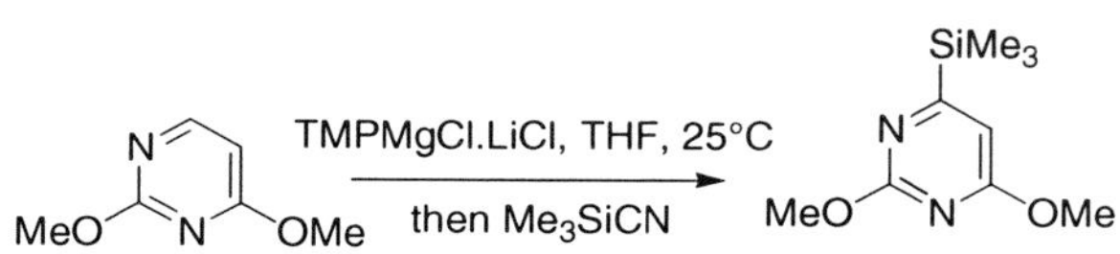

32, 70%

Interestingly, silicon can be introduced alpha to nitrogen of 2-methylsulfanyl-4-iodopyrimidine at $-100°c$ using a highly hindered base leading to silane **31** in acceptable yield (Scheme 15) [44].

Such an alpha silylation was also realized via magnesation using TMPMgCl.LiCl as reported by Knochel and coworkers giving **32** in good yield (Scheme 16) [45].

Other examples have been described by the same group on 2-bromopyrimidine. By decreasing the metalation temperature it was possible to tolerate the carbon–bromine bond [46]. This allowed to produce substituted bromopyrimidinyl silanes **33** and **34a–b** (Scheme 17).

Halogen metal exchange was also reported for the preparation of dipyrimidinylsilanes and germanes **35** (Scheme 18) [47]. These compounds were next used as platforms for building self-assembled molecules via coordination of palladium by the pyrimidine nitrogens.

Pyrimidinyl vinylsilane **36** has been prepared in acceptable yield via transmetalation of tributylstannane **15a** (Scheme 19) [48]. The azinyl vinyl silane is a powerful removable directing group in intermolecular Pauson–Khand

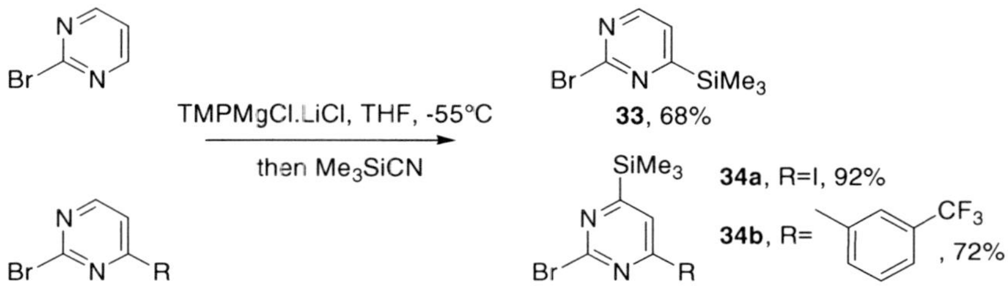

Scheme 17 Silylation via magnesation of bromopyrimidines

Scheme 18 Silylation via bromine–lithium exchange

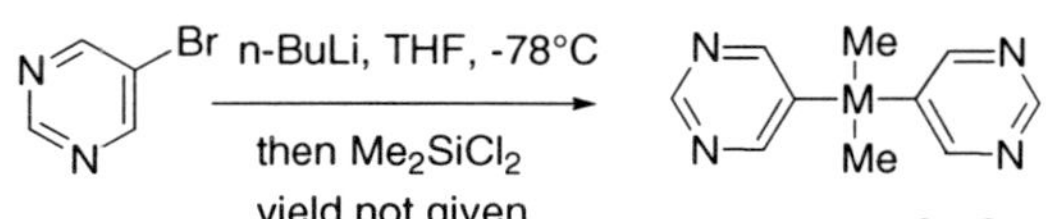

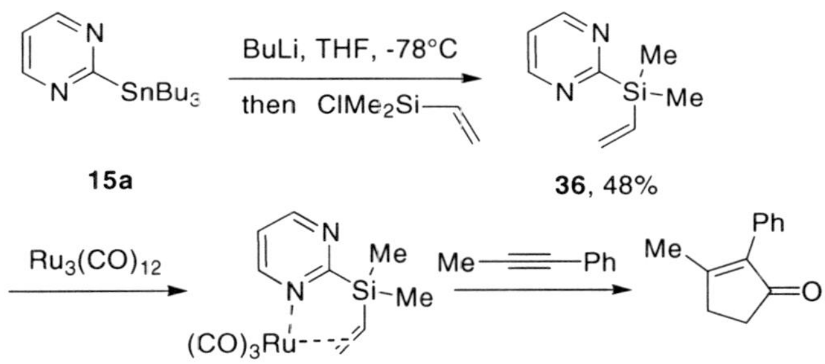

Scheme 19 Silylation via tin–lithium exchange and use of **36** in Pauson–Khand reaction

reactions. Compound **36** plays a double role as metal catalyst chelator and activator of the double bond via electron withdrawing effects.

An interesting synthesis of electroluminescent silyl pyrimidine-based compound **37** involving a nucleophilic silylation has been patented recently. A first double bromine–lithium exchange on 2,2′-dibromobiphenyl followed by trapping with trichloromethyl silane gave a tricyclic monochloro silane. Lithiation afforded an intermediate silyllithium that was added cleanly on a 2-chloropyrimidine derivative (Scheme 20) [49].

2.4 *Zincation*

2.4.1 Deprotonative Zincation

The pyrimidine ring was also subjected to zincation via direct metalation using organic or organometallic bases.

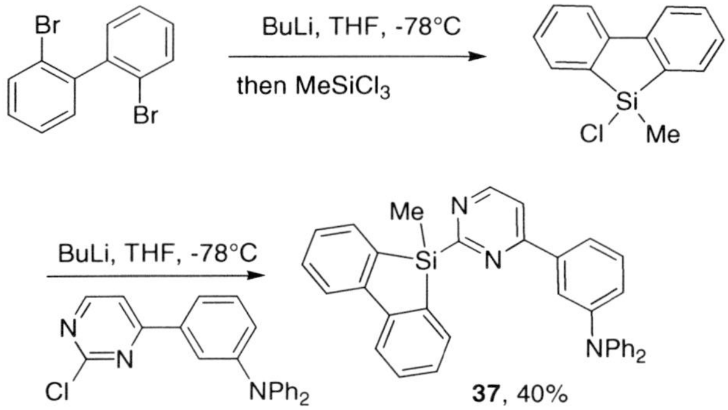

Scheme 20 Nucleophilic silylation of chloropyrimidine

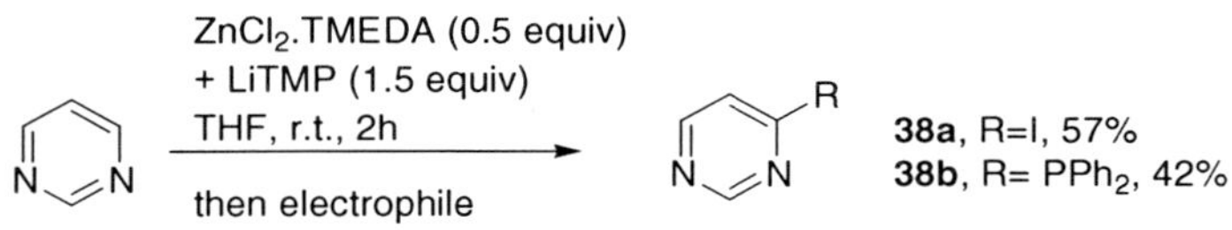

Scheme 21 Deprotonative zincation using homoleptic Li–Zn base

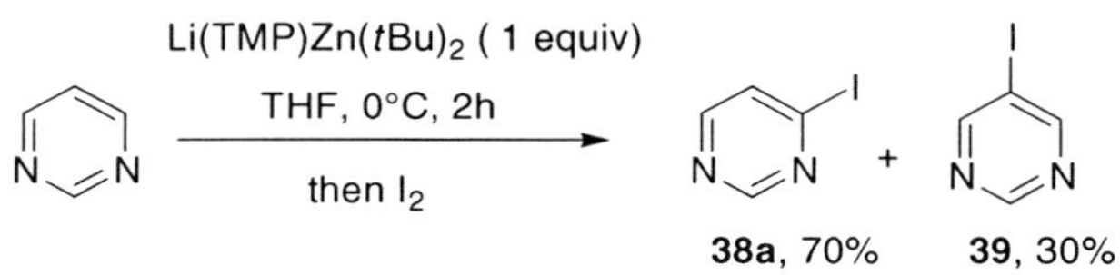

Scheme 22 Deprotonative zincation using heteroleptic Li–Zn base

Alkali-metal mediated zincations have been reported by several groups. Mongin and coworkers [50] have reported the zincation of pyrimidine using the homoleptic zincate prepared from $ZnCl_2$ and 3 equiv of LiTMP. Various conditions have been explored and it was possible to direct the reaction at C-4 when the reaction was performed at room temperature. Compounds **38a** and **38b** were obtained in fair yield after trapping the zinc species with appropriate electrophiles (Scheme 21). Note that LiTMP was unable to affect the reaction underlining the synergistic effect brought by the lithium–zinc reagent.

Hevia and Blair [51] have prepared and characterized the heteroleptic Li(TMP) Zn(tBu)$_2$ (TMP = 2,2,6,6-tetramethylpiperidide) and used it in zincation of N-containing heteroaromatic compounds. The reaction applied to pyrimidine gave a 70:30 mixture (determined by ^{1}H NMR) of 4- and 5-iodopyrimidine **38a** and **39** (Scheme 22). As for the case of the previously discussed homoleptic lithium–zinc reagent, C-4 zincation was the main process.

Knochel [52] reported the use of $TMP_2Zn.2MgCl_2.2LiCl$ for zincation of polyhalogenated pyrimidines in various positions (Scheme 23). The reaction was efficient at room temperature, and several substituted chloropyridines **40–43** were

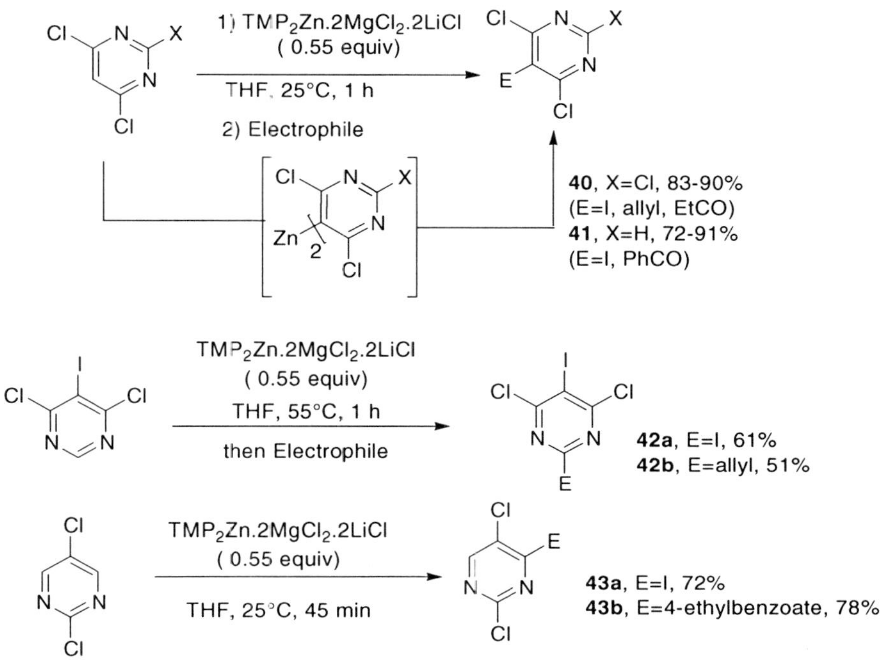

Scheme 23 Zincation of chloropyrimidines

obtained in acceptable to good yields. On 4,6-dichloropyrimidine the metalation occurred exclusively at C-5 and not at C-2 at room temperature (product **41**). The metalation of the latter position was achieved on 4,5,6-trisubstituted pyrimidine providing an increase of temperature up to 55°C was applied (products **42a–b**). The reaction performed on 2,5-dichloropyrimidine gave exclusively the expected orthometallation products **43a–b** without any formation of products resulting of potential 4,6-dimetallation.

As an alternative to organometallic bases, Kondo [53] has reported the use of highly basic Schwesinger phosphazene tBu-P_4 [54] for regioselective deprotonation of pyrimidine at C-5 (Scheme 24). In situ trapping by ZnI_2 and reaction with I_2 led to 5-iodopyrimidine **39** in 66% yield. The stoichiometry was found to be critical to ensure an efficient iodination. 3 equiv of base and 6 equiv of ZnI_2 were required. As an additional proof of formation of the zinc intermediate, an efficient Negishi-type coupling was realized leading to 5-phenylpyrimidine **44** in 73% yield.

2.4.2 Dehalogenative Zincation

Zincation via bromine–lithium exchange was realized on 4,6-methoxy-5-bromopyrimidine by treating the lithiated pyrimidine by $ZnCl_2$ (Scheme 25). The zincated heterocycle **45** was then directly coupled under palladium catalysis [55].

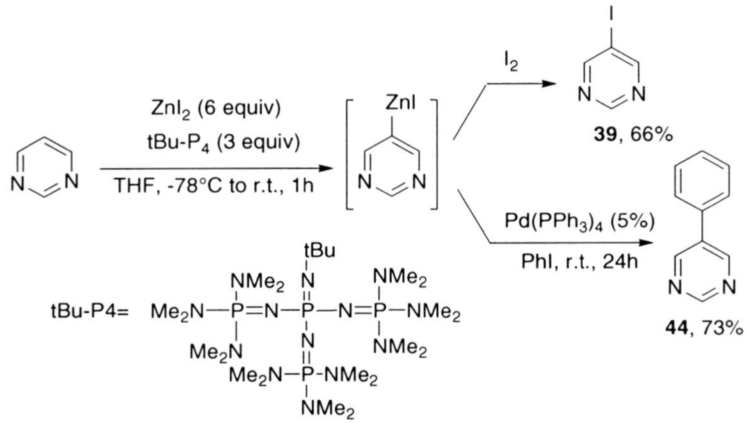

Scheme 24 tBu-P_4 induced deprotonative zincation

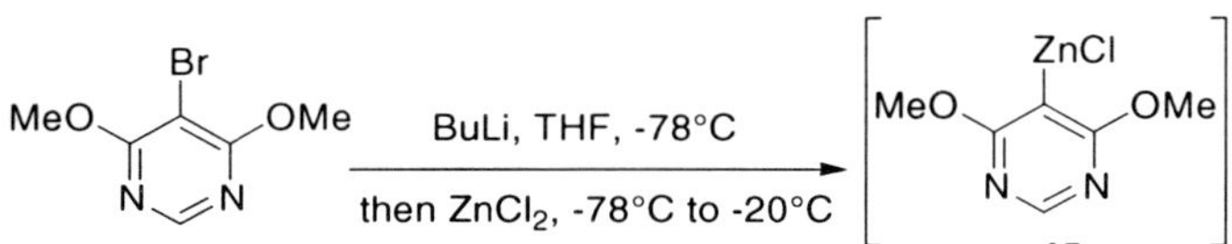

Scheme 25 Orthozincation of dimethoxypyrimidine

The direct zincation was also performed using commercially available zinc metal activated by a mixture of LiCl/Br$(CH_2)_2$Br and Me_3SiCl. The functionalization at the C-4 and/or C-5 positions of 2,6-dimethoxypyrimidine derivatives was found selective from C–I and C–Br bonds of pyrimidines (Table 2). A range of zincated pyrimidines **46a–f** was prepared almost quantitatively. Even an ester moiety was tolerated by the reagent. This procedure represents a new method for the polyfunctionalization of uracil derivatives [56].

More recently, the preparation of solid salt stabilized zincated dimethoxypyrimidine **47** has been reported [57]. The idea is to use $Zn(OPiv)_2$ (OPiv=Pivalate) as a source of zinc that is introduced via bromine magnesium exchange (Scheme 26).

2.5 *Cadmation*

Other metals can be introduced on the pyrimidine ring leading to improvement of reactivity of the metallated heterocycle and different selectivities.

Mongin has reported interesting deprotonative cadmation using a mixture of $CdCl_2$.TMEDA (TMEDA = tetramethylethylenediamine) and LiTMP as base [58,

Table 2 Zincation of halogenopyrimidines using activated Zn metal

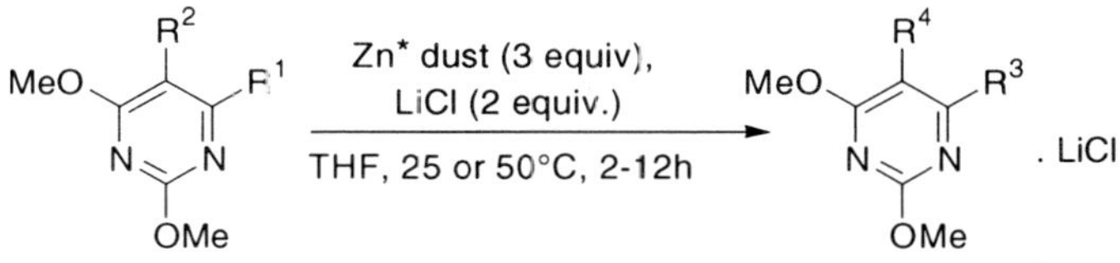

46a-f

R^1	R^2	R^3	R^4	Yield (%)
H	I	H	ZnI	**46a**, 95
I	H	ZnBr	H	**46b**, 96
I	CO_2Et	ZnBr	CO_2Et	**46c**, 98
Cl	I	Cl	ZnI	**46d**, 96
I	I	I	ZnI	**46e**, 98
Br	Br	Br	ZnBr	**46f**, 95

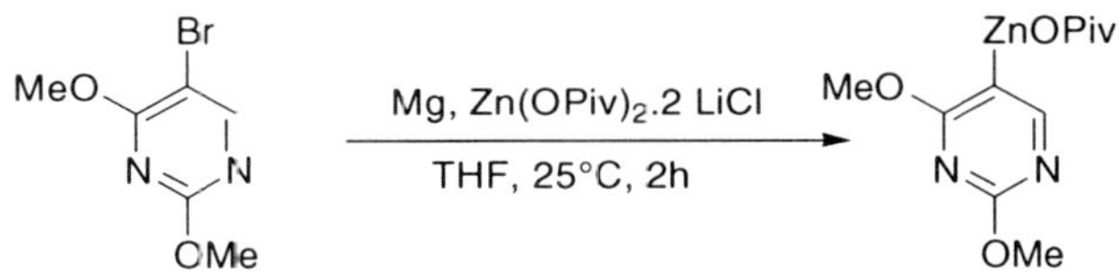

47, 65%

Scheme 26 Solid salt zincated pyrimidine

Scheme 27 Cadmation and zincation of pyrimidine

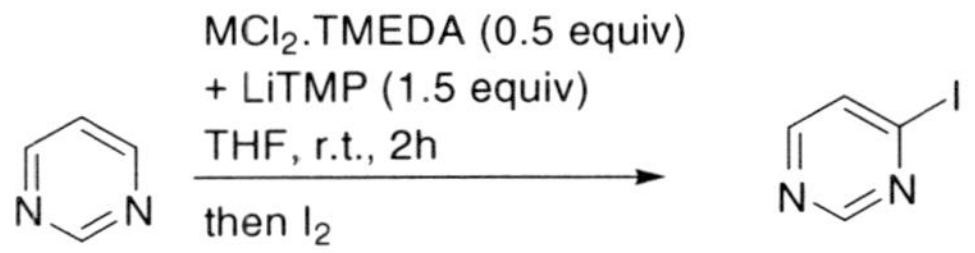

38a, M=Zn, 57%
38a, M=Cd, 71%

59]. The intermediate cadmate was trapped in good yield by iodine. It is noteworthy that the yield was much better than those obtained using the corresponding zinc reagent (Scheme 27).

2.6 *Cupration*

Cupration was also performed using ate complex TMP_2CuLi [60]. The pyrimidinyl cuprate was trapped with various electrophiles yielding compounds **48a–b** while oxidation gave the homocoupling product **49** (Scheme 28).

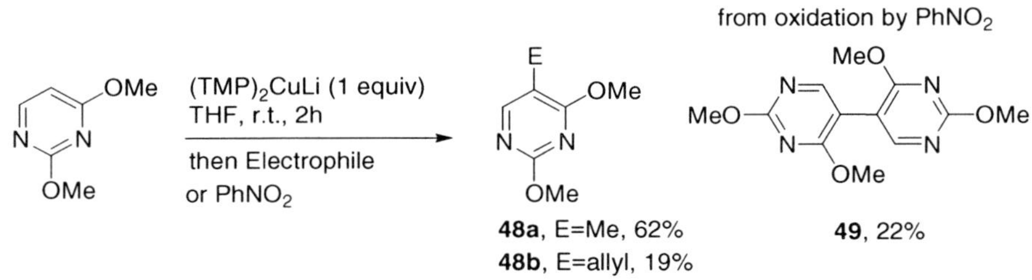

Scheme 28 Cupration of 2,4-dimethoxypyrimidine

Fig. 3 Quinazoline borane and boronic acid

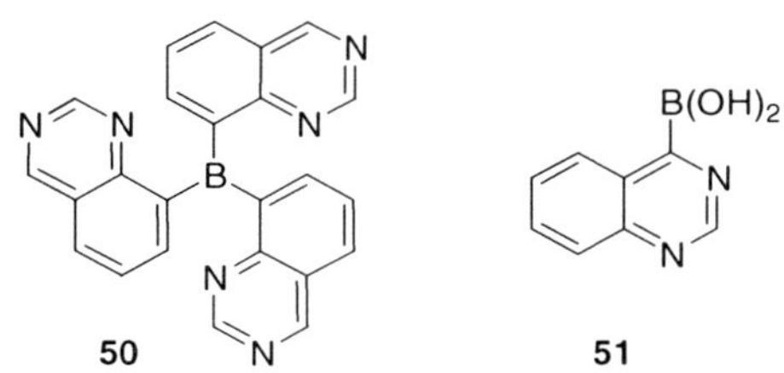

3 Quinazoline

In contrast to pyrimidine examples of metallated quinazoline are scarce.

3.1 Borylation

An example of trisquinazolinyl borane **50** has been patented as electroluminescent material [61], while a boronic acid has been introduced at the C-4 position to give compound **51** (Fig. 3) [62].

3.2 Stannylation

Stannylation was realized at C-2 of the quinazoline ring via a bromine–lithium exchange leading to stannane **52** in moderate yield (Scheme 29) [63].

Plé and coworkers stannylated the aromatic ring in excellent yield via deprotonation using LiTMP as basic reagent (Scheme 30) [64]. Compound **53** was found useful for nonlinear optics.

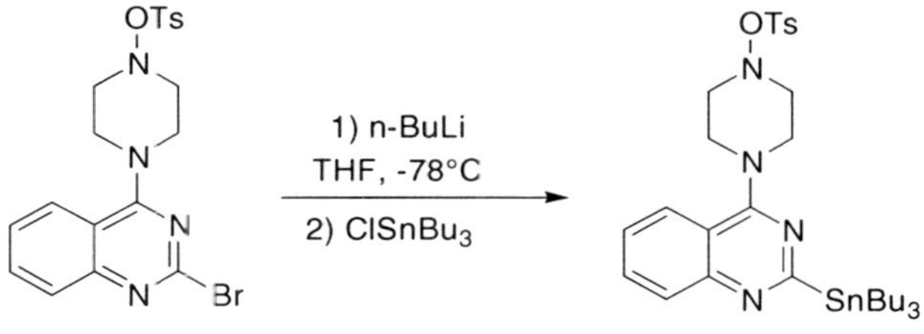

52, 29%

Scheme 29 Stannylation of piperazinyl quinazoline

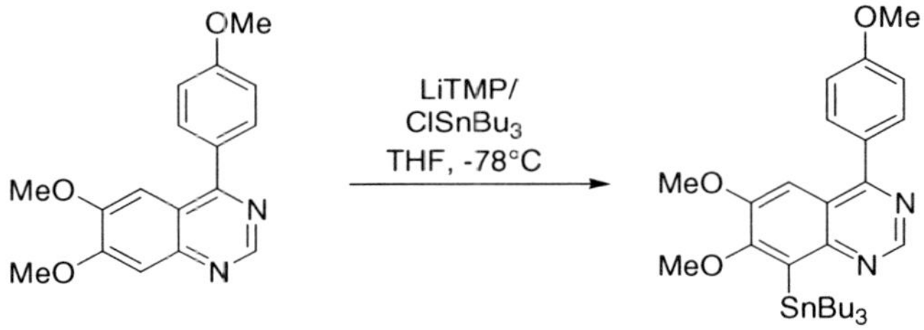

53, 92%

Scheme 30 Deprotonative stannylation of the aromatic ring of quinazoline

3.3 Silylation

Quéguiner and coworkers [65] have studied the lithiation of the non-heteroaromatic part of the molecule using LiTMP as a base. Some silyl derivatives have been prepared using the in situ (product **54**) or sequential trapping (products **55–57**) of lithio intermediates by $ClSiMe_3$ (Scheme 31).

3.4 Zincation

During their works on pyrimidine, Hevia and Blair have also used Li(TMP)Zn $(tBu)_2$ in zincation of quinazoline (Scheme 32) [51]. Trapping with iodine revealed a regioselective zincation at C-2 of the heteroaromatic ring. The 2-iodoquinazoline **58** was obtained for the first time by a metalation methodology in a good 69% yield.

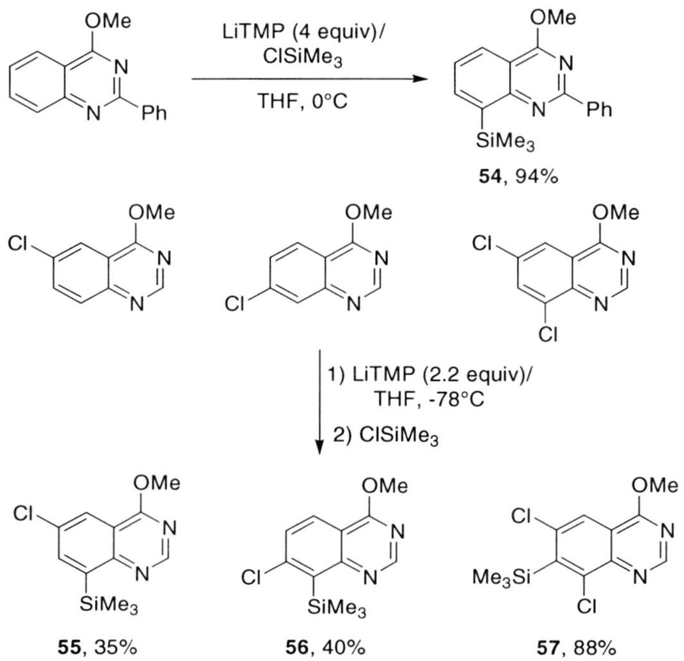

Scheme 31 Selective lithiation-silylation quinazolines

Scheme 32 First zincation-iodination of quinazolines

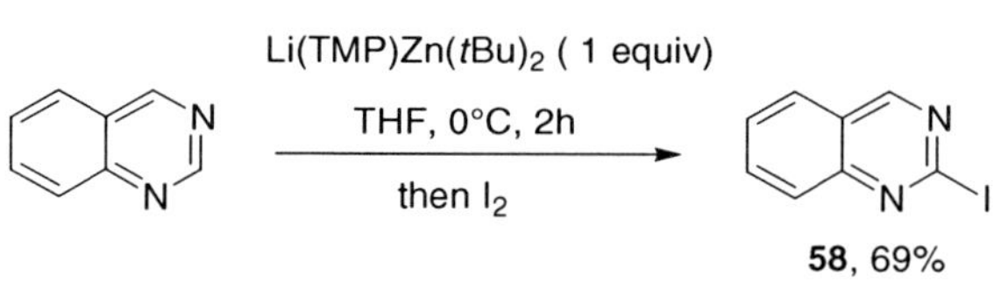

4 Conclusion

In this chapter it has been underlined how difficult it is to obtain general organometallic routes for functionalization of pyrimidine and quinazoline due to the sensitivity of these substrates and the instability of the corresponding intermediates. Especially, the incorporation of metal at the C-2 position (at the carbon between the two nitrogens) remains challenging. However, the recently developed bimetallic reagents based on zinc and magnesium seem to be promising in the field since they allow for the formation of stabilized organometallic azines. The broad spectrum of applications of pyrimidines ranging from pharmaceutical chemistry to materials chemistry (non-linear optics, luminescence, supramolecular assemblies) will urge chemists to pursue active research for development of selective and practical methodologies to functionalize these compounds.

References

1. Tyrrell E, Brookes P (2003) Synthesis 469
2. Kim CS, Eum SJ, Cho YJ, Kwon HJ, Kim BO, Kim SM, Yoon SS (2011) WO2011010840 A1
3. Surman MD, Freeman EE, Grabowski JF, Hadden M, Henderson AJ, Jiang G, Jiang X, Luche M, Khmelnitsky Y, Vickers S, Viggers J, Cheetham S, Guzzo PR (2010) Bioorg Med Chem Lett 20:7015
4. Raju BC, Saidachary G, Kumar JA, Sridhar B (2011) Helv Chim Acta 94:248
5. Cheng Y, White R, Amegadzie A, Brown J, Cheng AC, Dimauro EF, Dineen T, Epstein O, Gore VK, Human JB, Judd T, Kreiman C, Liu Q, Lopez P, Ma VV, Marx I, Minatti A, Nguyen HN, Paras NA, Patel VF, Qian W, Weiss M, Xue Q, Zheng XM, Zhong W (2011), WO 2011115938 A1
6. Dyck B, Tran JA, Tamiya J, Jovic F, Vickers T, Chen C, Harriott N, Coon T, Ashweek NJ (2010), WO 2010149684 A1
7. Machajewski TD, Gao Z, Levine BH, Antonios-Mccrea W, Bellamacina CR, Costales A, Doughan BM, Fong S, Hendrickson T, Lin X, Mcbride C, Mckenna M, Rico AC, Shafer CM, Wang XM, Zhou Y, Xia Y, Mendenhall KG (2006) WO 2006113498 A2
8. Bruton G, Orlek BS, Rana KK (2004), WO 2004101546 A1
9. Rodriguez Ropero S, Huguet Clotet J (2010) WO 2010012637 A1
10. Dick GR, Knapp DM, Gillis EP, Burke MD (2010) Org Lett 12:2314
11. Gillis EP, Lee SJ, Gray K, Burke MD, Knapp DM (2009) WO2009014550 A1
12. Berg S, Burrows J, Chessari G, Congreve MS, Hedstroem J, Hellberg S, Hoegdin K, Kihlstroem J, Kolmodin K, Lindstroem J, Murray C, Patel S (2007) WO 2007058602 A2
13. Ogawa A, Ok HO, Ondeyka D, Ujjainwalla F, Vande Bunte E (2010) WO 2010077624 A1 20100708
14. Boyle RG, Walker DW (2010), WO 2010136755 A1
15. Li X, Chu S, Feher VA, Khalili M, Nie Z, Margosiak S, Nikulin V, Levin J, Sprankle Kelly G, Tedder ME, Almassy R, Appelt K, Yager KM (2003) J Med Chem 46:5663
16. Sessions HE, Yin Y, Bannister TD, Weiser A, Griffin E, Pocas J, Cameron MD, Ruiz C, Lin L, Schürer SC, Schröter T, LoGrasso P, Feng Y (2008) Bioorg Med Chem Lett 18:6390
17. Feng Y, Lograsso P, Bannister T, Schroeter T, Chen YT, Yin Y, Sessions H, Smolinski MP, Yao L, Wang B (2009) WO 2009079008 A1
18. Bernotas RC, Inghaus RR, Ullrich JW, Kaufamn DH, Morris RL, Wrobel JE, Hu B, Jetter JW, Collini MD, Travins JM, Magolda RL (2008) WO 2008049047 A2
19. Wu Y, Iserloh U, Cumming JN, Liu X, Mazzola RD, Sun ZY, Huang Y, Stamford A, McKittrick B, Zhu Z (2007) US 20070287692 A1
20. Li W, Nelson DP, Jensen MS, Hoerrner RS, Cai D, Larsen RD, Reider PJ (2002) J Org Chem 67:5394
21. Saygili N, Batsanov AS, Bryce MR (2004) Org Biomol Chem 2:852
22. Clapham KM, Smith AE, Batsanov AS, McIntyre L, Pountney A, Bryce MR, Tarbit B (2007) Eur J Org Chem 2007:5712
23. Molander GA, Cavalcanti LN, Canturk B, Pan PS, Kennedy LE (2009) J Org Chem 74:7364
24. Molander GA, Canturk B, Kennedy LE (2008) J Org Chem 74:973
25. Molander GA, Beaumard F (2010) Org Lett 12:4022
26. Sandoshan J, Undheim K (1994) Tetrahedron 50:275
27. Darabantu M, Boully L, Turck A, Plé N (2005) Tetrahedron 61:2897
28. Hirano Y, Kojima S, Yamamoto Y (2011) J Org Chem 76:2123
29. Yoon YA, Park CS, Cha MH, Choi H, Sim JY, Kim JG (2010) Bioorg Med Chem Lett 20:5735
30. Turck A, Plé N, Mongin F, Quéguiner G (2001) Tetrahedron 57:4489
31. Arukwe J, Benneche T, Undheim K (1989) J Chem Soc, Perkin Trans 1 1989:255
32. Sandosham J, Benneche T, Moeller BS, Undheim K (1988) Acta Chem Scand B42:455
33. Parkanyi C, Cho NS, Yoo GS (1988) J Organomet Chem 342:1
34. Fiandor Roman JM, Bueno Calderon JM, Mallo Rubio A (2006) WO 2006094799 A2

35. Plé N, Turck A, Fiquet E, Queguiner G (1991) J Heterocycl Chem 28:283
36. Chang HF, Chapdelaine M, Dembofsky BT, Herzog KJ, Horchler C, Schmiesing RJ (2008) WO 2008155572 A2
37. Denmark SE, Sweiss RF (2002) Acc Chem Res 35:835
38. Hiyama T (2002) J Organomet Chem 653:58
39. Spivey AC, Gripton CJG, Hannah JP (2004) Curr Org Synth 1:211
40. Seganish WM, DeShong P (2004) J Org Chem 69:1137
41. Pierrat P, Gros PC, Fort Y (2005) Org Lett 7:697
42. Louërat F, Tye H, Napier S, Garrigou M, Whittaker M, Gros PC (2011) Org Biomol Chem 9:1768
43. Radinov R, Chanev K, Khaimova M (1991) J Org Chem 56:4793
44. Turck A, Plé N, Tallon V, Queguiner G (1993) J Heterocycl Chem 30:1491
45. Mosrin M, Boudet N, Knochel P (2008) Org Biomol Chem 6:3237
46. Mosrin M, Knochel P (2008) Org Lett 10:2497
47. Miura Y, Nanjo M, Mochida K, Iwamoto T, Kira M (2004) Inorg Chem Commun 7:1017
48. Itami K, Mitsudo K, Fujita K, Ohashi Y, Yoshida J-i (2004) J Am Chem Soc 126:11058
49. Lee SY, Cho YJ, Kwon HJ, Kim BO, Kim SM (2011) WO 2011055912 A1
50. Seggio A, Chevallier F, Vaultier M, Mongin F (2007) J Org Chem 72:6602
51. Blair VL, Blakemore DC, Hay D, Hevia E, Pryde DC (2011) Tetrahedron Lett 52:4590
52. Mosrin M, Knochel P (2009) Chem Eur J 15:1468
53. Imahori T, Suzuwa K, Kondo Y (2008) Heterocycles 76:1057
54. Schwesinger R, Schlemper H (1987) Angew Chem Int Ed 26:1167
55. D'Alarcao M, Bakthavachalam V, Leonard NJ (1985) J Org Chem 50:2456
56. Soorukram D, Boudet N, Malakhov V, Knochel P (2007) Synthesis 2007:3915
57. Bernhardt S, Manolikakes G, Kunz T, Knochel P (2011) Angew Chem Int Ed 50:9205
58. L'Helgoual'ch JM, Bentabed-Ababsa G, Chevallier F, Yonehara M, Uchiyama M, Derdour A, Mongin F (2008) Chem Commun 2008:5375
59. L'Helgoual'ch JM, Bentabed-Ababsa G, Chevallier F, Derdour A, Mongin F (2008) Synthesis 2008:4033
60. Nguyen TT, Marquise N, Chevallier F, Mongin F (2011) Chem Eur J 17:10405
61. Otsu S, Oshiyama T, Katoh E, Kita H (2005) WO 2005062676 A1
62. Johnson DE, Sarma JARP, Subrahmanyam D, Sudarsanam S, Grant F (2008) WO 2008042867 A2
63. Marugan JJ, Zheng W, Motabar O, Southall N, Goldin E, Westbroek W, Stubblefield BK, Sidransky E, Aungst RA, Lea WA, Simeonov A, Leister W, Austin CP (2011) J Med Chem 54:1033
64. Busch A, Turck A, Nowicka K, Barsella A, Andraud C, Plé N (2007) Heterocycles 71:1723
65. Plé N, Turck A, Chapoulaud V, Quéguiner G (1997) Tetrahedron 53:2871

Top Heterocycl Chem (2013) 31: 21–64
DOI: 10.1007/7081_2012_94

Published online: 8 November 2012

Lithiations and Grignard Reactions on Pyrimidine and Quinazoline

Andrej Kolarovič

Abstract Pyrimidine and quinazoline structural motifs are abundantly found in naturally occurring compounds, with uracil, thymine, and cytosine being the most prominent representatives. Furthermore, the pyrimidine core constitutes a subunit of many compounds rendering substantial medicinal or economic benefits. This review focuses on lithiations and Grignard reactions as tools for selective yet variable purposive derivatizations of pyrimidine and quinazoline, excluding stochiometric transmetalations. A brief introduction into the topic is followed by a detailed summary of what has been done in the field since 2005. The chemistry is discussed in several sections based on the structure of the substrates, which is expected to simplify orientation in the text. Deprotonative and halogen–metal exchange metalations, respectively, are reported separately for each major group of the substrates.

Keywords Deprotonation · Halogen–metal exchange · Pyrimidine functionalization · Pyrimidine nucleobases ·

Contents

A. Kolarovič (✉)
Institute of Organic Chemistry, Catalysis and Petrochemistry, Slovak University of Technology, Bratislava, Slovakia
e-mail: andrej.kolarovic@stuba.sk

Abbreviations

Ac	Acetyl
aq	Aqueous
Bn	Benzyl
BOM	Benzyloxymethyl
Bu	Butyl
Bz	Benzoyl
Cbz	Benzyloxycarbonyl
concd	Concentrated
dba	Dibenzylideneacetone
DME	Dimethoxyethane
DMF	Dimethylformamide
DMSO	Dimethyl sulfoxide
E+	Electrophile
equiv	Equivalent(s)
h	Hour(s)
HMPA	Hexamethylphosphoric triamide
i-Pr	Isopropyl
LDA	Lithium diisopropylamide
LHMDS	Lithium bis(trimethylsilyl)amide
LTMP	Lithium 2,2,6,6-tetramethylpiperidide
LUMO	Lowest unoccupied molecular orbital
min	Minute(s)
MOM	Methoxymethyl
NIS	*N*-Iodosuccinimide
NMP	*N*-Methyl-2-pyrrolidinone
PMB	4-Methoxyphenyl
rt	Room temperature
s	Second(s)
satd	Saturated
s-Bu	*sec*-Butyl
TBAI	Tetrabutylammonium iodide
TBDMS	*tert*-Butyldimethylsilyl
t-Bu	*tert*-Butyl
TDS	Dimethyl-(2,3-dimethylbut-2-yl)silyl (thexyl-dimethylsilyl)

THF	Tetrahydrofuran
THP	Tetrahydropyran-2-yl
TIPS	Triisopropylsilyl
TMEDA	*N,N,N′,N′*-Tetramethyl-1,2-ethylenediamine
TMP	2,2,6,6-Tetramethylpiperidyl
TMS	Trimethylsilyl
Tr	Triphenylmethyl

1 Introduction

Pyrimidines comprise an important class of heterocyclic compounds, with numerous applications in medicinal and agricultural chemistry [1]. A structural diversity in decoration of pyrimidines is highly desired, and carefully elaborated metalations of the pyrimidine core are expected to provide valuable synthetic shortcuts. Lithiations and magnesiations of pyrimidines can be either deprotonative or accessible by halogen–metal exchange. The synthesis of derivatized diazines via deprotonative metalation was reviewed several times, most recently by Quéguiner et al. [2] and Mongin and Chevallier [3].[1] Based on the values calculated in DMSO [5], the most acidic hydrogens are at C-5 in pyrimidine **1** and at C-4 in quinazoline **2**, respectively (Fig. 1).

However, coordination of a ring nitrogen to a metal usually favors deprotonations at adjacent positions. In general, the regioselective introduction of a metal requires the presence of directing metalation groups (DMGs; typically halo, methoxy, or methylthio). The electron-withdrawing properties of the two nitrogens are reflected in lower LUMO energy levels and make pyrimidines and quinazolines susceptible to nucleophilic additions. Thus alkyllithiums, being good nucleophiles, are in deprotonative lithiations usually avoided and non-nucleophilic alkylamides proved to be more reliable. LDA (p*K*a = 35.7) and LTMP (p*K*a = 37.3) are the most common choices. Importantly, due to relatively low basicity of alkylamides, the deprotonations of some pyrimidines are equilibria and as such are under thermodynamic control. Hence an excess of an amide may be required for satisfactory conversions. On the other hand, as halogen–lithium exchange reactions in arenes are known to be faster than deprotonations [6], alkyl lithiums are particularly suitable for dehalogenative lithiations at low temperatures. Over the last several years, mixed Mg/Li amides TMPMgCl·LiCl and $TMP_2MgCl·LiCl$ [7–10] have proved to be highly efficient both in deprotonative and in dehalogenative magnesiations and have become a valuable alternative to lithiations.

[1] A nice concise survey of pyrimidine lithiations can be found in the introduction part of a work published by Schlosser et al. [4].

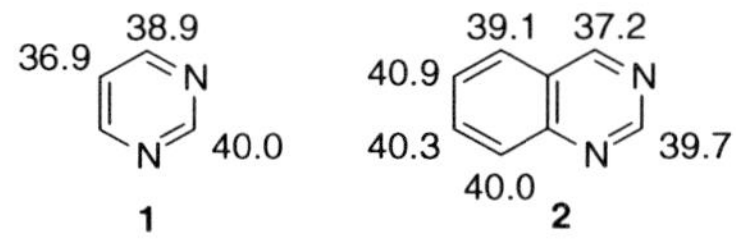

Fig. 1 Calculated p*K*a values for pyrimidine **1** and quinazoline **2**

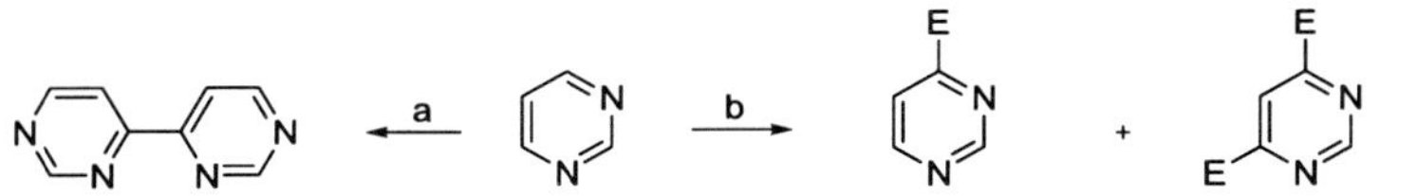

Scheme 1 (**a**) (i) 1.3 equiv LTMP, THF, −75°C, 1 h; (ii) HCl, EtOH, THF, −75°C; (**b**) (i) 1.2 equiv LTMP, 1.3 equiv E+ (TMSCl, PhCHO or Ph_2CO), THF, −75°C, 2 h; (ii) HCl, EtOH, THF, −75°C [12]

This review covers literature on lithiations and magnesiations published mainly since 2005. Stoichiometric transmetalations of metallated pyrimidines are a subject of the next chapter within this book and herein are included merely in a context. Also, lateral metalations are outside the scope of this account and are not presented.[2]

2 Unsubstituted Pyrimidines and Pyrimidines with Uncommon Directing Metalation Groups

Lithiated pyrimidines lacking a DMG are reported to be unstable even at low temperatures (−75°C) and prone to a quick dimerization (Scheme 1, conditions a) [12, 13].[3]

Significantly better results were achieved when trapping of lithiated species was performed in situ. LTMP proved to be compatible with some electrophiles and 4- and 4,6-disubstituted pyrimidines (**4a–c**, **5c**) were isolated in moderate to low yields.

Tacke et al. [15] studied co-condensation reactions of nitrogen-containing heterocycles with lithium atoms, in the presence of THF at −196°C. The reaction of pyrimidine **1** afforded the dimer **3** in 32% yield, supposedly through a radical intermediate.

The role of pyridine as DMG in pyridin-2-yl substituted diazines was examined by Plé et al. [16]. The regioselectivity of lithiations with an excess of LTMP was

[2] An interesting example of a lateral lithiation is discussed in Sect. 6.1, Scheme 32 [11].

[3] A radical coupling of pyrimidines promoted by 3 equiv of sodium and followed by air oxidation was reported to provide 6,6′-disubstituted 4,4′-bipyrimidines in yields of 56–93% [14].

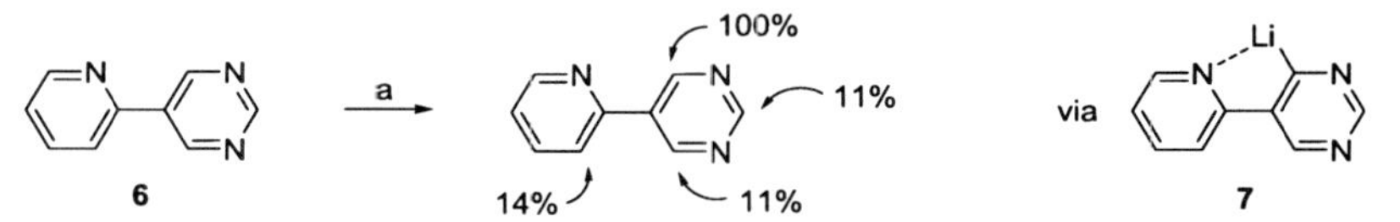

Scheme 2 (**a**) (i) 4 equiv LTMP, THF, −78°C, 15 min; (ii) 6 equiv DCl, EtOD, THF, −78°C, then warming to rt

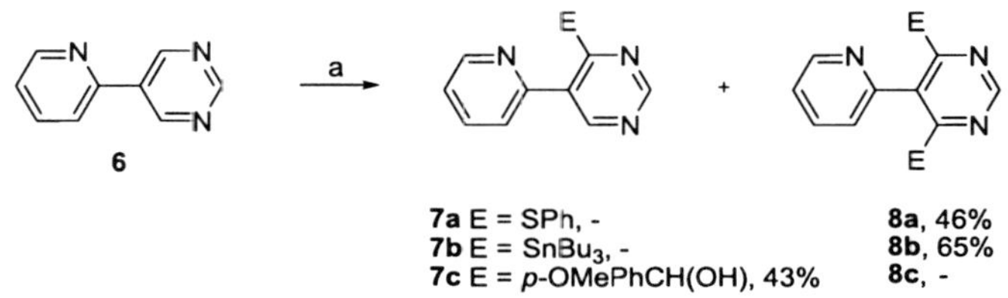

Scheme 3 (**a**) (i) 4 equiv LTMP, THF, −78°C, 15 min; (ii) 4 equiv E+ (PhSSPh, Bu_3SnCl or *p*-OMePhCHO), THF, −100°C, 30 min; (iii) THF/EtOH/H_2O

determined by quenching with DCl/EtOD and following examination of ^{1}H-NMR spectra of the mixtures (Scheme 2).

The *ortho*-position of the pyridine–pyrimidine linkage was completely deuterated via supposed intermediate **7**, along with minor deuterations at other sites. Nevertheless, two of the three other electrophiles exhibited a strong preference for repetitive substitutions at the *ortho*-position (Scheme 3).

Darabantu et al. [17] described lithiations of pyrimidines bearing two DMGs in positions 4 and 6. The system was rather complex, as each of the DMGs, that is α-(3,7-dioxa-r-1-azabicyklo[3.3.0]oct-*c*-5-ylmethoxy groups (DOABO-CH_2O), could exist in two chiral forms interconverting through a *meso* form (Scheme 4).

The CH_2O bridges were expected to coordinate the lithium atom and assist in hydrogen removal. The pyrimidine **9a** bearing (2H)DOABO units was unreactive at −78°C and the reaction temperature had to be raised to rt, which was rationalized by DOABO systems being locked in *meso* form. In case of pyrimidine **9b**, the (2Ph) DOABO conformers were still flipping at −78°C and a complete conversion was observed at this temperature. However, mechanistic details explaining the observed differences in reactivity remain unclear.

3 Alkoxy and Alkylthiopyrimidines

Lithiations of pyrimidines bearing methoxy group(s) have been known for a longer time and are well covered in several reviews [2, 3, 18]. Over the last several years research activity in this area has been relatively low. An example illustrating standard protocols is depicted in Table 1 [19].

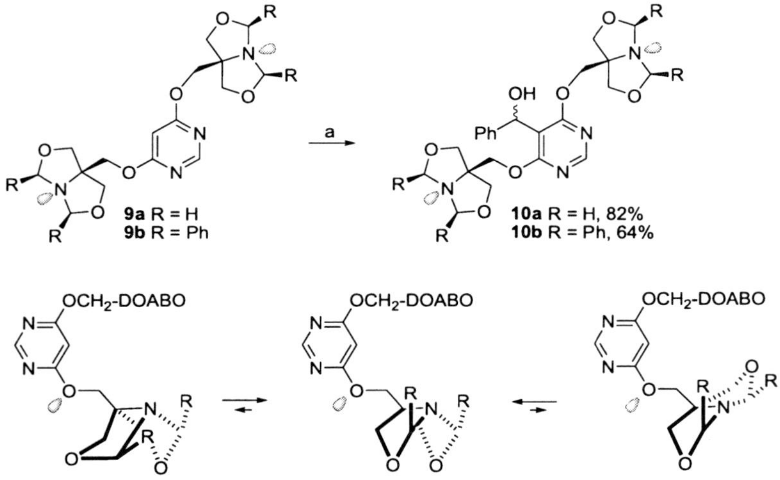

Scheme 4 (**a**) (i) 4 equiv LTMP, THF, −78°C, 2 h; (ii) 4 equiv PhCHO, THF, −78°C to rt, 12 h; (iii) CH_2Cl_2/H_2O, rt

Table 1 Functionalization of methoxy pyrimidines by means of *ortho*-lithiation

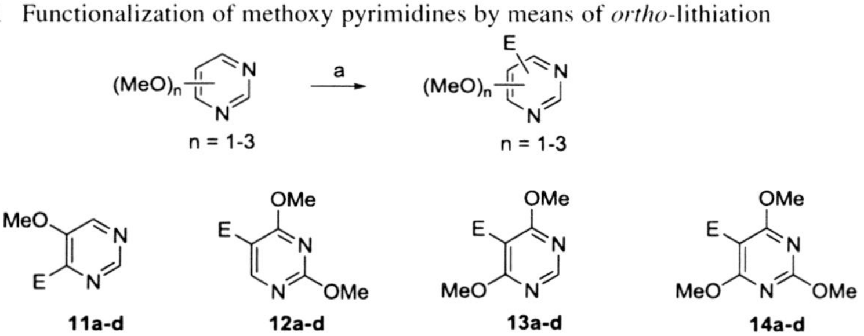

(**a**) (i) 1.3 equiv LTMP, THF, −78°C, 15 min; (ii) 1.5 equiv E+ (TMSCl, MeI, PhCHO or PhCOCl), THF, −78°C to rt, overnight

E	Products (%)			
TMS	**11a** (47)	**12a** (68)	**13a** (76)	**14a** (99)
Me	**11b** (47)	**12b** (83)	**13b** (75)	**14b** (92)
PhCHOH	**11c** (49)	**12c** (98)	**13c** (97)	**14c** (95)
PhCO	**11d**[a]	**12d**[a]	**13d** (84)	**14d** (91)

[a]Formation of a complex mixture was observed

In general, lithiated alkoxy substituted pyrimidines exhibit a better thermal stability and are less susceptible to dimerizations. Acidity of *ortho* hydrogens is increased by the inductive electron-withdrawing properties of the substituents. Moreover, coordination of the DMG to a metal makes *ortho* metalation more likely

Scheme 5 (**a**) (i) 1.2 equiv *n*-BuLi, Et_2O, −30°C, 1.5 h; (ii) 1.0 equiv HCO_2Et, −70°C, 1 h

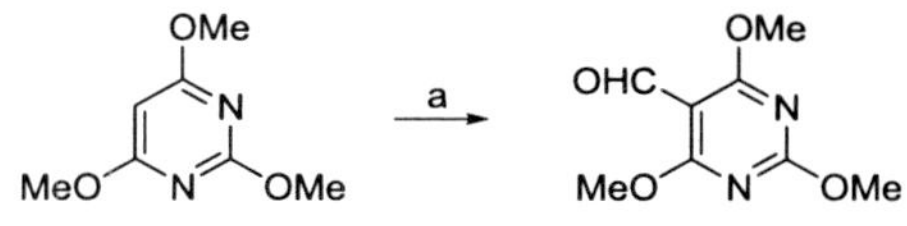

15 **16**, 40%

Table 2 A selective magnesiation utilized in derivatization of 2,4-dimethoxypyrimidine **17**

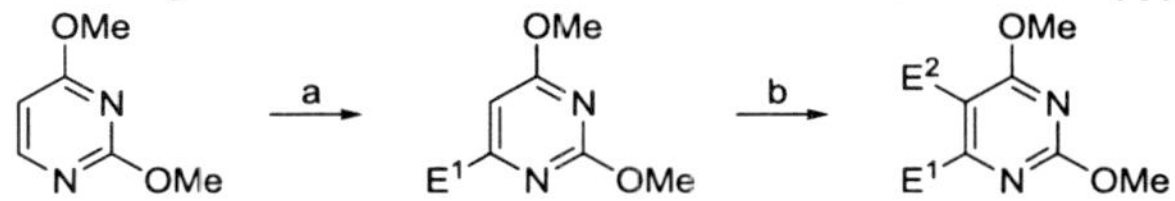

17 **18a-e** **19a-b**

(**a**) (i) 1.1 equiv TMPMgCl·LiCl, THF, 25°C, 15 min; (ii) E+; (**b**) (i) 1.1 equiv TMPMgCl·LiCl, THF, 0°C (**18a**) or −40°C (**18b**), 1 h; (ii) E+

Entry	Substrate	E+	Product	Yield (%)
1	**17**	I_2	**18a**, E^1 = I	74
2	**17**	TMS-CN	**18b**, E^1 = TMS	70
3	**17**	NC-CO_2Et	**18c**, E^1 = CO_2Et	71
4	**17**	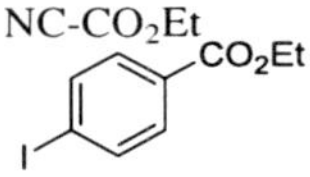	**18d**, E^1 = *p*-$PhCO_2Et$	75[a]
5	**17**	*t*-BuCOCl	**18e**, E^1 = *t*-BuCO	72[b]
6	**18a**, E^1 = I	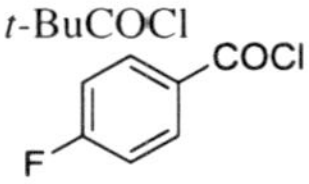	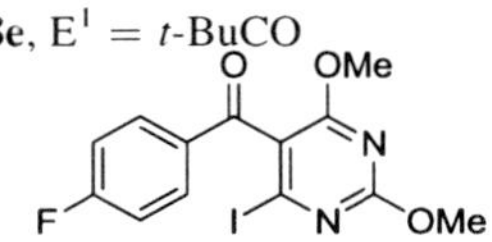 **19a**	84[b]
7	**18b**, E^1 = CO_2Et	PhCOCl	**19b**, E^1 = CO_2Et, E^2 = COPh	78[b]

[a]Transmetalation with 1.2 equiv $ZnCl_2$, followed by coupling catalyzed by 3 mol% $Pd(dba)_2$ and 6 mol% $P(o\text{-furyl})_3$

[b]1 equiv of CuCN·2LiCl was added [22]

due to the so-called "complex-induced proximity effect of the base" [20]. Methoxylated pyrimidines are less prone to nucleophilic additions, as illustrated by a recently reported lithiation of **15** with *n*-BuLi at −30°C (Scheme 5) [21].

A fresh blow to the field was brought by Knochel et al. [7]. They disclosed that 2,4-dimethoxypyrimidine **17** can be magnesiated by TMPMgCl·LiCl exclusively in position 6 (Table 2, **18**), thus establishing a complementary metalation procedure to lithiations by LTMP in position 5 (Table 1, **12a–d**).

Magnesiated intermediate can be directly reacted with an electrophile (entries 1–3) or transmetallated for a subsequent coupling reaction (entries 4, 5). Repeated magnesiation enables a successive functionalization of positions 6 and 5, respectively (entries 6, 7). A tolerance of relatively high reaction temperatures is

Table 3 A synthesis of 6- and 5,6-substituted 2,4-bis(methylthio)pyrimidines via magnesiation

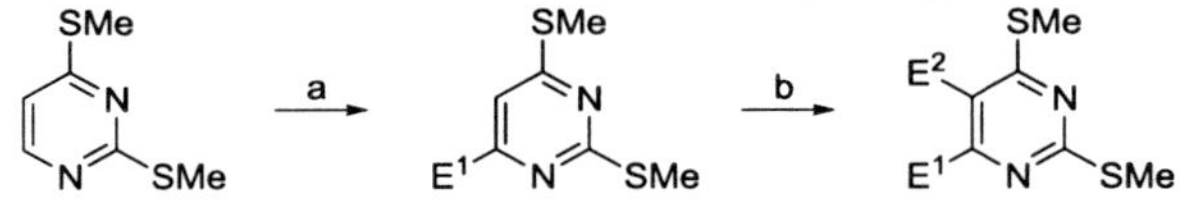

20 **21a-e** **22a-b**

(**a**) (i) 1.1 equiv $TMP_2Mg{\cdot}2LiCl$, THF, −20°C, 1 h; (ii) E+; (**b**) (i) 1.1 equiv $TMP_2Mg{\cdot}2LiCl$, THF, −5°C, 45 min; (ii) E+

Entry	Substrate	E+	Product	Yield (%)
1	**20**	I_2	**21a**, E^1 = I	76
2	**20**	$(BrCCl_2)_2$	**21b**, E^1 = Br	81
3	**20**	$FCCl_2CClF_2$	**21c**, E^1 = Cl	78
4	**20**	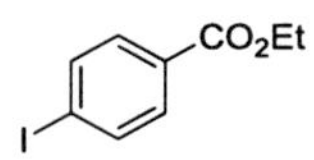	**21d**, E^1 = *p*-$PhCO_2Et$	71[a]
5	**20**	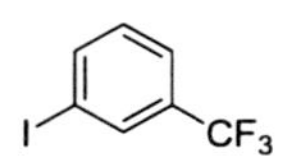	**21e**, E^1 = *m*-$PhCF_3$	80[a]
6	**21c**, E^1 = Cl	I_2	**22a**, E^1 = Cl, E^2 = I	61
7	**21c**, E^1 = Cl	PhCOCl	**22b**, E^1 = Cl, E^2 = COPh	65[b]
8	**21c**, E^1 = Cl	PhCHO	**22c**, E^1 = Cl, E^2 = CH(OH)Ph	66

[a]Transmetalation with 1.2 equiv $ZnCl_2$, followed by coupling catalyzed by 3 mol% $Pd(dba)_2$ and 6 mol% P(*o*-furyl)$_3$

[b]1 equiv of CuCN·2LiCl was added [22]

remarkable. The same approach was successfully extended to analogous 2,4-bis (methylthio)pyrimidine **20** (Table 3). The reagent switch for $TMP_2Mg{\cdot}2LiCl$ was not commented on.

4 Halopyrimidines

4.1 Deprotonative Lithiations and Grignard Reactions

Similar to unsubstituted pyrimidines, monohalogenated pyrimidines upon deprotonation tend to dimerise [4, 12, 23]. However, when lithiations are performed in the presence of an electrophile, 4-substituted-5-bromopyrimidines **24a–c** are accessible in moderate yields (Scheme 6). A decrease in reaction temperatures to −65°C was reported to affect the reaction outcome just marginally (a yield improvement of 5% for **24c**).

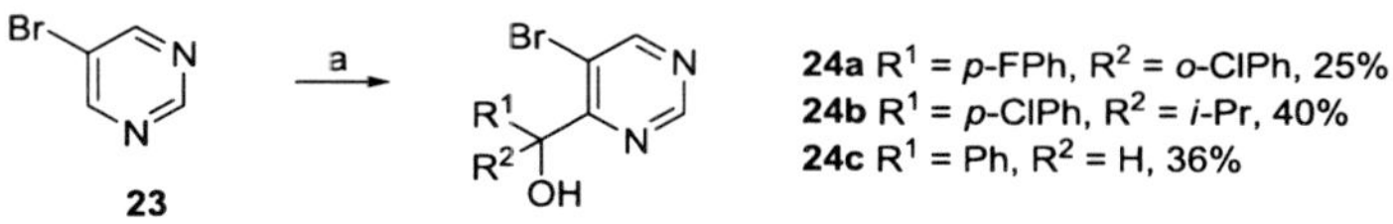

Scheme 6 (a) (i) 1 equiv LDA, 1 equiv E+, Et_2O, −10°C to 0°C, 2 h; (ii) 7% aq HCl [23]

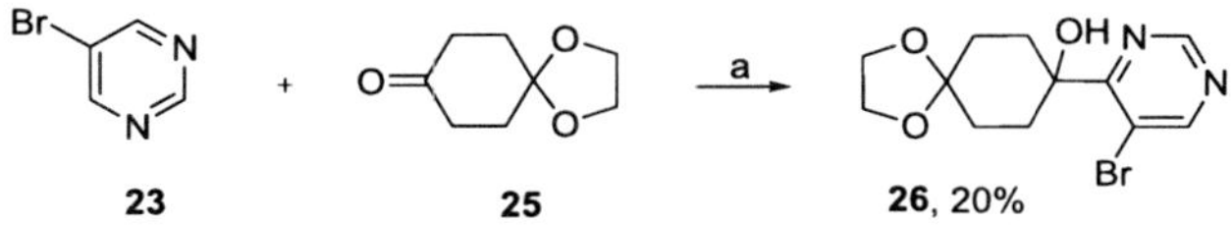

Scheme 7 (a) (i) **23**, LDA, THF, −78°C; (ii) **25**

Lately, a lithiation of 4-bromopyrimidine **23** and a quench with ketone **25** were performed in a stepwise manner with success; however, the target ketal **26** was isolated in only 20% yield (Scheme 7) [24].

In fluoro, chloro, and bromo 2,4-dihalopyrimidine series, similar trends in regioselectivity were observed. LDA turned out to be suitable for generation of the 5-lithio derivatives [25–27], while LTMP can preferentially deprotonate the position next to the nitrogen (Table 4, illustrative examples) [27, 28].

In general, the obtained yields were decreasing in the order of F $\approx$ Cl $>$ Br, and the regioselectivity was closely related to the reaction conditions, particularly to the reaction temperature.

A significantly better outcome can be expected if lithiated 2,4-dihalopyrimidine species are stabilized by an additional electron-withdrawing group (e.g., CF_3). Schlosser et al. described syntheses of 2,4-dihalo-6-(trifluoromethyl)pyrimidine-5-carboxylic acids **31a–b** in excellent yields (Scheme 8) [4].

A deprotonative lithiation of 2,4-dichloropyrimidine **27b** with LDA was recently applied in a synthesis of a novel p38α inhibitor **34**, exhibiting potent activity in treatment of arthritic diseases (Scheme 9) [29].

In addition to 2,4-dihalopyrimidines, only few reports exist on lithiations of other dihalo and trihalopyrimidines. Radinov et al. selectively converted 4,6-dichloro and 2,4,6-trichloropyrimidines to 5-lithio species, being flanked by two activating halogen atoms [30, 31].

In the field of magnesiations, an introduction of mixed Mg/Li amides by Knochel et al. represented a remarkable breakthrough [8]. TMPMgCl·LiCl has very good solubility (ca. 1.2 M) and stability (over 6 months) in THF and improved kinetic activity in comparison with TMPMgX. Upon inverse addition to THF solution of TMPMgCl·LiCl and subsequent reaction with an electrophile, monohalopyrimidines **23** and **37** furnished functionalized derivatives **36** and **39a–b** in a completely regioselective fashion (Scheme 10).

A remarkable synthetic flexibility of magnesiations effectuated by TMPMgCl·LiCl was convincingly demonstrated on successive functionalizations of 2-bromopyrimidine **40** (Table 5) [9, 10].

Table 4 Examples of 2,4-dihalopyrimidine lithiations

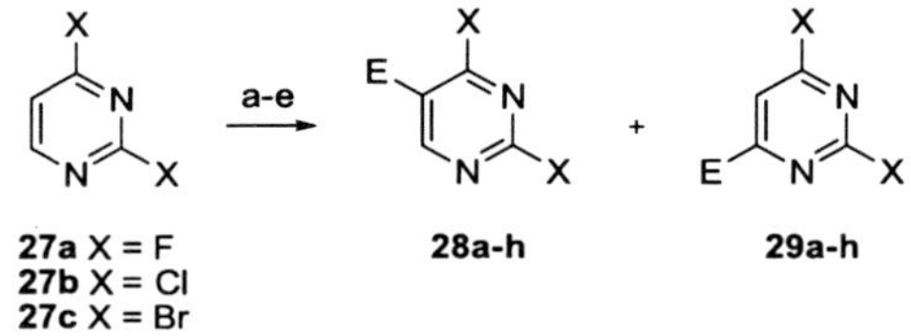

(**a**) (i) 2.3 equiv LDA, THF, −70°C, 30 min; (ii) E+, −70°C, 1 h; (iii) 35% aq HCl/EtOH/THF, −70°C; (**b**) (i) 2.3 equiv LDA, THF, −78°C, 1 h; (ii) E+, −78°C, 1 h; (iii) hydrolysis, −78°C; (**c**) (i) 1.4 equiv LTMP, 2.7 equiv HMPA, THF, −70°C, 1.5 h; (ii) E+, −70°C, 1 h; (iii) 35% aq HCl/EtOH/THF, −70°C; (**d**) (i) 3.2 equiv LDA, THF, −100°C, 30 min; (ii) E+, −100°C, 1.5 h; (iii) concd HCl/EtOH/THF, −100°C; (**e**) (i) 3.2 equiv LTMP, THF, −100°C, 30 min; (ii) E+, −100°C, 1.5 h; (iii) concd HCl/EtOH/THF, −100°C

Entry	Subst/Cond	E+	Product	Yield (%)
1	**27a**/a [26]	MeCHO	**28a**, X = F, E = MeCH(OH)	61
2	**27a**/a [26]	3,4,5-(OMe)$_3$PhCHO	**28b**, X = F, E = 3,4,5-(OMe)$_3$PhCH(OH)	47
3	**27b**/b [25]	DCl/CD$_3$OD	**28c**/**29c** = 94/6, X = Cl, E = D	58[a]
4	**27b**/b [25]	MeCHO	**28d**, X = Cl, E = MeCH(OH)	57
5	**27b**/b [25]	3,4,5-(OMe)$_3$PhCHO	**28e**/**29e** = 90/10, X = Cl, E = 3,4,5-(OMe)$_3$PhCH(OH)	58[a]
6	**27b**/c [28]	DCl/CD$_3$OD	**29f**, X = Cl, E = D	43
7	**27b**/c [28]	MeCHO	**29g**, X = Cl, E = MeCH(OH)	19
8	**27c**/d [27]	MeCHO	**28h**/**29h** = 80/20, X = Br, E = MeCH(OH)	25[a]
9	**27c**/e [27]	MeCHO	**29h**, X = Br, E = MeCH(OH)	21

[a]Overall yield

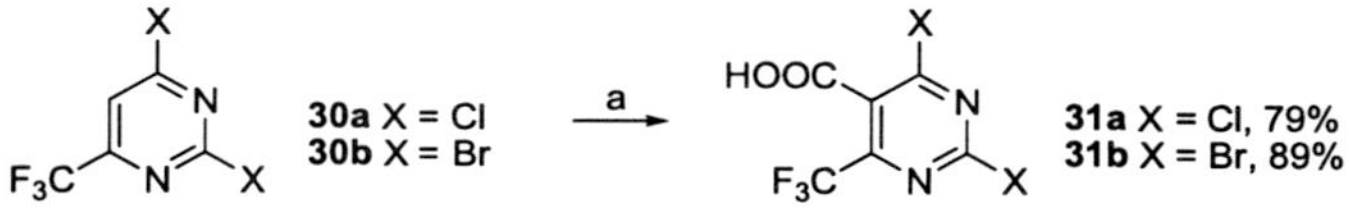

Scheme 8 (**a**) (i) 1 equiv LDA, THF, −75°C, 45 min; (ii) solid CO_2; (iii) 2M HCl

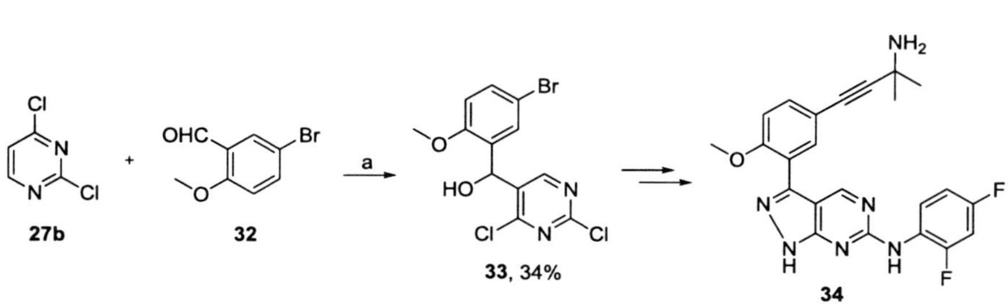

Scheme 9 (**a**) (i) **27b**, LDA, THF, −78°C; (ii) **32**

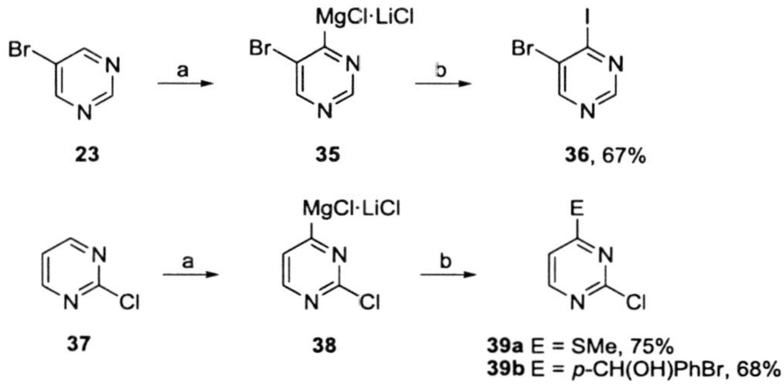

Scheme 10 (**a**) 1.2 equiv TMPMgCl·LiCl, THF, −55 to −40°C, 2 h; (**b**) (i) E+ (I_2, $MeSO_2SMe$, *p*-BrPhCHO); (ii) aq NH_4Cl

All positions of the pyrimidine ring could be regioselectively metalated in a stepwise fashion, in a sequence of C-4, C-6, and C-5. The magnesiated intermediates were either directly trapped by suitable electrophiles or transmetalated to broaden the coupling options. The method was extended to concise syntheses of p38 kinase inhibitor **46** and sPLA2 inhibitor **50**, both having anti-inflammatory properties, starting from commercially available pyrimidine **44** (Scheme 11).[4]

4.2 Lithiations and Grignard Reactions by Halogen–Metal Exchange

In cases when structural features of a pyrimidine substrate do not permit a region- and chemoselective deprotonative functionalization, a halogen–metal exchange is an alternative enabling a safe permutation at a desired position of the ring [6]. Availability of some starting halopyrimidines may nevertheless be cumbersome and if having an option, direct deprotonative metalations are usually preferred. A characteristic situation when halogen–metal permutation is the method of choice comprises lithiations at position C-5. Typically, lithiations are executed with *n*-BuLi at very low temperatures (−100°C) to suppress competitive alkylations, followed by a rapid quench with an electrophile (Table 6) [32–36].

The halogen–metal exchange can be extremely fast, as evident from reaction conditions reported for entries 4 and 5. However, depending on both the halogen itself and the flanking substituents, significant differences in reactivity between two halogen atoms can be observed and a highly chemo- and regioselective monometalation of dihalogenated pyrimidines is achievable (Table 7) [4, 37].

[4] Metalations of other alkylthiohalopyrimidines are discussed in Sect. 5.

Table 5 Repetitive functionalization of 2-bromopyrimidine **40** using TMPMgCl·LiCl

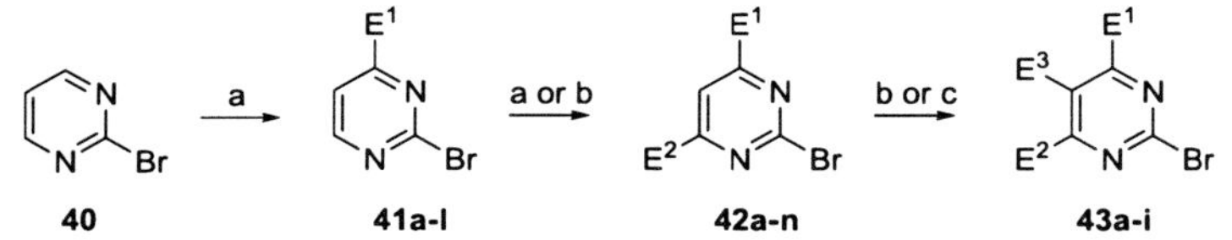

(**a**) (i) 1.1 equiv TMPMgCl·LiCl, THF, −60 to −55°C; (ii) E+; (**b**) (i) 1.1 equiv TMPMgCl·LiCl, THF, rt; (ii) E+; (**c**) (i) 1.1 equiv $TMP_2Mg{\cdot}2LiCl$, THF, −10°C; (ii) E+

Entry	Subst/Cond	E+	Product	Yield (%)
1	**40**/a [9]	$MeSO_2SMe$	**41a**, E^1 = SMe	81
2	**40**/a [10]	$PhSO_2SPh$	**41b**, E^1 = SPh	77
3	**40**/a [9]	I_2	**41c**, E^1 = I	85[a]
4	**40**/a [9]	$BrCCl_2CCl_2Br$	**41d**, E^1 = Br	71
5	**40**/a [10]	$ClCF_2CCl_2F$	**41e**, E^1 = Cl	71
6	**40**/a [9]	TMSCN	**41f**, E^1 = TMS	68
7	**40**/a [10]	*p*-IPhCl	**41g**, E^1 = *p*-PhCl	67[b]
8	**40**/a [9]	*m*-$IPhCF_3$	**41h**, E^1 = *m*-$PhCF_3$	72[b]
9	**40**/a [9]	*p*-$IPhCO_2Et$	**41i**, E^1 = *p*-$PhCO_2Et$	81[b]
10	**40**/a [10]	2-iodothiophene	**41j**, E^1 = thiophen-2-yl	62[b]
11	**41c** [10]	hex-1-yne	**41k**, E^1 = hex-1-ynyl	56[c]
12	**41c** [9]	phenylacetylene	**41l**, E^1 = phenylethynyl	71[c]
13	**41a**/b [9]	$ClCF_2CCl_2F$	**42a**, E^1 = SMe, E^2 = Cl	76
14	**41a**/b [9]	$BrCCl_2CCl_2Br$	**42b**, E^1 = SMe, E^2 = Br	81
15	**41a**/b [9]	I_2	**42c**, E^1 = SMe, E^2 = Br	78
16	**41b**/a [10]	$ClCF_2CCl_2F$	**42d**, E^1 = SPh, E^2 = Cl	90
17	**41b**/a [10]	TsCN	**42e**, E^1 = SPh, E^2 = CN	67
18	**41b**/a [10]	$BrCCl_2CCl_2Br$	**42f**, E^1 = SPh, E^2 = Br	85
19	**41c**/a [9]	TMS-CN	**42g**, E^1 = I, E^2 = TMS	93
20	**41c**/a [10]	I_2	**42h**, E^1 = I, E^2 = I	82[a]
21	**41c**/a [10]	$BrCCl_2CCl_2Br$	**42i**, E^1 = I, E^2 = Br	67
22	**41c**/a [10]	TMS-CN	**42j**, E^1 = I, E^2 = TMS	89
23	**41h**/a[d] [9]	$ClCF_2CCl_2F$	**42k**, E^1 = *m*-$PhCF_3$, E^2 = Cl	91
24	**41h**/a[d] [9]	TMS-CN	**42l**, E^1 = *m*-$PhCF_3$, E^2 = TMS	72
25	**41h**/a[d] [9]	$MeSO_2SMe$	**42m**, E^1 = *m*-$PhCF_3$, E^2 = SMe	76
26	**41l**/a[e] [9]	$ClCF_2CCl_2F$	**42n**, E^1 = phenylethynyl, E^2 = Cl	84
27	**42a**/b [9]	PhCOCl	**43a**, E^1 = SMe, E^2 = Cl, E^3 = PhCO	81[f]
28	**42a**/b [9]	PhCHO	**43b**, E^1 = SMe, E^2 = Cl, E^3 = PhCH (OH)	75
29	**42a**/b [9]	$MeSO_2SMe$	**43c**, E^1 = SMe, E^2 = Cl, E^3 = SMe	92
30	**42d**/c [10]	*o*-ClPhCOCl	**43d**, E^1 = SPh, E^2 = Cl, E^3 = *o*-ClPhCO	77[f]
31	**42d**/c [10]	Allyl bromide	**43e**, E^1 = SPh, E^2 = Cl, E^3 = allyl	89[a,f]

(continued)

Table 5 (continued)

Entry	Subst/Cond	E+	Product	Yield (%)
32	**42d**/c [10]	$MeSO_2SMe$	**43f**, E^1 = SPh, E^2 = Cl, E^3 = SMe	80
33	**42n**/b[g] [9]	I_2	**43g**, E^1 = phenylethynyl, E^2 = Cl, E^3 = I	71
34	**42n**/b[g] [9]	PhCOCl	**43h**, E^1 = phenylethynyl, E^2 = Cl, E^3 = PhCO	69[f]
35	**42n**/b[g] [9]	allyl bromide	**43i**, E^1 = phenylethynyl, E^2 = Cl, E^3 = allyl	67[f]

[a]The magnesiated pyrimidine derivative was transmetalated with 1.1 equiv $ZnCl_2$
[b]Transmetalation with $ZnCl_2$, followed by coupling catalyzed by $Pd(dba)_2/P(o\text{-furyl})_3$
[c]2-Bromo-4-iodopyrimidine **41c** was generated in situ and reacted with $Pd(dba)_2/P(o\text{-furyl})_3$, CuI, Et_3N, and alkyne
[d]Magnesiations at −40°C
[e]Magnesiations at −20°C
[f]Transmetalation with CuCN·2LiCl [22]
[g]Magnesiations at −5°C

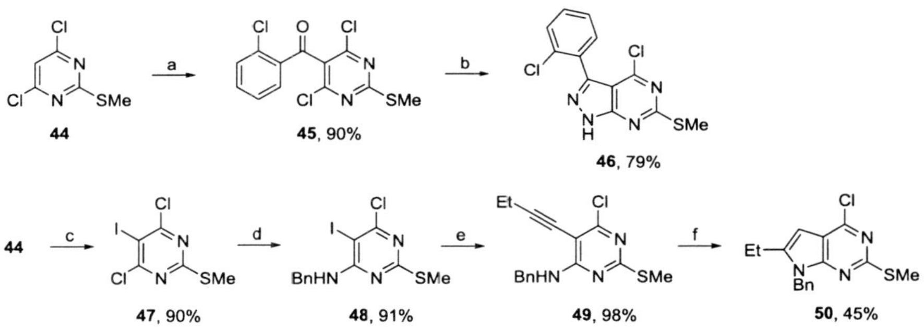

Scheme 11 (**a**) (i) 1.1 equiv TMPMgCl·LiCl, THF, rt, 20 min; (ii) 1.1 equiv CuCN·2LiCl, 2.0 equiv *o*-ClPhCOCl, rt, 1 h; (**b**) $NH_2NH_2·H_2O$, THF, rt, 10 min; (**c**) (i) 1.1 equiv TMPMgCl·LiCl, THF, rt, 20 min; (ii) 1.5 equiv I_2, THF, rt, 20 min; (**d**) 1.7 equiv $BnNH_2$, THF, rt, 20 min; (**e**) (i) 3 mol% $Pd(dba)_2$, 6 mol% $P(o\text{-furyl})_3$, 4 mol% CuI, Et_3N; (ii) 2.0 equiv but-1-yne, 50°C, 1 h; (**f**) 2.0 equiv *t*-BuOK, NMP, rt, 1 h

5 Pyrimidines Bearing a Combination of Halo, Alkoxy, and Alkylthio Substituents

5.1 Deprotonative Lithiations and Grignard Reactions

Unsurprisingly, metalations of pyrimidines decorated with miscellaneous arrangements of halo, alkoxy, and alkylthio substituents are often found to exhibit features similar to those of halogenated or alkoxylated pyrimidines. Analogous with 2,4-dihalopyrimidines **27a–c** (Table 4), 4-halo-2-methylthiopyrimidines were

Table 6 Recent examples of dehalogenative lithiations on 5-bromopyrimidines **23** and **51a–c**, respectively

23 (R = H)
51a-c

52a-e

(**a**) (i) 1.4 equiv *n*-BuLi, THF/Et_2O 1:1, −100°C, 30 min; (ii) 1.4 equiv E+, −100 to 25°C, 16 h; (**b**) (i) 1.0 equiv *n*-BuLi, THF, −100°C, 20 min; (ii) 1.0 equiv E+, −100°C, 20 min; (iii) 2 M HCl/Et_2O, <−85°C; (**c**) (i) 2 equiv *n*-BuLi, 1 equiv TMEDA, THF, −110°C, 30 min; (ii) 1.0 equiv E+, −110°C, 20 min; (iii) HCl, dioxane, −110°C; (**d**) (i) 1.2 equiv *n*-BuLi, THF, −95°C, 30 s; (ii) excess of E+, −90°C, 20 min; (iii) aq NH_4Cl; (**e**) (i) 1.0 equiv *n*-BuLi, THF, −100°C, 2 min; (ii) 1.0 equiv E+, −100°C to rt

Entry	Subst/Cond	E+	Product	Yield (%)
1	**23**/a [32]	2-F-6-CF_3PhCHO	**52a**, R = H, E = 2-F-6-CF_3PhCH(OH)	65
2	**23**/b [33]	HCO_2Et	**52b**, R = H, E = CHO	52
3	**51a**/c [34]	HCO_2Et	**52c**, E = CHO	54
	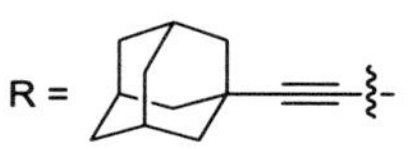		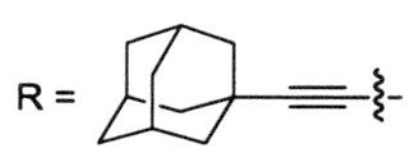	
4	**51b**/d [35] R = *p*-CF_3OPh	DMF	**52d**, R = *p*-CF_3OPh, E = CHO	74
5	**51c**/e [36]	$(CH_3)_2CHCHO$	**52e**, E = $(CH_3)_2CHCH(OH)$	79
	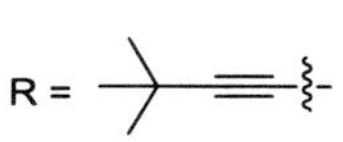		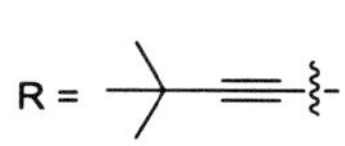	

regioselectively lithiated with LDA at C-5, while bulkier and more basic LTMS provided mainly the corresponding C-6-regioisomers [25, 26, 38]. The preference for C-6 was even more pronounced when a highly hindered lithium *N*,*N*-*t*-butyl-(1-*i*-propylpentyl)amide was employed as a base [39]. However, in the presence of a methoxy group at C-4, *ortho*-lithiations at C-5 are favored even with LTMS (Scheme 12) [40, 41].

Remarkably, on the contrary to the other electrophiles, trapping of lithiated pyrimidine **55** with iodine at −70°C provided six-substituted derivative **57**. This has been attributed to a possible isomerization of 5-iodo derivative **56d**, promoted by either LTMP or lithiated intermediates ("halogen-dance" [42, 43]). Indeed, iodination at −100°C furnished 6-iodo pyrimidine **56d** in 42% yield, along with 18% of **57**.

Table 7 Monometalations of dihalogenated pyrimidines **53a–d**

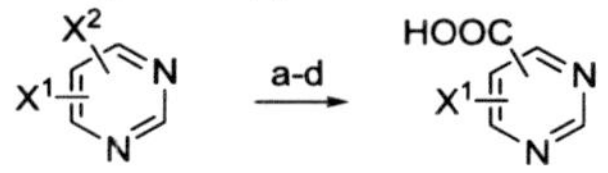

53a-d **54a-d**

(**a**) (i) 1.05 equiv *n*-BuLi, THF, −78°C, 1 h; (ii) CO_2 (s); (iii) AcOH; (**b**) (i) *n*-BuLi, THF, −75°C, 15 min; (ii) CO_2 (s); (iii) HCl; (**c**) (i) 1.0 equiv *i*-PrMgCl, THF/Et_2O, −10°C, 2 h; (ii) CO_2 (s); (iii) 2 M HCl; (**d**) (i) 1.0 equiv *n*-BuLi, hexanes/toluene, −75°C, 45 min; (ii) CO_2 (s); (iii) 2 M HCl

Entry	Substrate	Cond	Product
1	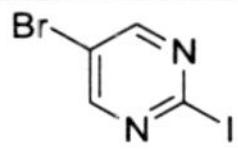 **53a**	a [37]	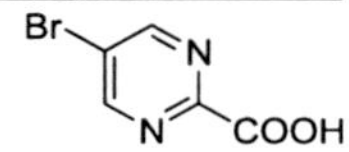 **54a**, 62%
2	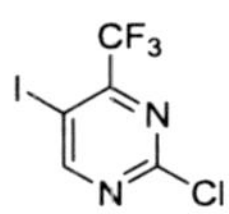 **53b**	b [4]	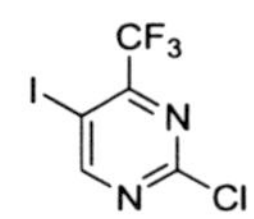 **54b**, 29%
3	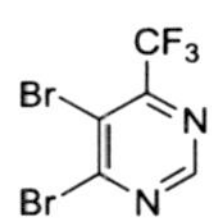 **53c**	c [4]	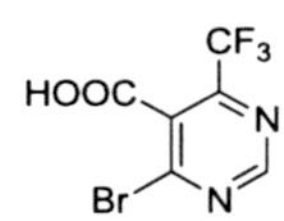 **54c**, 54%
4	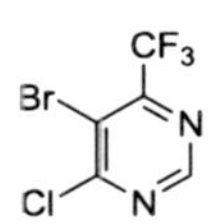 **53d**	d [4]	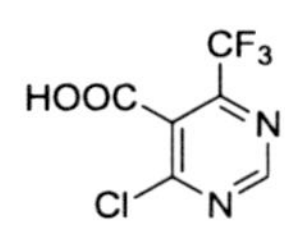 **54d**, 73%

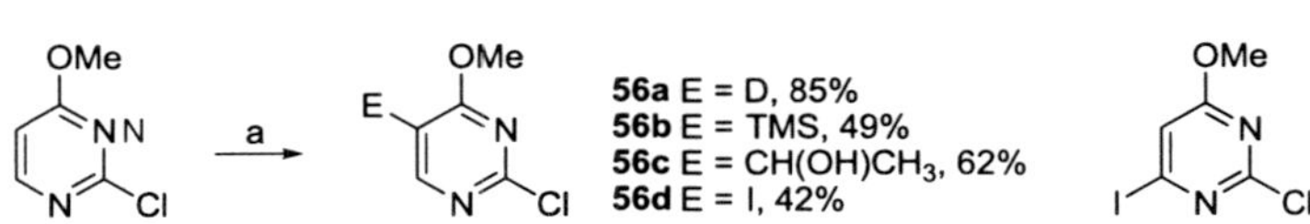

55

56a E = D, 85%
56b E = TMS, 49%
56c E = CH(OH)CH_3, 62%
56d E = I, 42%

57, 85%

Scheme 12 (**a**) (i) 2.3 equiv LTMP, THF, −70°C, 1 h; (ii) E+ (DCl/CD_3OD, TMSCl, CH_3CHO or I_2), −70°C (−100°C for **56d**), 1–1.5 h; (iii) 35% HCl/EtOH/THF

The protocol describing preparation of pyrimidine **57** [41], as depicted in Scheme 12, was later exploited by Hecht et al. in synthesis of analogues of anticancer agent camptothecin **59a** (Scheme 13, X = CH), 14-azacamptothecin

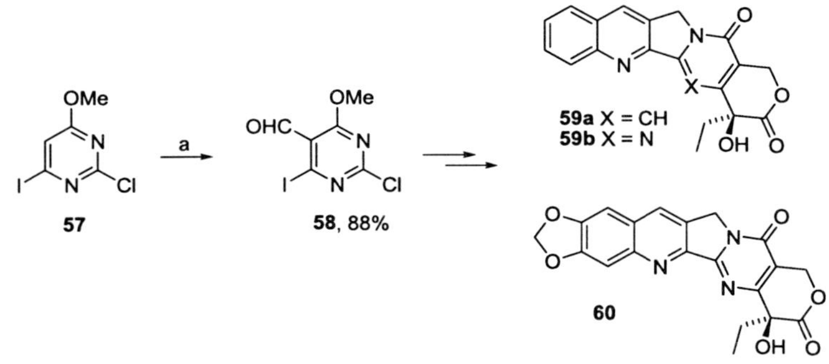

Scheme 13 (**a**) (i) 2.5 equiv LTMP, THF, −70°C, 1 h; (ii) 13 equiv HCO_2Et, −70°C, 1 h; (iii) HCl/EtOH/THF

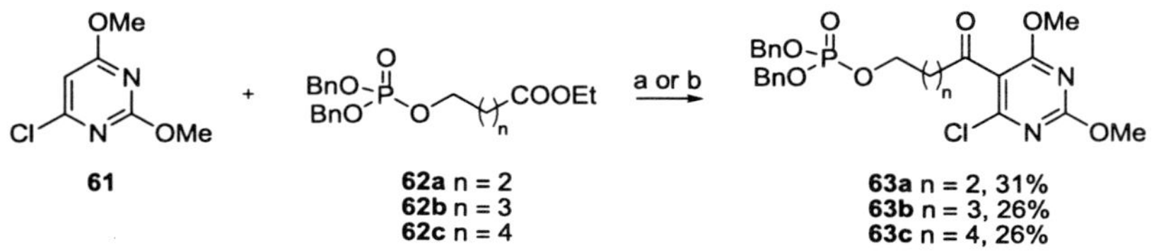

Scheme 14 (**a**) (i) 1.0 equiv *n*-BuLi, THF, −30°C, 1.5 h; (ii) 1.0 equiv **62a**, −78°C, 1 h; (iii) 35% HCl, THF; (**b**) (i) 1.0 equiv *n*-BuLi, THF, −78 to −70°C, 20 min; (ii) 1.0 equiv **62b** or **62c**, −40°C, 1–1.5 h; (iii) brine

59b [44], and 10,11-methylenedioxy-14-azacamptothecin **60** [45], respectively. A follow-up deprotonative lithiation of **57** enabled a smooth introduction of a formyl group to the remaining position at C-5.

On the other hand, acylations of lithiated pyrimidine **61** were observed to give only low yields of **63a–c** (Scheme 14) [46]. Interestingly, preparation of homologue **63a** turned out to be problematic under conditions leading to **63b–c**, and a deprotonative lithiation at relatively high temperature (−30°C; −70°C for **63b–c**) followed by a quench with an ester at low temperature (−78°C; −40°C for **63b–c**) was found to be essential to a successful synthesis.

Hřebabecký et al. [47] described a simple, yet informative study of thermal stability of lithiated pyrimidine **64**, based on the regular treatment of samples with 3 equiv of benzaldehyde and GC-MS analysis after a reaction time of 1 h at −75°C (Table 8).

A control experiment at −75°C with exclusion of a prolonged stirring of **64** yielded **65** in 88%. Thus, the deterioration of lithiated intermediate **64** at −75°C with a drop of 4% in 8.5 h can be considered to be relatively slow. However, the obtained data indicate a dramatic decrease in stability of **64** at temperatures above −25°C. This know-how was exploited in synthesis of derivatives **68** and **69**, containing a masked uracil subunit (Scheme 15).

Table 8 Yields of pyrimidine **65** reflecting a time-dependent thermal stability of intermediate **64**

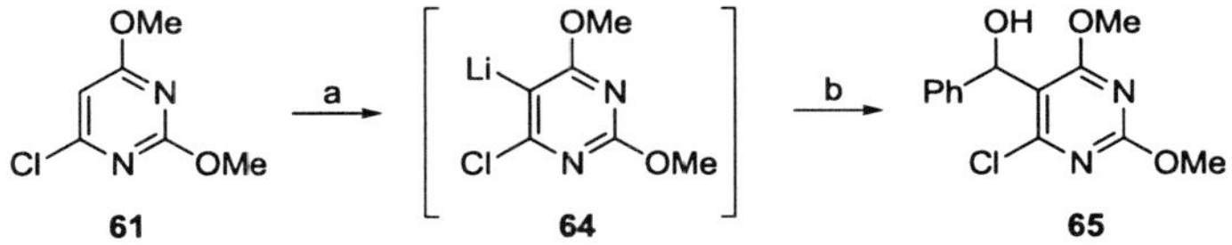

(**a**) (i) 1.1 equiv *n*-BuLi, THF, −75°C, 30 min; (ii) stirring at defined temperature (**b**) (i) 3 equiv PhCHO, THF, −75°C, 1 h; (ii) aq NH_4Cl

	Temperature (°C)		
Time of stirring (h)	−75	−25	0
2	87%	78%	16%
4.5	86%	70%	0%
8.5	84%	43%	0%

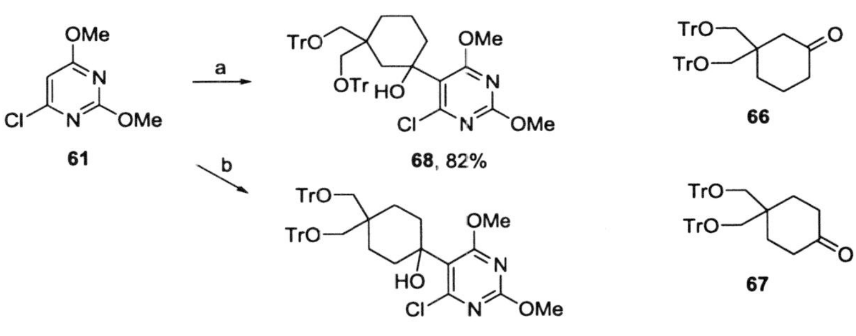

Scheme 15 (**a**) (i) 1.1 equiv *n*-BuLi, THF, −75°C, 25 min; (ii) 0.77 equiv **66**, 35 min at −75°C, 2 h at −30°C, 16 h at 0°C; (iii) aq NH_4Cl; (**b**) (i) 1.1 equiv *n*-BuLi, THF, −75°C, 30 min; (ii) 0.77 equiv **67**, 35 min at −75°C, 4 h at −30°C, 16 h at −10°C; (iii) aq NH_4Cl

A related 4-bromo-2,6-dimethoxypyrimidine **70** was reported to be prone to a nucleophilic displacement of the halogen upon formylation with DMF (Scheme 16) [48]. Presumably, the initial introduction of formyl group activated the pyrimidine ring towards the released dimethylamide and a substitution occurred via addition–elimination mechanism. Formation of **72** was partially suppressed by quenching the reaction mixture at −78°C (conditions b).

Lithiated pyrimidines are eligible for high-yielding oxidations, as demonstrated on 2-chloro-4,6-dimethoxypyrimidine **73** (Scheme 17) [49]. Alternative to *t*-BuOOH, dioxygen can serve as a source of the hydroxyl group, affording **74** in a less appealing 44% yield.

A series of 2-alkylthio-4,6-dichloropyrimidines **44**, **75a–b** (Scheme 18) can be safely deprotonated with LDA at the only free position left and quenched with an

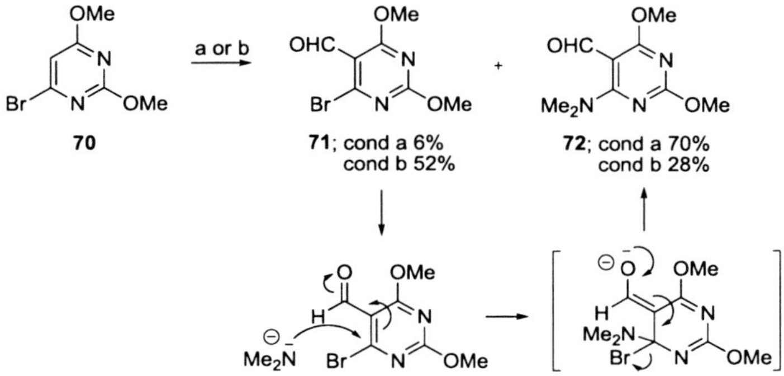

Scheme 16 (**a**) (i) 2.5 equiv LDA, THF, −78°C, 10 min; (ii) 2.0 equiv DMF, −78°C, 1 h; (iii) aq NH_4Cl, rt; (**b**) (i) 2.5 equiv LDA, THF, −78°C, 2 h; (ii) 2.0 equiv DMF, −78°C, 2 h; (iii) aq NH_4Cl, -78°C

Scheme 17 (**a**) (i) 2.0 equiv LDA, THF, −78°C, 1 h; (ii) 2.1 equiv *t*-BuOOLi, −78 to 0°C, 3 h; (iii) 6M HCl

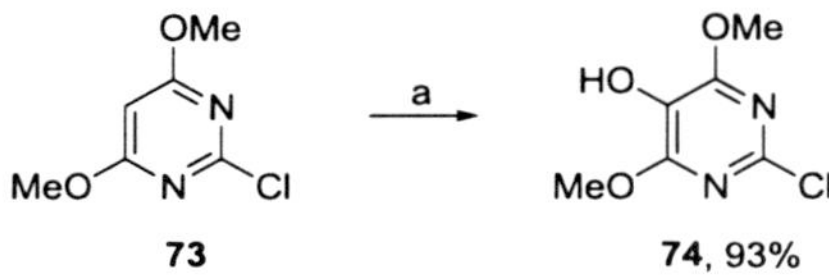

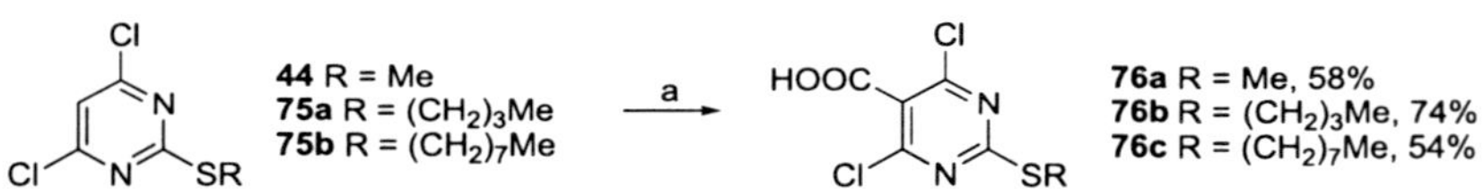

Scheme 18 (**a**) (i) 1.5 equiv LDA, THF, −78°C, 3–5 h; (ii) CO_2 (g), −78°C, 10 min; (iii) HCl

electrophile (e.g., CO_2) [50–52]. Irie et al. [53] reported 98% yield for **76a** under analogous reaction conditions, but with no experimental details.

In comparison with lithiated 2-methylthio-4,6-dichloropyrimidine **44**, its magnesiated counterpart **77** has a remarkable thermal stability and can be generated and handled at ambient temperature (Table 9) [54].[5] The same magnesiation protocol, with somewhat shorter metalation time, is applicable to 2-chloro-4-methylthiopyrimidine **39a**. The intermediates **77** and **79** reliably react with a variety of electrophiles, with no substantial deviation in the reaction outcome.

A synthetic potential of the described magnesiations was demonstrated in preparation of the fungicide Mepanipyrim **83** (Scheme 19) [54].

Additional modes of functionalization of various halopyrimidines are enabled by zincation with $TMP_2Zn{\cdot}2MgCl_2{\cdot}2LiCl$ [54].

[5] For related magnesiations see Tables 2, 3, and 5.

Table 9 Selective magnesiations of pyrimidines **44** and **39a** using TMPMgCl·LiCl, followed by a quench with an electrophile

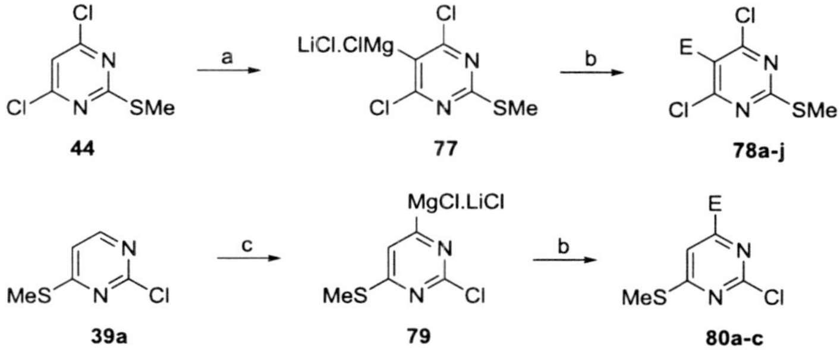

(**a**) 1.1 equiv TMPMgCl·LiCl, THF, 25°C, 30 min; (**b**) E+; (**c**) as for (**a**), 5 min

Entry	Subst	E+	Product	Yield (%)
1	**44**	PhCHO	**78a**, E = CH(OH)Ph	90
2	**44**	$NCCO_2Et$	**78b**, E = CO_2Et	86
3	**44**	MeI	**78c**, E = Me	92
4	**44**	$BrCCl_2CCl_2Br$	**78d**, E = Br	89
5	**44**	TMS-CN	**78e**, E = TMS	79
6	**44**	t-$BuCOOCH_2I$	**78f**, E = t-$BuCOOCH_2$	76
7	**44**	PhCOCl	**78g**, E = PhCO	90[a]
8	**44**	p-FPhCOCl	**78h**, E = p-FPhCO	93[a]
9	**44**	furan-2-carbonyl chloride	**78i**, E = furan-2-carbonyl	86[a]
10	**44**	t-$BuCH_2COCl$	**78j**, E = t-$BuCH_2CO$	84[a]
11	**39a**	I_2	**80a**, E = I	71
12	**39a**	$BrCCl_2CCl_2Br$	**80b**, E = Br	79
13	**39a**	$ClCF_2CCl_2F$	**80c**, E = Cl	72

[a]The magnesiated pyrimidine derivative was transmetalated with 1.1 equiv CuCN·2LiCl [22]

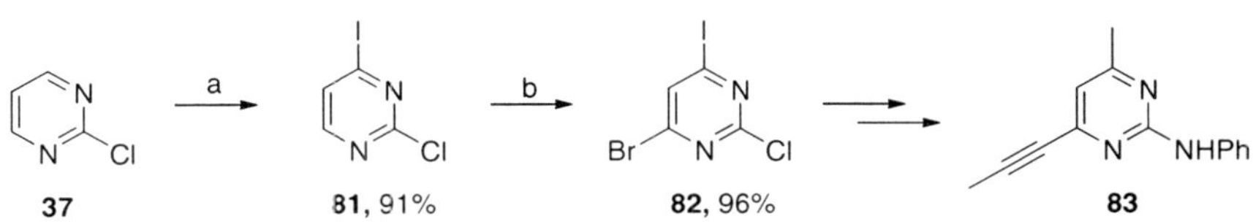

Scheme 19 (**a**) (i) 1.1 equiv TMPMgCl·LiCl, THF, −60°C, 2 h; (ii) 1.1 equiv $ZnCl_2$, −60°C; (iii) 1.5 equiv I_2, THF, rt, 20 min; (**b**) (i) 1.1 equiv TMPMgCl·LiCl, THF, −60°C, 1 h; (ii) 1.5 equiv $BrCCl_2CCl_2Br$, −78 to −65°C, 3 h

5.2 *Lithiations and Grignard Reactions by Halogen–Metal Exchange*

Halogen–metal exchange has been established as a reliable method for a regioselective metalation of challenging substrates. As such can be regarded, for

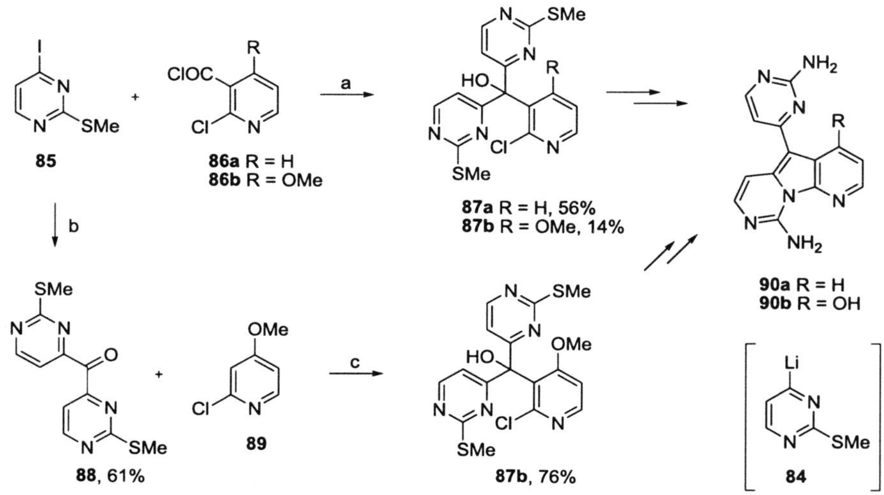

Scheme 20 (**a**) (i) **85**, 1.0 equiv *n*-BuLi, THF, −95°C, 40 min; (ii) 0.34 equiv **86**, −95°C, 3.5 h; (iii) MeOH; (**b**) (i) **85**, 1.0 equiv *n*-BuLi, THF, −97°C, 50 min; (ii) 0.5 equiv $(EtO)_2CO$, −97°C to rt; (iii) aq NH_4Cl; (**c**) (i) **89**, 1.0 equiv *n*-BuLi, THF, −90 to −78°C, 80 min; (ii) 0.93 equiv **88**, −90 to −78°C, 3.5 h; (iii) MeOH, aq NH_4Cl

instance, a generation of 4-lithio-2-methylthiopyrimidine **84** species, readily decomposing at temperatures above −95°C. Morris et al. [55] employed iodine–lithium permutation of 4-iodo-2-methylthiopyrimidine **85** in total syntheses of a marine alkaloid Variolin B **90b** and its analogue Deoxyvariolin B **90a** (Scheme 20).

Addition of *n*-BuLi to pyrimidine **85** (conditions a) through a special glasware set[6] secured efficient precooling of the reagent solution and suppressed decomposition of the lithiated intermediate. The Deoxyvariolin B building block **87a**, as a result of double-addition, was isolated in a good yield of 56%. Synthesis of **87a** was as well attempted under Barbier-type conditions,[7] however, the outcome was less satisfactory (40% yield). The analogous pathway to **87b** (conditions a) proved to be troublesome, and the yield dropped to unacceptable 14%. The authors took advantage of the product symmetry and proposed a two-step alternative (conditions b, c). The ketone **88** was readily available by reaction of lithiated pyrimidine **84** with diethyl carbonate. Thanks to the directing effects of the chlorine and the methoxy group, pyrimidine **89** was lithiated in position 3 and upon addition to **88** provided the Variolin B building block **87b** in a yield of 76%.

Interestingly, pyrimidine **85** can be metalated and further functionalized under significantly less stringent conditions using lithium tri-*n*-butylmagnesate

[6] The *n*-BuLi solution was precooled to −95°C when added through a spiral port. The glassware was first described by Suzuki and Noyori [56].

[7] The *n*-BuLi solution was carefully added to a mixture of **85** and **86a** in THF, at −95°C.

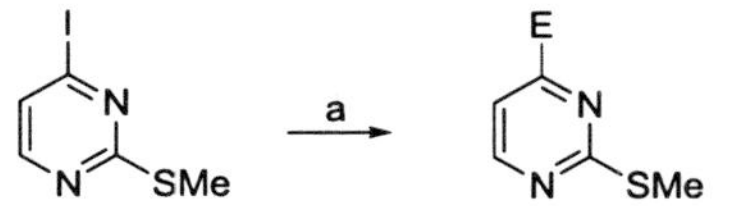

85

91a E = *p*-MeOPhCH(OH), 86%
91b E = MeCH(OH), 79%
91c E = PhC(OH)Ph, 78%
91d E = PhS, 85%

Scheme 21 (**a**) (i) 0.35 equiv *n*-Bu_3MgLi, THF, −10°C, 2.5 h; (ii) E+ (*p*-MeOPhCHO, MeCHO, PhCOPh and PhSSPh, respectively), −10°C to rt, 18 h; (iii) H_2O

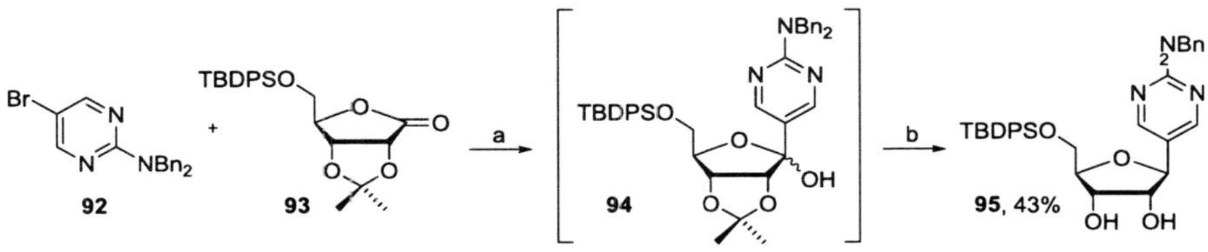

Scheme 22 (**a**) (i) **92**, 1.2 equiv *n*-BuLi, THF, −78°C, 1 h; (ii) 0.83 equiv **93**, −78°C to rt, 5 h; (iii) H_2O, work-up; (**b**) (i) 2.5 equiv Et_3SiH, 2.5 equiv $BF_3{\cdot}Et_2O$, −78°C to rt, overnight; (ii) satd $NaHCO_3$

and *n*-Bu_3MgLi [57]. Prior to metalation, *n*-Bu_3MgLi was readily prepared in situ by reacting 2 equiv *n*-BuLi with *n*-BuMgCl at −10°C in THF. Metalated pyrimidine smoothly reacted with four electrophiles and afforded products **91a–d** in good yields (Scheme 21).

For not quite obvious reasons, metalations of *N*-protected aminopyrimidines are rarely reported and a lithiation of *N*,*N*-dibenzyl-2-amino-5-bromopyrimidine **92** constitutes the only recent example [58]. Strictly considered, a pyrimidine bearing an amino group does not fall into scope of this section. Nevertheless, based on the structural features of **92**, discussing its chemistry here may be considered to be appropriate. Upon bromide–lithium exchange of **92** with *n*-BuLi at −78°C, the lithiated intermediate underwent addition to lactone **93**. The crude product **94** was consumed in a following reduction–deprotection sequence, providing C-nucleoside **95** in an overall yield of 43% (Scheme 22). On that account, the efficiency of initial metalation-addition steps remains unclear.

Halogenated 2,4-dialkoxypyrimidines have proven to be a useful tool in syntheses of complex targets containing a uracil subunit (Scheme 23). A bromine–lithium permutation at 2,4-bis(benzyloxy)-6-bromopyrimidine **96** by the action of *n*-BuLi and a subsequent formylation furnished **97**. This served as a building block in synthesis of the putative structure of 7-deoxycylindrospermopsin **98**, an analogue of hepatotoxic metabolites of the cyanobacterium *Cylindrospermopsis raciborskii* [59]. In analogy, 5-bromo-2,4-dimethoxypyrimidine **99** was upon lithiation added to lactone **100** giving **101** as a mixture of anomers in an α/β ratio of 1:9. Both of them were incorporated into structure of a potent fungicide malayamycin A **102** [60].

In a study published by Koszytkowska-Stawińska et al. [61], pyrimidine **99** was lithiated and magnesiated in parallel with the aim to compare the efficiency of both metalated reagents **103** and **105** in addition to reactions with a nitrone **104** (Table 10).

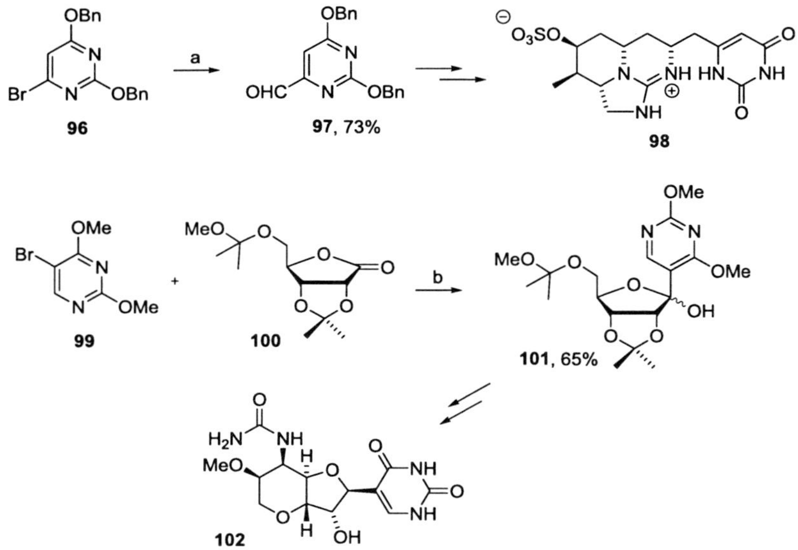

Scheme 23 (**a**) (i) 1.3 equiv *n*-BuLi, Et_2O, −100°C, 20 min; (ii) 5 equiv DMF, −100°C to reflux; (iii) 10% HCl; (**b**) (i) **99**, 2.1 equiv *t*-BuLi, THF, −78°C, 30 min; (ii) 0.77 equiv **100**, −78°C, 1.5 h; (iii) brine

Due to reported additions to ketonitrones, metalations with *n*-BuLi were not attempted. Pyrimidine **99** was successfully lithiated by the action of either 1.2 equiv or 2 equiv of *t*-BuLi (entries 1 and 2, respectively) and upon addition to nitrone **104** oxidized with $Cu(OAc)_2/NH_3(aq)$. However, only the latter conditions (entry 2) led to the desired nitrone **106**, in a low yield of 17%. The accompanying dimer **107** presumably arises from deprotonative α-lithiation of the nitrone **104**, either by **103** or residual *t*-BuLi, followed by a cannibalistic addition to another nitrone **104**, and the oxidation. On the other hand, in the experiment with magnesiated pyrimidine **105** (entry 3), formation of **107** was completely suppressed and nitrone **106** was isolated in a somewhat better, though still unsatisfactory yield (36%).

Halogen–magnesium permutations pose a powerful tool in synthesis of 4,5-difunctionalized-2,6-dimethoxypyrimidines **65**, **112a–f**, and **113a–e**, which are easily convertible to the corresponding valuable uracil derivatives (Table 11) [62].

In the first step, starting from either 5-bromo-4-chloro or 4,5-dibromosubstituted pyrimidines **108** and **109**, respectively, magnesiation occurred selectively at C5. Subsequent reactions of the Grignard intermediates **110** (entries 1–7) or **111** (entries 8–13) with various electrophiles afforded the target pyrimidines in excellent yields. The method permits a successive introduction of two different electrophiles at C5 and C4 and was successfully executed even in a "one-pot" version (two examples). A considerable potential of stepwise magnesiations is apparent from the synthesis of HIV-1 inhibitor Emivirine **119**, described in the same paper (Scheme 24, steps a and c).

Table 10 Differences in reactivity of pyrimidines **103** and **105** with nitrone **104**

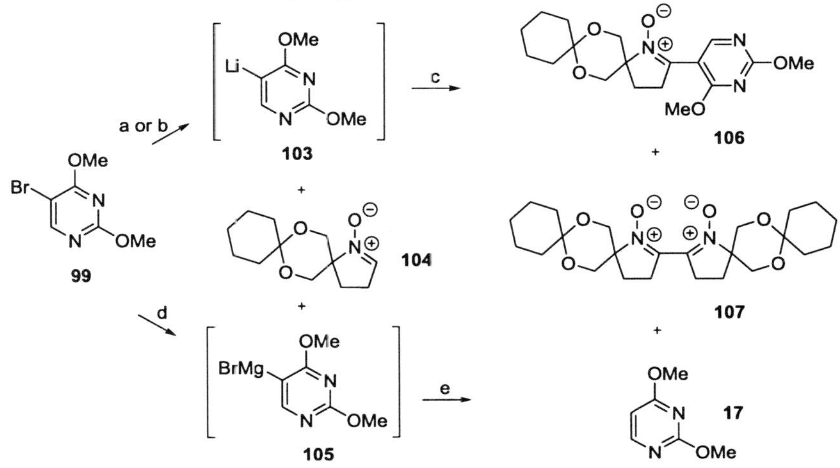

(a) 1.2 equiv *t*-BuLi, THF, −78°C, 30 min; (b) 2 equiv *t*-BuLi, THF, −78°C, 30 min; (c) (i) 0.89 equiv **104**, −78°C to rt, 3 h; (ii) a work-up; (iii) $Cu(OAc)_2$, NH_3 (aq), 1,4-dioxane, CH_2Cl_2, rt, 1 day; (d) 3 equiv Mg, 1.5 equiv $Br(CH_2)_2Br$, THF, reflux, 3 h; (e) (i) **104**, −10°C, 2 h; (ii) a work-up; (iii) $Cu(OAc)_2$, NH_3 (aq), 1,4-dioxane, CH_2Cl_2, rt, 1 day

Entry	Conditions	**106/107/17** (yield in %)
1	a + c	-/34/71
2	b + c	17/39/70
3	d + e	36/-/39

McCaleb [38] and Soth [63] reported closely related examples of halogen–magnesium exchange at 5-bromo-2,4-bis-(2,4-difluorophenoxypyrimidine), performed with *i*-PrMgCl. Thereafter the metalated intermediate was reacted with either acid chloride or CO_2, to furnish the corresponding ketone and carboxylic acid, respectively.

Interesting synthetic applications can arise from a so far little explored area of transition-metal-catalyzed cross-couplings of magnesiated pyrimidines. In the synthesis of antibiotic Trimethoprim, Kofink, and Knochel [64] employed a copper-catalyzed coupling of magnesiated pyrimidine **120** with phosphate **121** as the key step (Scheme 25).

Triethyl phosphite $P(OEt)_3$ is presumed to stabilize the intermediate arylcopper species and along with tetrabutylammonium iodide were essential for the coupling reaction (conditions a). The protocol was expanded to various magnesiated aryl and indole derivatives.

Table 11 Functionalization of pyrimidines **108** and **109** at C5 via Br/Mg exchange

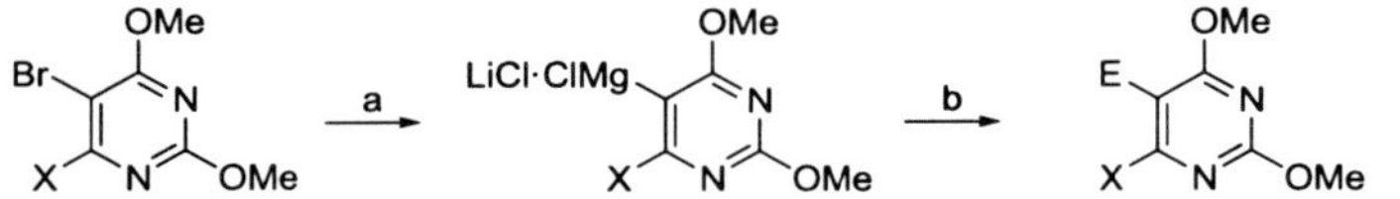

108 X = Cl
109 X = Br

110 X = Cl
111 X = Br

65, **112a-f** X = Cl
113a-e X = Br

(**a**) 1.05 equiv *i*-PrMgCl·LiCl, THF, rt, 15 min; (**b**) (i) E+; (ii) aq NH_4Cl

Entry	Subst	E+	Product	Yield (%)
1	**108**	PhCHO	**65**, E = PhCH(OH)	91
2	**108**	*o*-MeOPhCHO	**112a**, E = *o*-MeOPhCH(OH)	83
3	**108**	PhCOCl	**112b**, E = PhCO	86
4	**108**	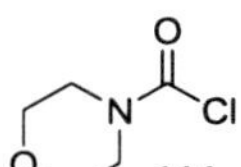 **114**	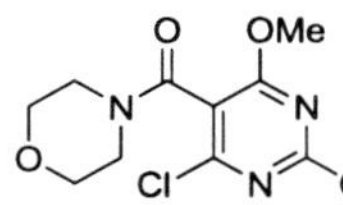 **112c**	85
5	**108**	TsCN	**112d**, E = CN	89
6	**108**	$NCCO_2Et$	**112e**, E = CO_2Et	87
7	**108**	BnBr	**112f**, E = Bn	75
8	**109**	PhCHO	**113a**, E = PhCH(OH)	95
9	**109**	$NCCO_2Et$	**113b**, E = CO_2Et	81
10	**109**	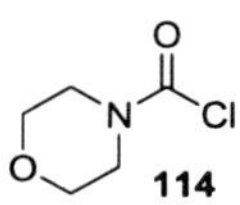 **114**	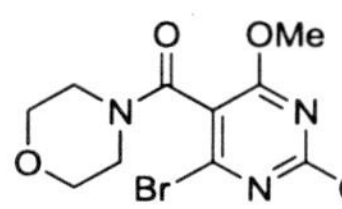 **113c**	95
12	**109**	TMSCl	**113d**, E = TMS	91
13	**109**	Allyl bromide	**113e**, E = allyl	91

In a quest for novel carbocyclic C-nucleosides, Hřebabecký et al. [47] studied cross-coupling reactions of metallated pyrimidines with allyl chlorides (Scheme 26).

The attempts to employ lithium–tin or lithium–zinc exchange were unsuccessful, as well as coupling with an allyl acetate instead of allyl chlorides (**128**, **129**). After a switch to Grignard reagents [62, 65] and transmetalation with copper under catalytic conditions [66],[8] protected uracil (**130**, **132**) and thymine (**131**, **133**) derivatives were furnished in good yields. Allylic rearrangement leading to isomeric products was not observed. Comparable results were achievable with allyl iodide related to **129**; however, it should be avoided due to its problematic stability.

[8] Attempted palladium- or nickel-catalyzed couplings failed.

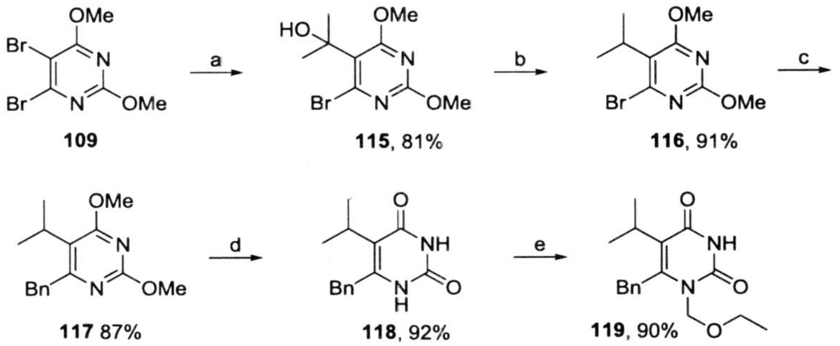

Scheme 24 (**a**) (i) 1.05 equiv *i*-PrMgCl·LiCl, THF, rt, 15 min; (ii) 1.0 equiv $LaCl_3$·LiCl, 1.0 equiv acetone, 0°C, 4 h; (iii) aq NH_4Cl; (**b**) (i) 3 equiv Et_3SiH, 3.5 equiv TFA, CH_2Cl_2, rt, 12 h; (ii) H_2/PtO_2, 1 bar, EtOH, 30 min; (**c**) (i) 1.05 equiv *i*-PrMgCl·LiCl, THF, rt, 5 h; (ii) 2 equiv BnBr, 20 h; (iii) aq NH_4Cl; (**d**) concd HCl, MeOH, reflux, 4 h; (**e**) (i) 3.5 equiv *N*,*O*-bis(trimethylsilyl) acetamide, MeCN, rt, 10 min; (ii) 1.07 equiv TMS-triflate, 2 equiv $CH_2(OEt)_2$, MeCN, −45°C to rt, 3.5 h; (iii) aq $NaHCO_3$

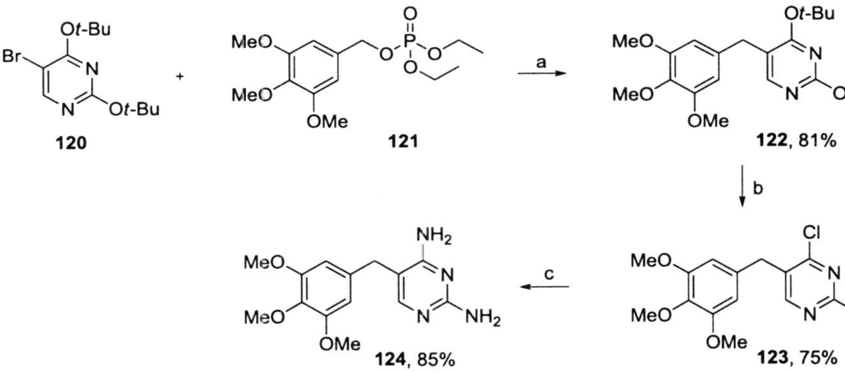

Scheme 25 (**a**) (i) **120**, 1.07 equiv *i*-PrMgCl·LiCl, THF, 0°C, 4 h; (ii) 10 mol% CuBr, 20 mol% P$(OEt)_3$; (iii) 0.67 equiv **121**, 10 mol% TBAI, DME, 60°C, 1 h; (**b**) (i) concd HCl, MeOH, rt, 15 min; (ii) $POCl_3$, reflux, 1 h; (**c**) (i) 7N NH_3/MeOH, autoclave, 175°C, 7 h

The palladium-catalyzed cross-coupling of organomagnesium reagents and unsaturated halides (Kumada coupling) [67, 68] avoids transmetalation steps and, thus, represents an attractive tool of coupling chemistry. Knochel and Manolikakes [69] have recently described coupling of 4-bromobenzonitrile **136** with pyrimidylmagnesium reagent **135**, prepared from pyrimidine **134** by I/Mg exchange (Scheme 27).

Interestingly, an accelerating effect of *i*-PrI, a sideproduct of I/Mg exchange, was observed and the desired diaryl product **137** could be cleanly prepared within several minutes at ambient temperature. A similar rate acceleration was found with various alkyl iodides, and after a brief investigation a radical mechanism was proposed (Scheme 28). In the initiation step, the palladium catalyst [LPd^0] reacts with R-I by a radical path, affording the Pd^I intermediate [LPdX]. This combines with Ar^1-Br generating an aryl radical Ar^1, which is trapped by [$LPdX_2$] to form

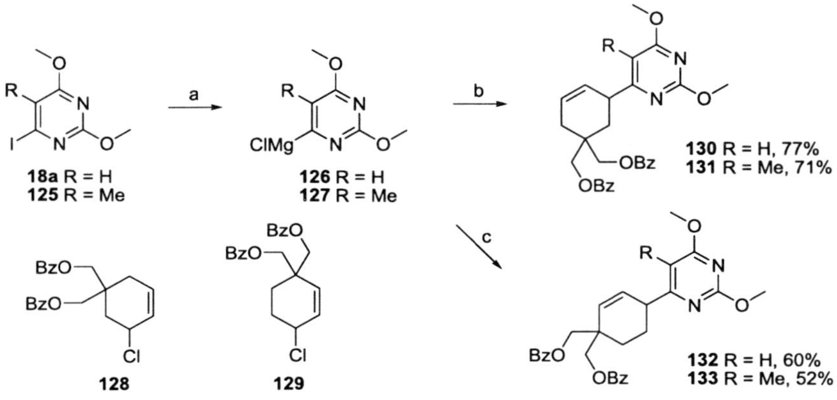

Scheme 26 (**a**) *i*-PrMgCl·LiCl, THF, −25°C; (**b**) (i) 13 mol% CuI·2LiCl, **128**, THF, −25°C, 36 h; (ii) aq NH_4Cl; (**c**) (i) 13 mol% CuI·2LiCl, **129**, THF, −25°C, 36 h; (ii) aq NH_4Cl

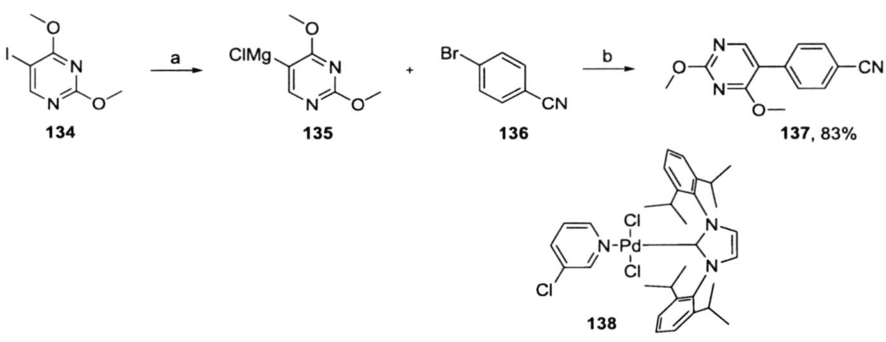

Scheme 27 (**a**) *i*-PrMgCl·LiCl, THF, −20°C, 1 h; **b** 3 mol% PEPPSI (**138**), THF, 25°C, 10 min

Pd^{III} intermediate **139**. Subsequent ligand exchange with Ar^2-MgX and reductive elimination provide the coupling product Ar^1-Ar^2 and the regenerated palladium radical [LPdX].

The method is applicable for a wide range of arylmagnesium reagents and aryl bromides.

6 Uracil, Thymine, and Cytosine Derivatives

6.1 C-Deprotonative Lithiations

Uracil (**140**) and cytosine (**142**) (Fig. 2) represent fundamental subunits of ribonucleic acid (RNA), while thymine (**141**) and cytosine (**142**) are found in deoxyribonucleic acid (DNA).

Scheme 28 Suggested radical mechanism of the Kumada cross-coupling

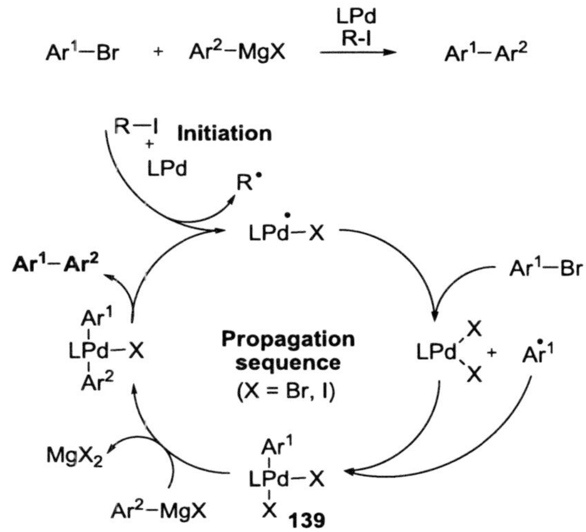

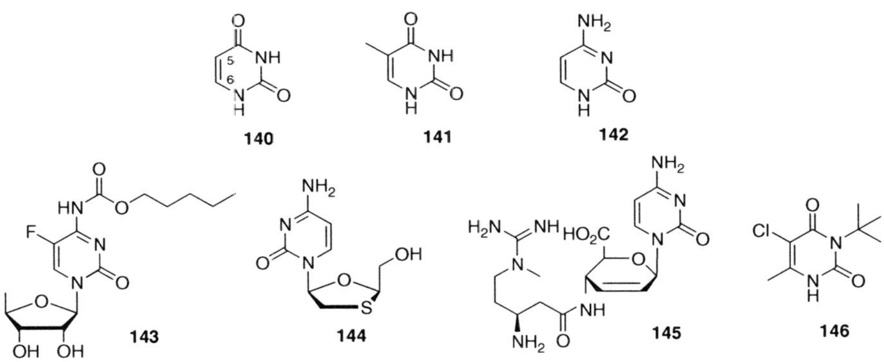

Fig. 2 Uracil (**140**), thymine (**141**), cytosine (**142**), capecitabine (**143**), lamivudine (**144**), blasticidin S (**145**), terbacil (**146**)

As such, therapeutic potential of numerous derivatives of uracil, thymine, and cytosine has been thoroughly investigated, particularly in antiviral and anticancer activity. Capecitabine (**143**) (breast and colorectal cancer) [70], lamivudine (**144**) (HIV and hepatitis B) [71], blasticidin S (**145**) [72] (fungicide), and terbacil (**146**) (herbicide), to mention just some of them, are illustrative representatives of the successful candidates, currently broadly used in medical practice and agriculture.

It is not of surprise that a lot of effort has been put into finding reliable synthetic methods enabling selective structural modifications of the pyrimidine bases. C-Lithiations are particularly suitable for introduction of substituents at C-5 and C-6.[9] In general, deprotonation at C-6 is thermodynamically favored. In uridine

[9] Occasionally, deprotonative lithiations are used for synthesis of *N*-alkylated or *N*,*N*-dialkylated pyrimidine bases. Illustrative examples and a discussion of chemoselectivity can be found in [73, 74].

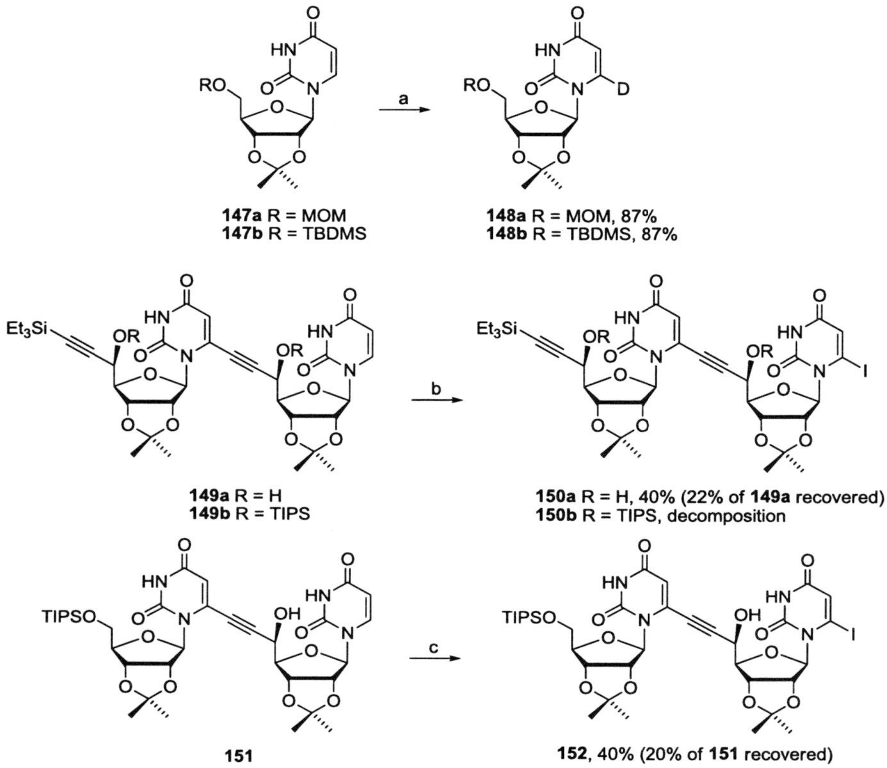

Scheme 29 (**a**) (i) 3 equiv LDA, THF, −70°C, 1 h; (ii) CD_3OD; (**b**) (i) 11.4 equiv LDA, THF, −78°C, 3 h; (ii) 11.9 equiv NIS, −78°C, 2 h; (**c**) (i) 12 equiv LDA, THF, −78°C, 2.5 h; (ii) 12 equiv NIS, −78°C, 1.5 h

[75, 76] and cytidine [77] series, C-6 can be selectively deprotonated under thermodynamic control (LDA). The assistance of chelating groups at C-5′ does not seem to play a substantial role (identical yields for **148a** and **148b**, Scheme 29) [78]. However, an example of a more complex substrate **149a** was reported, when unprotected propargylic OH group was essential for a moderate conversion and TIPS-*O*(5′) protected analogue **149b** decomposed under identical reaction conditions [79].

A selective deprotonation at C-5 usually requires a stronger base (*s*-BuLi) and a careful work at lower temperatures (Scheme 30) [80, 81]. Typically, C-5′ bears a nonchelating silyl group and sterical factors at C-2′ have to be attentively considered.

Over the last years, deprotonative lithiation followed by nucleophilic addition or substitution was routinely used for introduction of various substituents at C-6. Table 12 contains an overview of the reported structures and related reaction conditions.

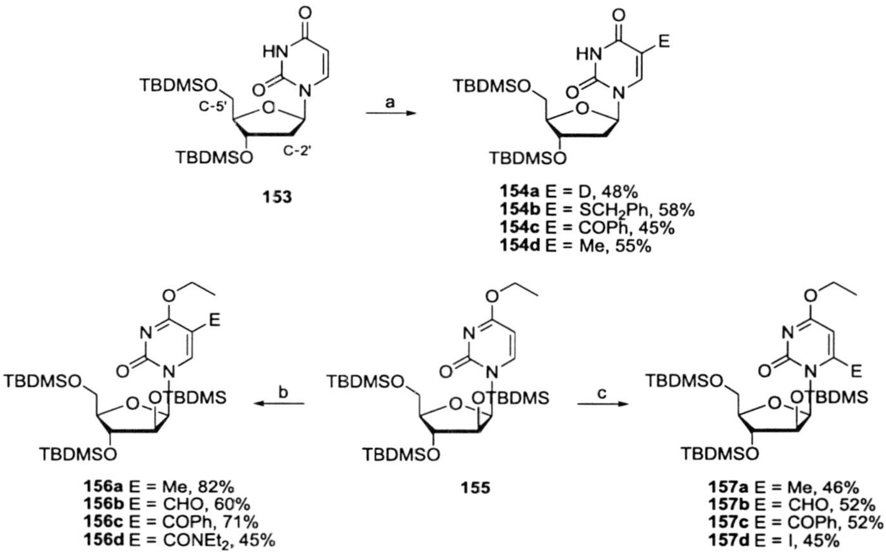

Scheme 30 (**a**) (i) 2.2 equiv *s*-BuLi, 2.2 equiv TMEDA, THF, −78°C, 30 min; (ii) E+, then warming to rt [80]; (**b**) (i) 4 equiv LDA, THF, −80°C, 5 min; (ii) E+ [77]; (**c**) (i) 1 equiv LDA, THF, −40°C, 5 min; (ii) E+

Miyasaka et al. [75] observed that deprotonative lithiation of uridines at C-6 followed by methylation can be accompanied by undesired α-methylation of the newly attached substituent. Bello et al. [86] turned the fact to good account and smoothly ethylated substrate **163** via a two-stage methylation (Scheme 31).

A monomethylation was not observed. The remarkable chemoselectivity was not discussed; however, a lateral lithiation with assistance of the neighboring fluorine (**190**) is likely behind the facile alkylation. In the absence of fluorine, the second methylation is a bit more tricky and under similar conditions (2.5 equiv LDA followed by 3.3 equiv MeI at −78°C) a mixture of 6-methyluridine **160a** (44%) and 6-ethyluridine **191** (17%) was isolated [11]. Double alkylation can be suppressed by a slow addition of the preformed dianion **192** to a solution of MeI (reverse-addition mode), furnishing 72% of uridine **160a**.

Utilization of a stepwise alkylation turned to be essential for a synthesis of 6-ω-alkenyl uridines **194a–b** (Scheme 32) [11].

Direct alkylation of lithiated 2′,3′-*O*-isopropylideneuridine of a putative structure **193** with 4-bromobut-1-ene failed to afford any C-6 alkylated product, in spite of literature precedent reporting a successful alkylation with *n*-butylbromide [75]. Starting from 6-methyluridine **160a**, the role of base was investigated in detail (Table 13).

A formation of a dilithiated reagent **198** is presumed, with negative charges being delocalized over the conjugated system. The resonance structures **199** and **200** give origin to regioisomers **195a–b** and **194a–b**, respectively. With respect to yields and regioselectivity, a performance of LDA or LTMP (entries 1–4) appears to be superior over LiHMDS or *s*-BuLi/TMEDA (entries 5–6). Moreover, LiHMDS

Table 12 A summary of recent protocols in deprotonative lithiations of pyrimidine bases

Substrate	Conditions	Product
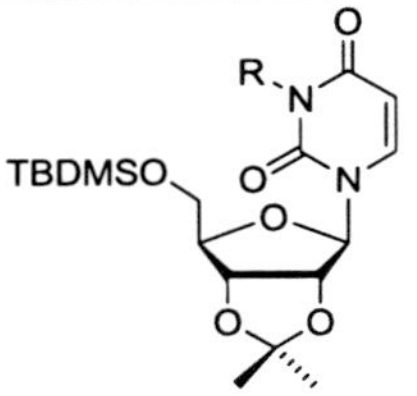 **147b** R = H **158a** R = Me **158b** R = Bn **158c** R = PMB	**159a–d** [82]: (i) 2 equiv LDA, THF, −78°C, 1 h (ii) 2 equiv I_2, −78°C, 3 h; **159a** 55%, **159b** 45%, **159c** 50%, **159d** 43% **159a** [83]: (i) **147b**, 2.2 equiv LDA, THF, −78°C, 1 h (ii) 1 equiv I_2, −78°C, 5 h; 68%	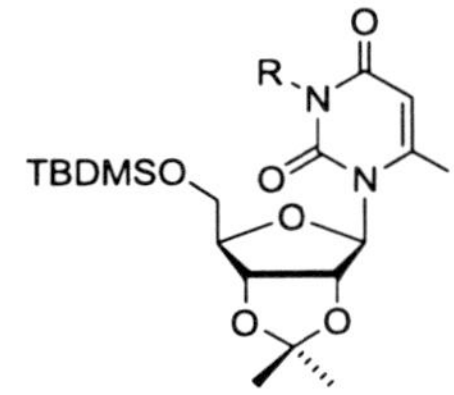 **159a** R = H **159b** R = Me **159c** R = Bn **159d** R = PMB
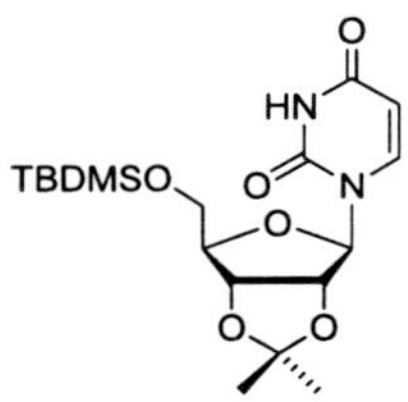 **147b**	**160a** [84]: (i) 1.9 equiv LDA, THF, −78°C, 1 h (ii) 1 equiv MeI, −78°C, 5 h; unspecified yield **160b** [85]: (i) 6.2 equiv LDA, THF, −76°C, 2.5 h (ii) DMF, −76°C, 2.5 h; 85%	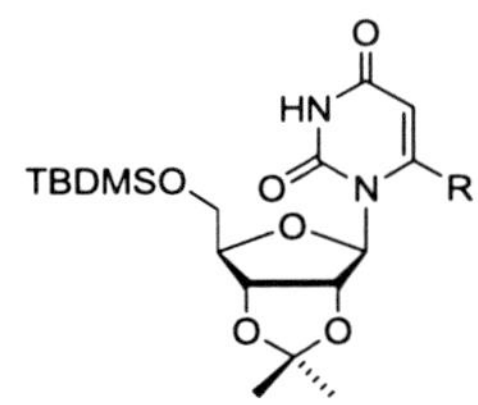 **160a** R = Me **160b** R = CH_2OH
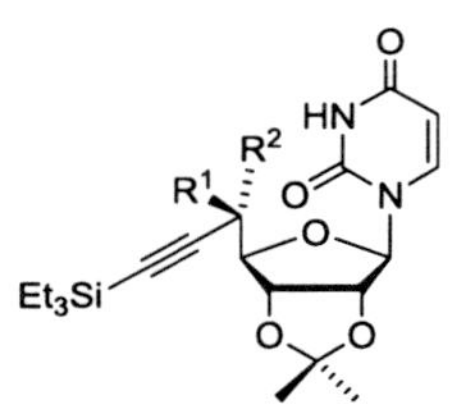 **161a** R^1 = OH, R^2 = H **161b** R^1 = H, R^2 = OH	**162a,b** [85]: (i) 8 equiv LDA, THF, −76°C, 3 h (ii) DMF, −76°C, 2.5 h (iii) AcOH, EtOH (iv) $NaBH_4$, rt; **162a** 60%, **162b** 54%	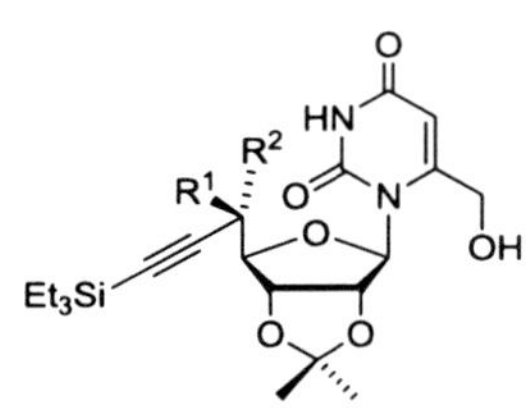 **162a** R^1 = OH, R^2 = H **162b** R^1 = H, R^2 = OH
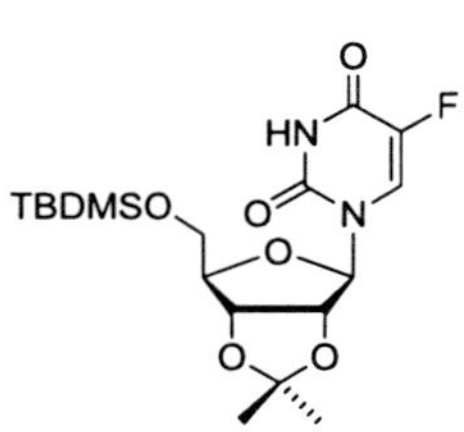 **163**	**164a** [86]: (i) 3 equiv LDA, THF, −78°C, 1 h (ii) 1 equiv I_2, −78°C, 5 h; 95% **164b** [86]: (i) 5 equiv LDA, THF, −78°C, 1 h (ii) 1.5 equiv HCO_2Me, −78°C, 5 h; 70%	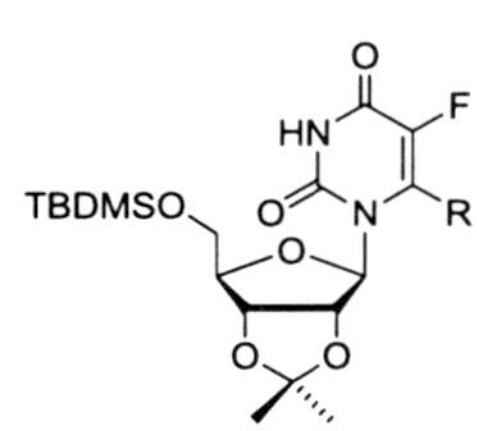 **164a** R = I **164b** R = CHO

(continued)

Table 12 (continued)

Substrate	Conditions	Product
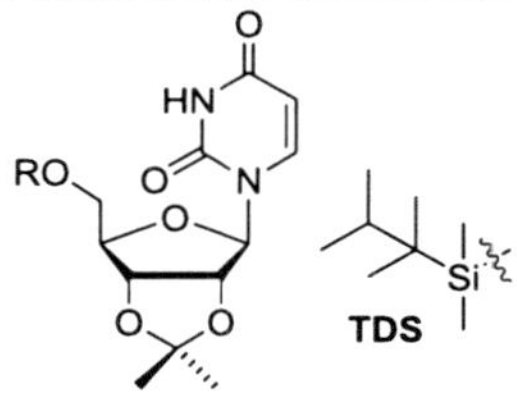 **165a** R = TDS **165b** R = TIPS	**166a** [87]: (i) **165a**, 5 equiv LDA, THF, −76°C, 1.5 h (ii) 25 equiv DMF, −76°C, 1 h, then warming to rt; 98% **166b** [88] (i) **165a**, 5 equiv LDA, THF, −70°C, 1 h (ii) 25 equiv DMF, −70°C, 1 h (iii) AcOH, EtOH (iv) $NaBH_4$, rt; 80% **166c** [89, 90] (i) **165b**, 5 equiv LDA, THF, −78°C, 1 h (ii) 25 equiv DMF, −78°C, 2.5 h (iii) AcOH, EtOH (iv) $NaBH_4$, rt; 70%	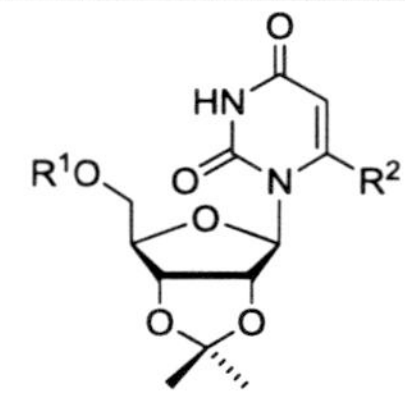 **166a** R^1 = TDS, R^2 = CHO **166b** R^1 = TDS, R^2 = CH_2OH **166c** R^1 = TIPS, R^2 = CH_2OH
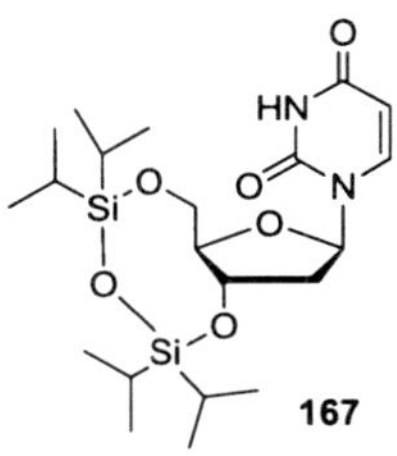 **167**	**168** [91]: (i) 5 equiv LDA, THF, −78°C, 1 h (ii) 1.5 equiv MeI, −78°C, 2 h; 58%	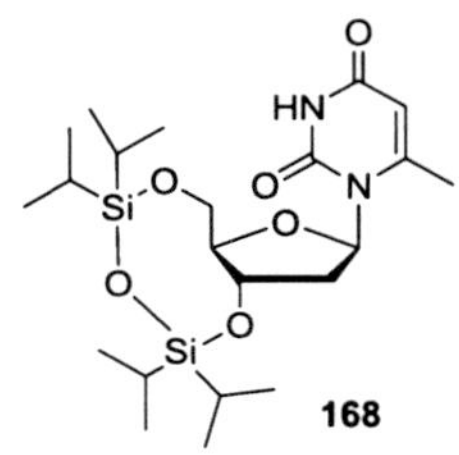 **168**
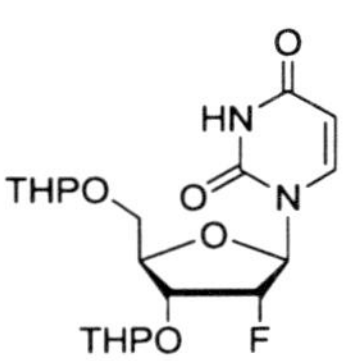 **169**	**170a** [92]: (i) 3.5 equiv LDA, THF, −78°C, 2 h (ii) 2 equiv I_2, −78°C, 6 h; 45% **170b** [92]: (i) 3.8 equiv LDA, THF, −78°C (ii) 4.2 equiv $ClCO_2Et$, −78°C, 2 days; 30%	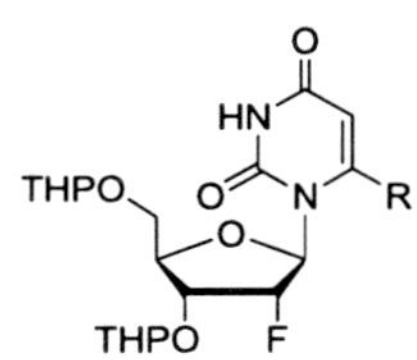 **170a** R = I **170b** R = CO_2Et
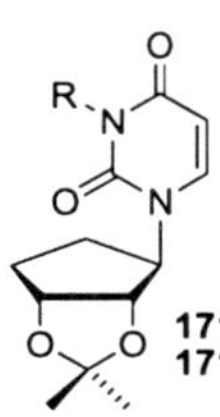 **171a** R = H **171b** R = PMB	**172a** [93, 94]: (i) **171a**, 2.2 equiv LDA, THF, −78°C, 30 min (ii) 2.2 equiv I_2, −78°C, 2 h; 60% **172b** [94]: (i) **171b**, 2 equiv LDA, THF, −78°C, 30 min (ii) 4 equiv I_2, −78°C, 3 h; 78%[a]	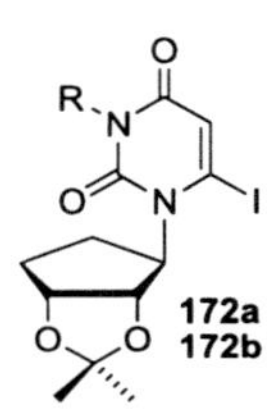 **172a** R = H **172b** R = PMB

(continued)

Table 12 (continued)

Substrate	Conditions	Product
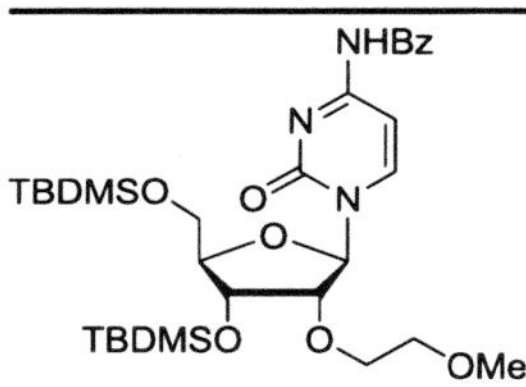 **173**	**174** [95]: (i) 13 equiv LDA, THF, −78°C, 4 h (ii) 26 equiv HCOOMe, −78°C, 3 h, then warming to rt; 4:6 mixture of **173** and **174**, unspecified yield	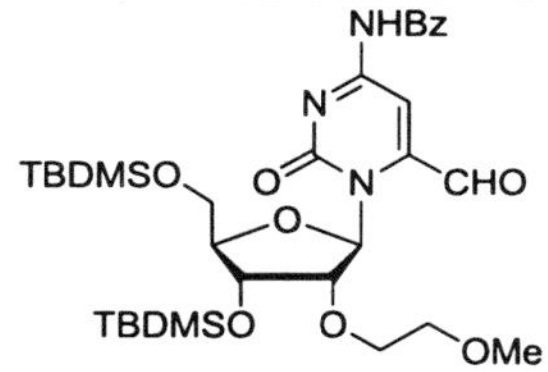 **174**
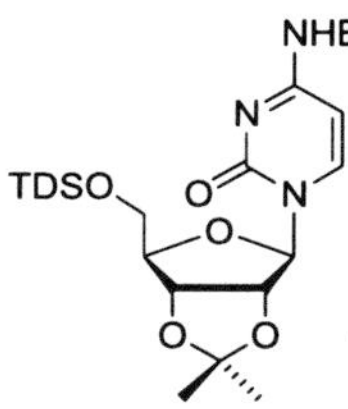 **175**	**176** [96]: (i) 5 equiv LDA, THF, −70°C, 1 h (ii) 10 equiv DMF, −60°C, 1 h, then warming to −20°C (iii) AcOH (iv) $NaBH_4$, EtOH, rt; 86%	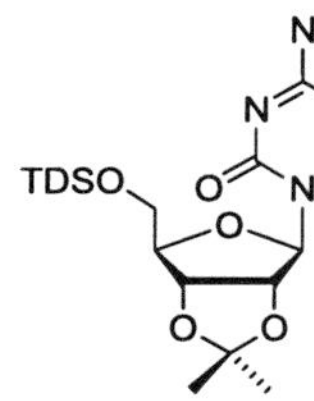 **176**
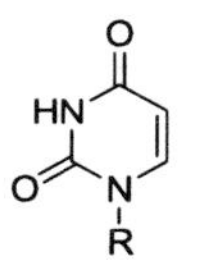 **177a** R = C_8H_{17} **177b** R = C_6H_{13}	**178a** [97]: (i) **177a**, 2.5 equiv LDA, 1.25 equiv *n*-BuLi, THF, −78°C, 1.5 h (ii) 2 equiv I_2, −78°C, 2 h; 61% **178b** [98, 99]: (i) **177b**, 5 equiv LDA, THF, −78°C, 1.5 h (ii) 5 equiv I_2, −78°C, 2 h; 72%	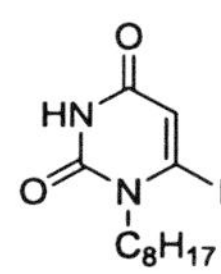 **178a** R = C_8H_{17} **178b** R = C_6H_{13}
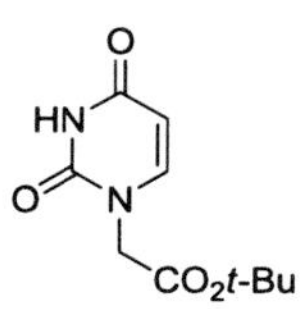 **179**	**180a** [100]: (i) 6 equiv LDA, THF, −76°C, 2 h (ii) 20 equiv DMF, −76°C, 2.5 h, then warming to rt; 37% **180b** [100]: (i) 5 equiv LDA, THF, −70°C, 2 h (ii) 20 equiv DMF, −70°C, 1.5 h (iii) AcOH, EtOH (iv) $NaBH_4$, 0°C; **180b** 45%, **179** 43%[b]	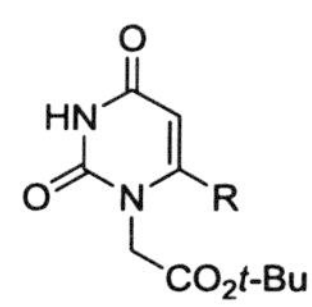 **180a** R = CHO **180b** R = CH_2OH
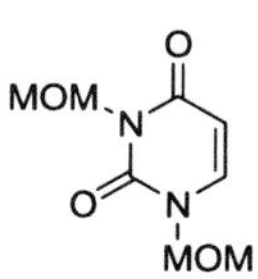 **181**	**182** [101]: (i) 3 equiv LDA, THF, −78°C, 1 h (ii) 4.2 equiv DMF, −78°C, 2.5 h; 66%	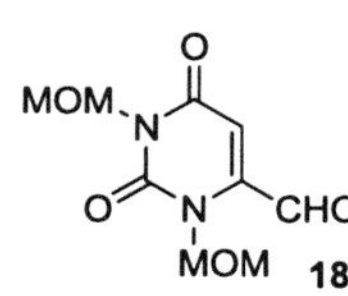 **182**
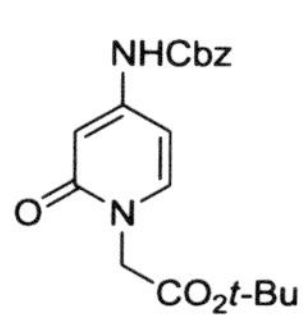 **183**	**184** [100]: (i) 5 equiv LDA, THF, −70°C, 2 h (ii) 20 equiv DMF, −70°C, 2 h, then warming to rt (iii) AcOH, EtOH (iv) $NaBH_4$, rt; **184** 51%, **183** 27%	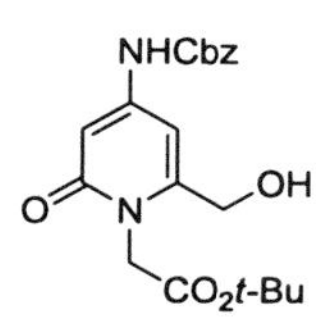 **184**

(continued)

Table 12 (continued)

Substrate	Conditions	Product
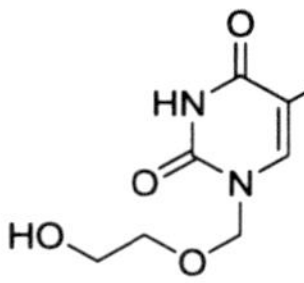 185	**186a,b** [102]: (i) 2.7 equiv LDA, THF, −78°C, 1 h (ii) 2 equiv aldehyde, −78°C, 2 h; **186a** 32%, **186b** 51%	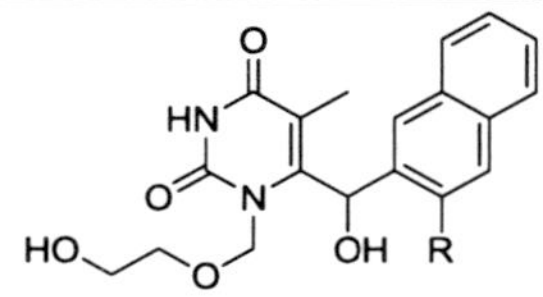 **186a** R = H **186b** R = OMe
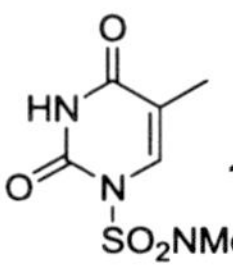 187	**188** [103]: (i) 2.2 equiv *n*-BuLi, THF, −78°C, 1 h (ii) 1.1 equiv aldehyde, −78°C, 1 h, then warming to rt (iii) 4 M HCl, rt, 1 h; 71%	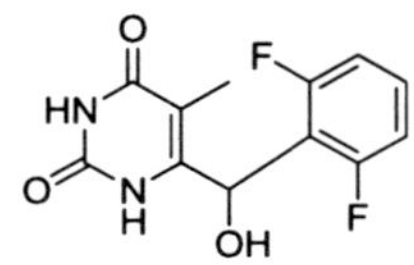 **188**

[a]With Bz-protecting group (R = Bz), iodination occurred at several sites, including Bz-group. On the other hand, substrates bearing Bn- and BOM-protection were successfully iodinated, though yields are not reported

[b]Attempted optimization using LTMP, LHMDS, or TrLi as a base did not provide better yields. As well, replacement of DMF for *N*-methylformanilide or *N*-formylmorpholine did not lead to an improvement

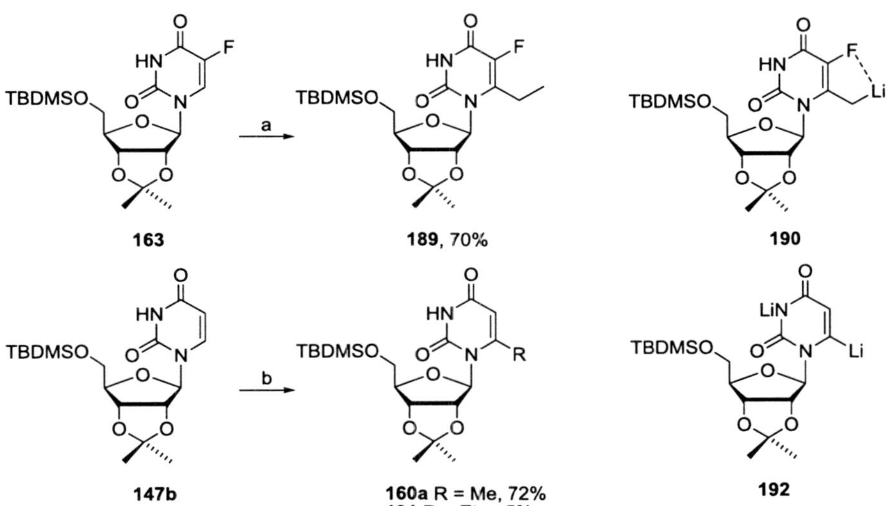

Scheme 31 (**a**) (i) 5 equiv LDA, THF, −78°C, 1 h; (ii) 3 equiv MeI, −78°C, 5 h [86]; (**b**) (i) 2.5 equiv LDA, THF, −78°C, 3 h; (ii) 3.3 equiv MeI, −78°C, 1 h, then warming to rt [11]

did not furnish any amount of uridines **194a** and/or **195a**, and bisallylated products **196**, **197** were obtained.

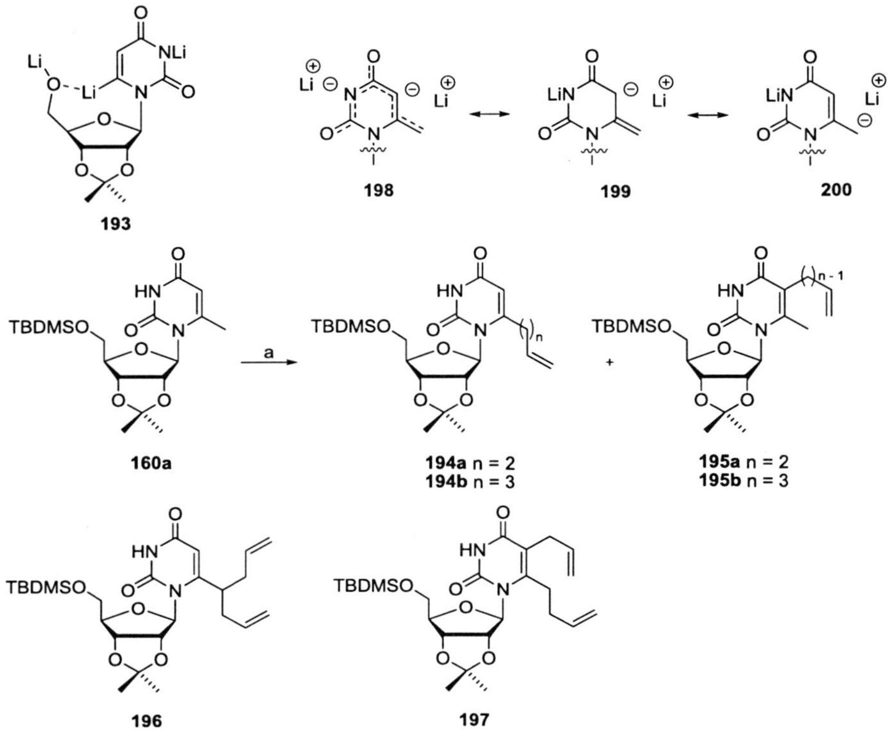

Scheme 32 (**a**) (i) 4 equiv base, THF, −70°C, 1 h; (ii) 8 equiv $Br(CH_2)_{n-1}$-$CH{=}CH_2$, −70°C, 30 min, then warming to rt

Table 13 Product distribution in lateral lithiation of uridine **160a**

Entry	Base	*n*	**194a/195a** (%)	**194b/195b** (%)
1	LDA	2	58/10	–
2	LTMP	2	65/20	–
3	LDA	3	–	44/0
4	LTMP	3	–	56/0
5	LiHMDS	2	0/0	–
6	*s*-BuLi/TMEDA	3	–	38/0

Mechanistically interesting is the reported synthesis of pentafluoropropenyl derivatives of pyrimidine [104]. In the first step, a lithiated pyrimidine **202** adds to hexafluoropropene (Scheme 33). Sequentially, the intermediate carbanion **203** undergoes a fluoride elimination to afford vinylated products **204** as a mixture of *E* and *Z* isomers.

In case of thymine derivative **205**, a competing 6-6′ dimerization was observed and **207** was isolated as a mixture of diastereisomers (dr = 65:35), bearing exclusively *E* configuration across the newly formed double bond. Attempted addition–elimination reactions of substrates **208**, **209** failed and **210** provided

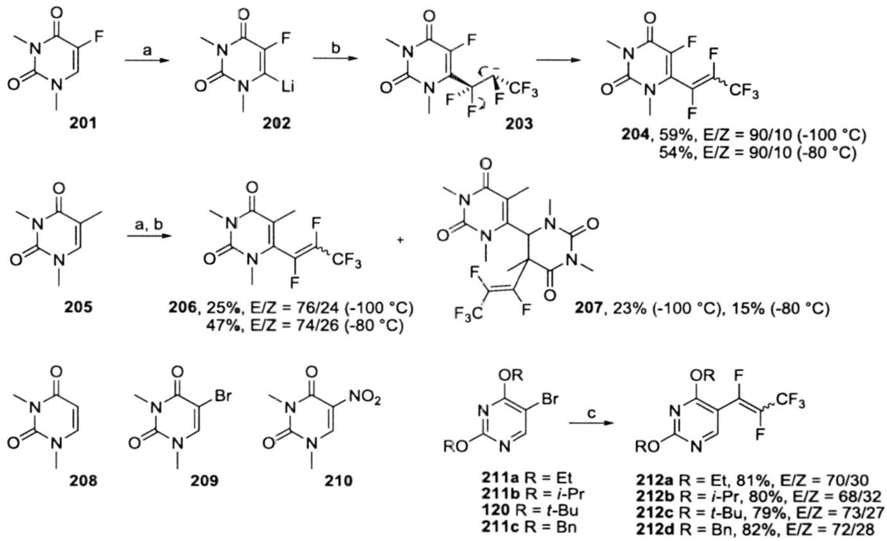

Scheme 33 (**a**) 1.2–2.5 equiv LDA, THF, −100°C or −80°C, 10–60 min; (**b**) 3 equiv hexafluoropropene, −100°C or −80°C, 30 min, then warming to rt; (**c**) (i) 1.5 equiv *n*-BuLi, THF, −100°C, 30 min; (ii) 3 equiv hexafluoropropene, −100°C, 30 min, then warming to rt

only the 6-6′ dimerization product (22%). However, the protocol was very well applicable for 5-bromo-2,4-dialkoxy-pyrimidines **120** and **211a–c** using a halogen–metal exchange reaction.

6.2 Lithiations and Grignard Reactions by Halogen–Metal Exchange

Halogen–metal exchange is typically used for introduction of substituents at the less reactive C-5 position of pyrimidine bases and is a useful alternative to deprotonative lithiations with *s*-BuLi [80]. An important contribution to the field was published by Aso et al. [105]. They observed that attempted reaction of 5-iodo-2′-deoxyuridine **213** with 2-methyl-2-nitrosopropane (MNP) gave mainly product of deiodination **153** (Scheme 34, Method A).

Presumably, imide deprotonation and iodo-lithium exchange were competitive at the early stage of the reaction and a proton scrambling between intermediates **215** and **216** or **215** itself yielded **217**, which was hard to deprotonate at C-5 with *n*-BuLi. Pretreatment of **213** with NaH (Method B) significantly limited formation of **153** and afforded 5-substituted uridines **214**, **154a, d**, and **220a–c** in good yields (Table 14).

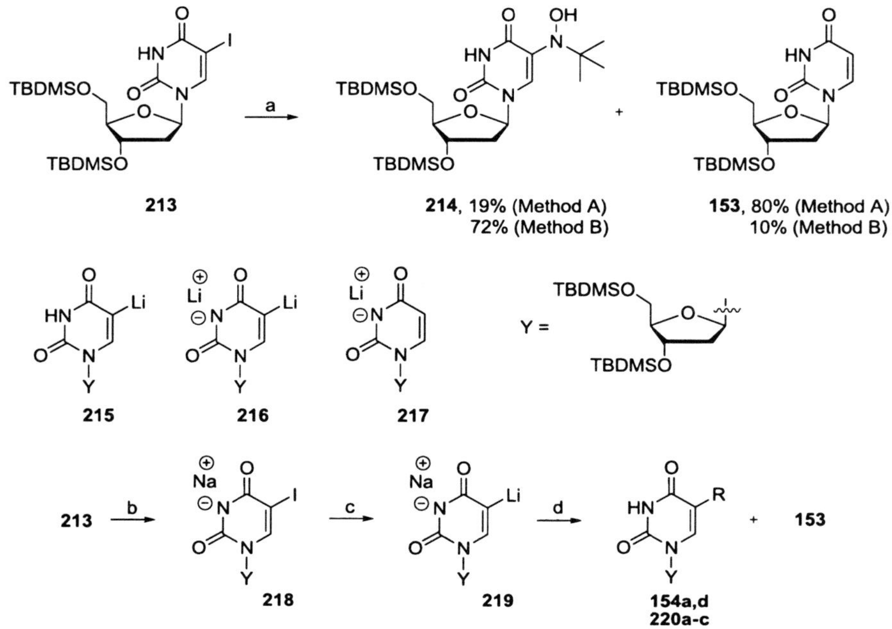

Scheme 34 (**a**) Method A: (i) 2 equiv *n*-BuLi, THF, −78°C; (ii) 6 equiv MNP, −78°C; Method B: (i) 1.5 equiv NaH, THF, 0°C; (ii) 1.2 equiv *n*-BuLi, −78°C; (iii) 6 equiv MNP, −78°C; (**b**) NaH; (**c**) *n*-BuLi; (**d**) E+

Table 14 5-Substituted 2′-deoxyuridines **154a, d** and **220a–c** prepared by Method B

n-BuLi (equiv)	E+	R	Product (%)	**153** (%)
1.2	CD_3OD	D	**154a** (64)	9
1.2	TMSCl	TMS	**220a** (62)	12
1.2	PhCHO	PhCH(OH)	**220b** (79)	0
3.0	MeI	Me	**154d** (81)	7
3.0	MeSSMe	SMe	**220c** (85)	0

Slightly altered conditions were reported to introduce an ethoxycarbonyl group [106]. The scope of the method was successfully expanded to alkylations of 5-bromosubstituted cyclonucleoside **221** (Scheme 35) [107].

The target compounds **222a–b** and **222d–g** were prepared in moderate yields along with unreported varying amounts of dehalogenated uridine **223**. Interestingly, when using PrI, the reaction did not yield any 5-substituted product **222c**. 4-Chlorobenzyl bromide afforded only the product of *N*-alkylation **224**.

Halogen–magnesium exchange reactions of nucleobase derivatives are convenient with respect to relatively high reaction temperatures and variability of reagents (Table 15) [65].

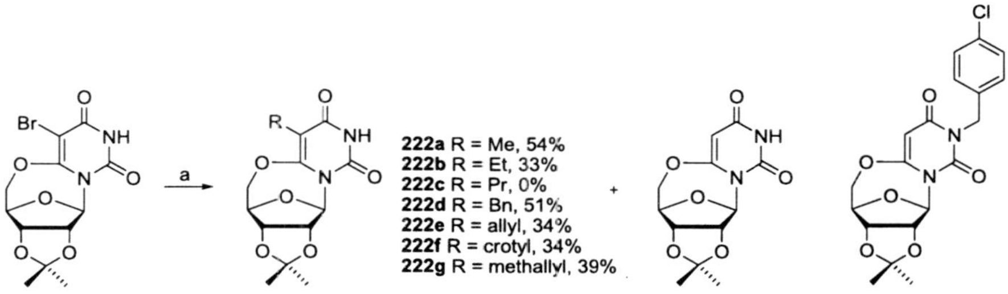

Scheme 35 (**a**) (i) 1 equiv NaH, THF, rt, 10 min; (ii) 3 equiv HMPA (for **222e–f**); (iii) 2 equiv *n*-BuLi, −78°C, 30 min; (iv) 5 equiv RI, −78°C, overnight

In comparison with *i*-PrMgCl, MeMgCl exhibits a low aptitude for halogen–magnesium exchange and a dehalogenative protonation of the substrates **225a–c** via proton scrambling can be avoided (conditions a). The use of LiCl is essential as it both increases the rate of the exchange reaction and improves the solubility. A versatility of the described method was demonstrated in synthesis of precursors of anti-HIV agents, HEPT and Emivirine.

7 Quinazolines

Unlike pyrimidines, the metalation chemistry of quinazolines has been relatively little explored. Smith et al. [108] disclosed that quinazolinones **230a–b**, upon litiation with LDA, could be selectively substituted at C2 (Scheme 36). The more nucleophilic MeLi or *t*-BuLi gave 1,2-addition products.

Introduction of a bulky *t*-Bu group to the C2-position effectively prevents from a nucleophilic attack and quinazolinone **232** can be regioselectively lithiated at C5 when treated with *s*-BuLi and TMEDA (Scheme 37) [109, 110].

When compared to other electrophiles, reaction of lithiated **232** with sterically hindered *t*-BuSS*t*-Bu required higher temperatures (conditions b) and along with **233e**, a product of a substitution at C8 was isolated in a yield of 18%.

Quinazolines bearing two DMGs at C6 and C7 of the benzene ring prefer to undergo deprotonative lithiations at C8, and not C5 (Scheme 38; a recent illustrative example) [111–113].

Interestingly, in case that the starting quinazoline **234** was bearing a fluorine atom on the phenyl substituent (X = F), iodination was observed exclusively at C3′, yielding 30% of the corresponding product.

Unexpectedly, high regioselectivity was observed in lithiations of 5-phenylsulfinyl substituted quinazoline **236** (Scheme 39) [114]. The sulfoxide could assist both in *ortho* (C6) and in *peri* (C4) metalations; however, the latter did not occur. This is surprising with respect to a high acidity of H4.

The attempts to employ aldehydic electrophiles were unsuccessful and gave tarry products. Lithiation experiments in presence of TMSCl revealed a competitive *ortho* metalation on the phenyl ring (Scheme 40).

Table 15 Synthesis of substituted uracil derivatives by means of halogen–magnesium exchange

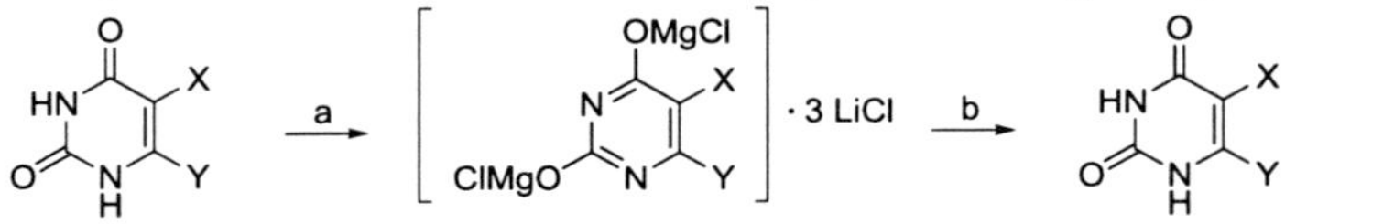

225a X = I, Y = H
225b X = Br, Y = H
225c X = H, Y = I

226a X = I, Y = H
226b X = Br, Y = H
226c X = H, Y = I

227a-g X = E, Y = H
228a-b X = H, Y = E

(**a**) (i) 2 equiv MeMgCl·LiCl, THF, −20°C, 30 min; (ii) 1.2 equiv *i*-PrMgCl·LiCl, −20°C to rt, 1 h; (**b**) E+, −20°C to rt

Entry	Substrate	E+	Product	Yield (%)
1	**225a**	*t*-BuCHO	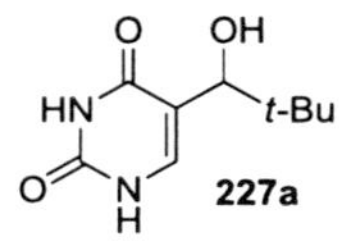 **227a**	77
2[a]	**225b**	*t*-BuCHO	**227a**	57
3[b]	**225c**	*t*-BuCHO	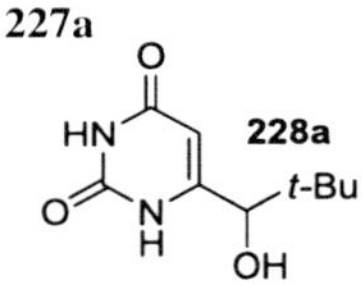 **228a**	69
4	**225a**	PhCHO	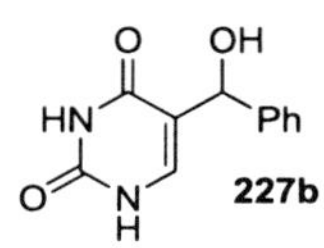 **227b**	78
5	**225a**	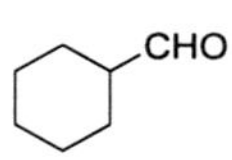	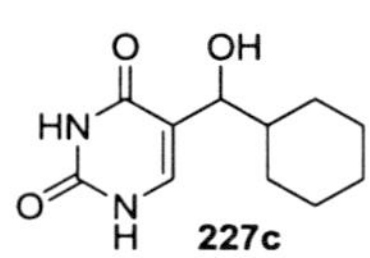 **227c**	70
6	**225a**	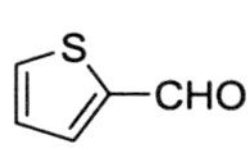	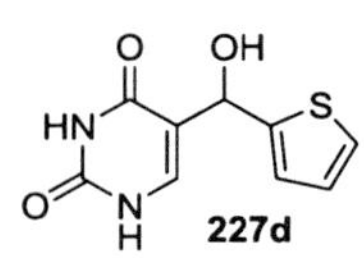 **227d**	55
7[c]	**225a**	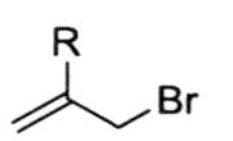 **229a** R = H	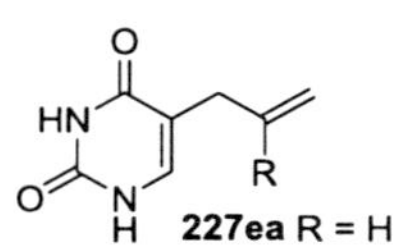 **227ea** R = H	84
8[a,c]	**225b**	**229b** R = Me	**227eb** R = Me	54

(continued)

Table 15 (continued)

Entry	Substrate	E+	Product	Yield (%)
9[b,c]	**225c**	Allyl bromide	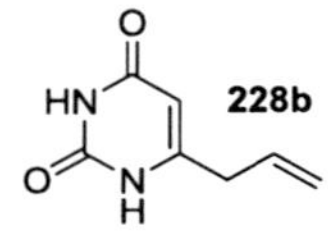 **228b**	64
10	**225a**	TMSCl	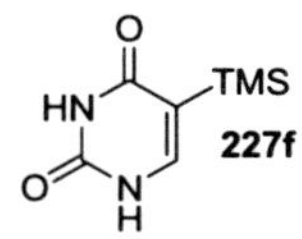 **227f**	72
11	**225a**	$PhSSO_2Ph$	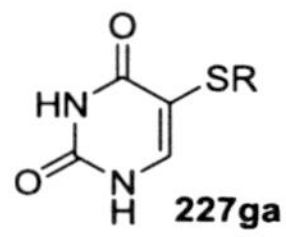 **227ga** R = Ph	77
12	**225a**	$MeSSO_2Me$	**227gb** R = Me	64

[a]1.1 Equiv $(i\text{-Pr})_2MgCl{\cdot}LiCl$, −20°C to rt, 2 h
[b]Step **a** carried out at −25°C
[c]Reaction in the presence of 1 mol% $CuCN{\cdot}2LiCl$ [22]

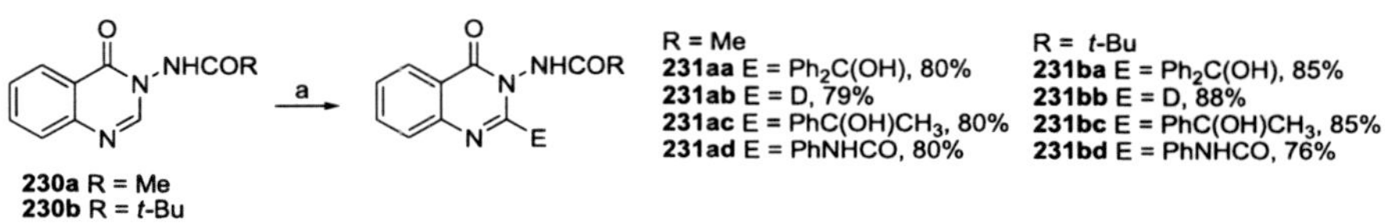

Scheme 36 (**a**) (i) 2.2 equiv LDA, THF, −78°C, 1 h; (ii) 1.1 equiv E+, −78°C, 4 h; (iii) aq NH_4Cl, rt

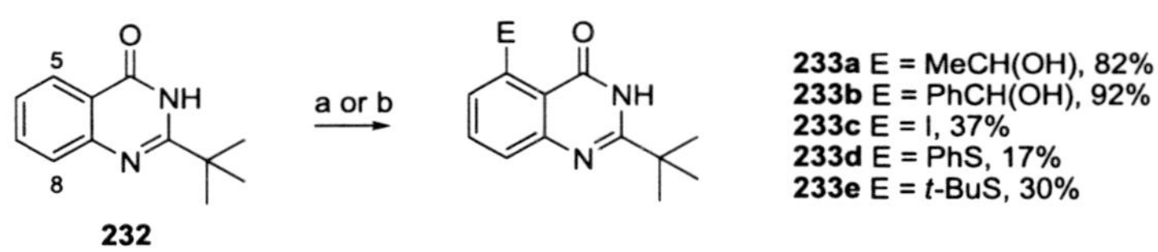

Scheme 37 (**a**) (**233a–d**) (i) 4.2 equiv *s*-BuLi, 4 equiv TMEDA, THF, −20°C, 1 h; (ii) 5 equiv E+, −78°C; (iii) $EtOH/H_2O$, −78°C [109]; (**b**) (**233e**) (i) 4 equiv *s*-BuLi, 4 equiv TMEDA, THF, −78 to 0°C, 1 h; (ii) 4 equiv *t*-BuSS*t*-Bu, 0°C, 3 h; (iii) H_2O, 0°C [110]

Presumably quinazoline **238** undergoes a sequence of a second lithiation, a nucleophilic addition at the diazine ring and an air oxidation upon work-up. Consequently, a side-product **239** was isolated along with the expected derivative **238** in identical yield.

Halogen–lithium exchange reactions with alkyl lithiums in good yields have been reported both at the diazine [115] and the benzene ring [116] of quinazolines. Attempts to lithiate quinazolin-4-one **242** (Scheme 41) using *n*-BuLi or *t*-BuLi were unsuccessful

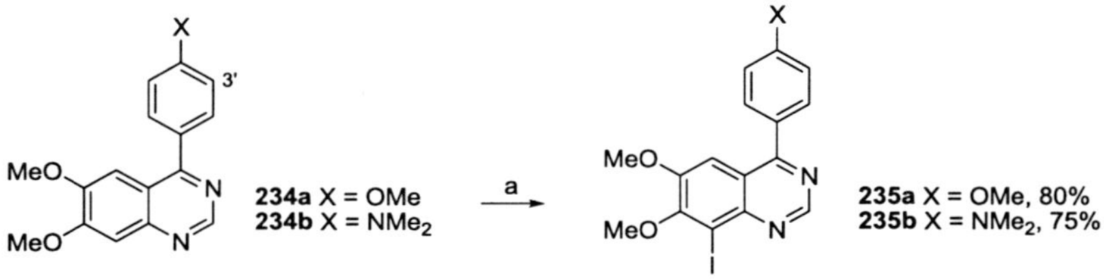

Scheme 38 (**a**) (i) 4 equiv LTMP, THF, −78°C; (ii) 2.5 equiv I_2, −78°C, 2 h; (iii) EtOH/H_2O [111]

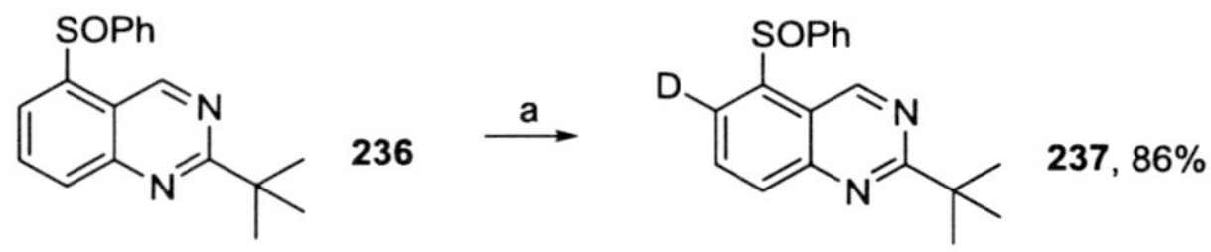

Scheme 39 (**a**) (i) 2.1 equiv LTMP, THF, −78°C, 5 min; (ii) DCl, −78°C, 5 min

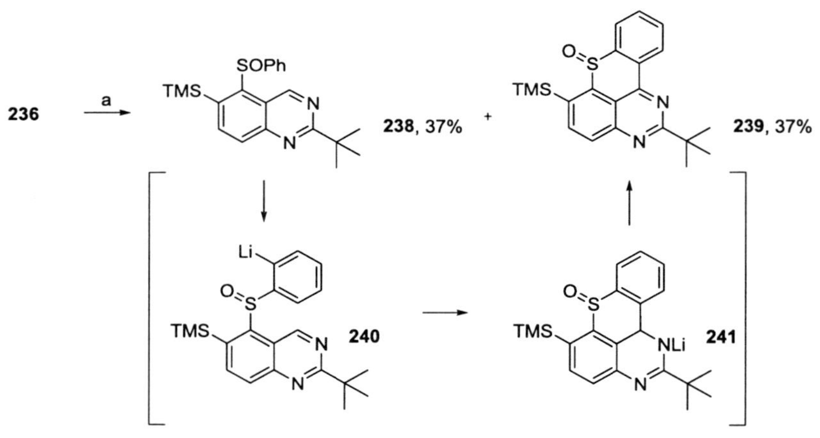

Scheme 40 (**a**) (i) 2.1 equiv LTMP, 2.1 equiv TMSCl, THF, −78°C, 90 min; (ii) H_2O/EtOH/THF

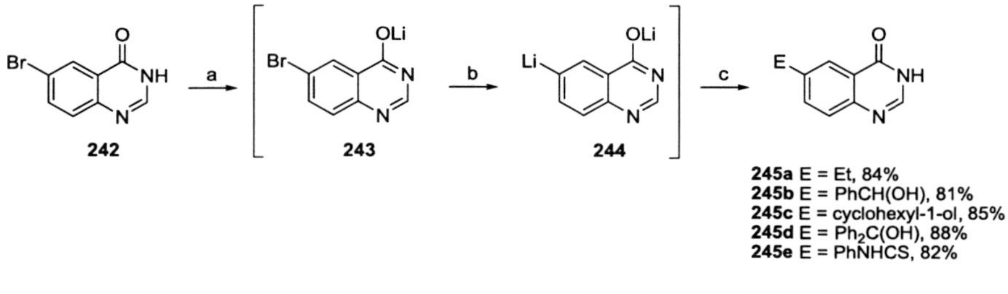

Scheme 41 (**a**) 1.1 equiv MeLi, THF, −78°C, 5 min; (**b**) 2.2 equiv *t*-BuLi, −78°C, 1 h; (**c**) (i) 1.1 equiv E+, −78°C, 2 h; (ii) aq NH_4Cl, rt

and competitive 1,2-additions were observed [117]. A stepwise *N*-deprotonation with MeLi followed by a bromine–lithium exchange performed with *t*-BuLi seemed to be essential for an efficient introduction of various electrophiles to the C6-position of **242**. In Scheme 41 are depicted selected examples **245a–e** of the prepared quinazolin-4-ones, isolated in somewhat surprisingly similar yields of 81–88%.

8 Conclusions

Both deprotonative and halogen–metal exchange lithiations and magnesiations provide a multitude of strategies for an aimed synthesis of variously modified compounds containing a pyrimidine subunit. Yet, the potential of catalytic transmetalations remains almost unrevealed and shows promise for a future dynamic development of the area.

Acknowledgment This work was supported by the Slovak Research and Development Agency under the contract No. APVV-0457-11.

References

1. Rewcastle GW (2008) Pyrimidines and their benzo derivatives. In: Katritzky AR, Ramsden CA, Scriven EFV, Taylor RJK (eds) Comprehensive heterocyclic chemistry III, vol 8. Elsevier Science Ltd. The Boulevard, Langford Lane, Kidlington, Oxford, UK
2. Turck A, Plé N, Mongin F, Quéguiner G (2001) Tetrahedron 57:4489
3. Chevallier F, Mongin F (2008) Chem Soc Rev 37:595
4. Schlosser M, Lefebvre O, Ondi L (2006) Eur J Org Chem 1593
5. Shen K, Fu Y, Li JN, Liu L, Guo QX (2007) Tetrahedron 63:1568
6. Nájera C, Sansano JM, Yus M (2003) Tetrahedron 59:9255
7. Mosrin M, Boudet N, Knochel P (2008) Org Biomol Chem 6:3237
8. Krasovskiy A, Krasovskaya V, Knochel P (2006) Angew Chem Int Ed 45:2958
9. Mosrin M, Knochel P (2008) Org Lett 10:2497
10. Mosrin M, Petrera M, Knochel P (2008) Synthesis 3697
11. Nguyen NH, Len C, Castanet AS, Mortier J (2011) Beilstein J Org Chem 7:1228
12. Plé N, Turck A, Couture K, Quéguiner G (1995) J Org Chem 60:3781
13. Clarke AJ, McNamara S, Meth-Cohn O (1974) Tetrahedron Lett 15:2373
14. Ioachim E, Medlycott EA, Polson MIJ, Hanan GS (2005) Eur J Org Chem 3775
15. Mendoza O, Müller-Bunz H, Tacke M (2005) J Organomet Chem 690:3357
16. Berghian C, Darabantu M, Turck A, Plé N (2005) Tetrahedron 61:9637
17. Berghian C, Condamine E, Plé N, Turck A, Silaghi-Dumitrescu I, Maiereanu C, Darabantu M (2006) Tetrahedron 62:7339
18. Turck A, Plé N, Quéguiner G (1994) Heterocycles 37:2149
19. Mattson RJ, Sloan CP (1990) J Org Chem 55:3410
20. Snieckus V (1990) Chem Rev 90:879
21. Arai S, Okamura T, Takeoka S (2010) Tetrahedron Lett 51:5177
22. Knochel P, Yeh MCP, Berk SC, Talbert J (1988) J Org Chem 53:2390
23. Kress TJ (1979) J Org Chem 44:2081
24. Xue CB, Feng H, Cao G, Huang T, Glenn J, Anand R, Meloni D, Zhang K, Kong L, Wang A, Zhang Y, Zheng C, Xia M, Chen L, Tanaka H, Han Q, Robinson DJ, Modi D, Storace L, Shao L, Sharief V, Li M, Galya LG, Covington M, Scherle P, Diamond S, Emm T,

Yeleswaram S, Contel N, Vaddi K, Newton R, Hollis G, Friedman S, Metcalf B (2011) Med Chem Lett 2:450
25. Plé N, Turck A, Martin P, Barbey S, Quéguiner G (1993) Tetrahedron Lett 34:1605
26. Plé N, Turck A, Heynderickx A, Quéguiner G (1994) J Heterocycl Chem 31:1311
27. Decrane L, Plé N, Turck A (2005) J Heterocycl Chem 42:509
28. Turck A, Plé N, Mojovic L, Quéguiner G (1990) J Heterocycl Chem 27:1377
29. Revesz L, Blum E, Di Padova FE, Buhl T, Feifel R, Gram H, Hiestand P, Manning U, Neumann U, Rucklin G (2006) Bioorg Med Chem Lett 16:262
30. Radinov R, Haimova M, Simova E (1986) Synthesis 886
31. Radinov R, Chanev C, Haimova M (1991) J Org Chem 56:4793
32. Buurman ET, Blodgett AE, Hull KG, Carcanague D (2004) Antimicrob Agents Chemother 48:313
33. Coleman PJ, Brashear KM, Askew BC, Hutchinson JH, McVean CA, Duong LT, Feuston BP, Fernandez-Metzler C, Gentile MA, Hartman GD, Kimmel DB, Leu CT, Lipfert L, Merkle K, Pennypacker B, Prueksaritanont T, Rodan GA, Wesolowski GA, Rodan SB, Duggan ME (2004) J Med Chem 47:4829
34. Busch M, Schlageter M, Weingand D, Gehring T (2009) Chem Eur J 15:8251
35. Kmentova I, Sutherland HS, Palmer BD, Blaser A, Franzblau SG, Wan B, Wang Y, Ma Z, Denny WA, Thompson AM (2010) J Med Chem 53:8421
36. Kawasaki T, Sato M, Ishiguro S, Saito T, Morishita Y, Sato I, Nishino H, Inoue Y, Soai K (2005) J Am Chem Soc 127:3274
37. Le Bourdonnec B, Windh RT, Leister LK, Zhou QJ, Ajello CW, Gu M, Chu GH, Tuthill PA, Barker WM, Koblish M, Wiant DD, Graczyk TM, Belanger S, Cassel JA, Feschenko MS, Brogdon BL, Smith SA, Derelanko MJ, Kutz S, Little PJ, DeHaven RN, DeHaven-Hudkins DL, Dolle RE (2009) J Med Chem 52:5685
38. Abbot SC, Billedeau RJ, Dewdney NJ, Gabriel T, Goldstein DM, McCaleb KL, Soth M, Trejo-Martin TA, Zecic H (2009) Heterocycles 78:2811
39. Plé N, Turck A, Heynderickx A, Quéguiner G (1998) Tetrahedron 54:9701
40. Plé N, Turck A, Bardin F, Quéguiner G (1992) J Heterocycl Chem 29:467
41. Plé N, Turck A, Couture K, Quéguiner G (1994) Tetrahedron 50:10299
42. Schnürch M, Spina M, Khan AF, Mihovilovic MD, Stanetty P (2007) Chem Soc Rev 36:1046
43. Schnürch M (2012) Top Heterocycl Chem 27:185
44. Rahier NJ, Cheng K, Gao R, Eisenhauer BM, Hecht SM (2005) Org Lett 7:835
45. Elban MA, Sun W, Eisenhauer BM, Gao R, Hecht SM (2006) Org Lett 8:3513
46. Zhang Y, Jin G, Illarionov B, Bacher A, Fischer M, Cushman M (2007) J Org Chem 72:7176
47. Nencka R, Šála M, Dejmek M, Dračínský M, Holý A, Hřebabecký H (2010) Synthesis 4119
48. Mukherjee S, Ghorai BK (2010) Synth Commun 40:1939
49. Jin H, Randazzo J, Zhang P, Kador PF (2010) J Med Chem 53:1117
50. Seto S, Tanioka A, Ikeda M, Izawa S (2005) Bioorg Med Chem Lett 15:1485
51. Seto S, Tanioka A, Ikeda M, Izawa S (2005) Bioorg Med Chem 13:5717
52. Seto S, Okada K, Kiyota K, Isogai S, Iwago M, Shinozaki T, Kitamura Y, Kohno Y, Murakami K (2010) J Med Chem 53:5012
53. Irie O, Yokokawa F, Ehara T, Iwasaki A, Iwaki Y, Hitomi Y, Konishi K, Kishida M, Toyao A, Masuya K, Gunji H, Sakaki J, Iwasaki G, Hirao H, Kanazawa T, Tanabe K, Kosaka T, Hart TW, Hallett A (2008) Bioorg Med Chem Lett 18:4642
54. Mosrin M, Knochel P (2009) Chem Eur J 15:1468
55. Anderson RJ, Hill JB, Morris JC (2005) J Org Chem 70:6204
56. Suzuki M, Noyori R (1994) Conjugate addition – enolate trapping reactions. In: Taylor RJK (ed) Organocopper reagents: a practical approach. Oxford University Press, New York
57. Buron F, Plé N, Turck A, Marsais F (2006) Synlett 1586
58. Lu J, Li NS, Koo SC, Piccirilli JA (2009) J Org Chem 74:8021
59. Looper RE, Runnegar MTC, Williams RM (2005) Angew Chem Int Ed 44:3879

60. Hanessian S, Marcotte S, Machaalani R, Huang G, Pierron J, Loiseleur O (2006) Tetrahedron 62:5201
61. Koszytkowska-Stawińska M, Mironiuk-Puchalska E, Sas W (2011) Tetrahedron Lett 52:1866
62. Boudet N, Knochel P (2006) Org Lett 8:3737
63. Soth M, Abbot S, Abubakari A, Arora N, Arzeno H, Billedeau R, Dewdney N, Durkin K, Frauchiger S, Ghate M, Goldstein DM, Hill RJ, Kuglstatter A, Li F, Loe B, McCaleb K, McIntosh J, Papp E, Park J, Stahl M, Sung ML, Suttman R, Swinney DC, Weller P, Wong B, Zecic H, Gabriel T (2011) Bioorg Med Chem Lett 21:3452
64. Kofink CC, Knochel P (2006) Org Lett 8:4121
65. Kopp F, Knochel P (2007) Org Lett 9:1639
66. Bäckvall JE, Persson ESM, Bombrun A (1994) J Org Chem 59:4126
67. Corriu RJP, Masse JP (1972) J Chem Soc Chem Commun 144
68. Tamao K, Sumitani K, Kumada M (1972) J Am Chem Soc 94:4374
69. Manolikakes G, Knochel P (2009) Angew Chem Int Ed 48:205
70. Shimma N, Umeda I, Arasaki M, Murasaki C, Masubachi K, Kohchi Y, Miwa M, Ura M, Sawada N, Tahara H, Kuruma I, Horii I, Ishitsuka H (2000) Bioorg Med Chem 8:1697
71. Goodyear MD, Hill ML, West JP, Whitehead AJ (2005) Tetrahedron Lett 46:8535
72. Lamberth C (2005) Heterocycles 65:667
73. Gardelli C, Nizi E, Muraglia E, Crescenzi B, Ferrara M, Orvieto F, Pace P, Pescatore G, Poma M, Ferreira MRR, Scarpelli R, Homnick CF, Ikemoto N, Alfieri A, Verdirame M, Bonelli F, Gonzalez Paz O, Taliani M, Monteagudo E, Pesci S, Laufer R, Felock P, Stillmock KA, Hazuda D, Rowley M, Summa V (2007) J Med Chem 50:4953
74. Gambacorta A, Tofani D, Loreto MA, Gasperi T, Bernini R (2006) Tetrahedron 62:6848
75. Tanaka H, Nasu I, Miyasaka T (1979) Tetrahedron Lett 20:4755
76. Tanaka H, Hayakawa H, Miyasaka T (1981) Chem Pharm Bull 29:3565
77. Shimizu M, Tanaka H, Hayakawa H, Miyasaka T (1990) Tetrahedron Lett 31:1295
78. Hayakawa H, Tanaka H, Maruyama Y, Miyasaka T (1985) Chem Lett 14:1401
79. Eppacher S, Solladié N, Vasella A (2004) Helv Chim Acta 87:2926
80. Armstrong RW, Gupta S, Whelihan F (1989) Tetrahedron Lett 30:2057
81. Hayakawa H, Tanaka H, Obi K, Itoh M, Miyasaka T (1987) Tetrahedron Lett 28:87
82. Shih YC, Chien TC (2011) Tetrahedron 67:3915
83. Bello AM, Poduch E, Fujihashi M, Amani M, Li Y, Crandall I, Hui R, Lee PI, Kain KC, Pai EF, Kotra LP (2007) J Med Chem 50:915
84. Bello AM, Poduch E, Liu Y, Wei L, Crandall I, Wang X, Dyanand C, Kain KC, Pai EF, Kotra LP (2008) J Med Chem 51:439
85. Zhang X, Bernet B, Vasella A (2006) Helv Chim Acta 89:2861
86. Bello AM, Konforte D, Poduch E, Furlonger C, Wei L, Liu Y, Lewis M, Pai EF, Paige CJ, Kotra LP (2009) J Med Chem 52:1648
87. Chiesa K, Shvoryna A, Bernet B, Vasella A (2010) Helv Chim Acta 93:668
88. Ritter A, Egli D, Bernet B, Vasella A (2008) Helv Chim Acta 91:673
89. Matthews AJ, Bhardwaj PK, Vasella A (2003) Chem Commun 950
90. Matthews AJ, Bhardwaj PK, Vasella A (2004) Helv Chim Acta 87:2273
91. Holmes SC, Gait MJ (2005) Eur J Org Chem 5171
92. Lewis M, Meza-Avina ME, Wei L, Crandall IE, Bello AM, Poduch E, Liu Y, Paige CJ, Kain KC, Pai EF, Kotra LP (2011) J Med Chem 54:2891
93. Sunkara NK, Mosley SL, Seley-Radtke KL (2006) Collect Czech Chem Commun 71:1161
94. Sadler JM, Mosley SL, Dorgan KM, Zhou ZS, Seley-Radtke KL (2009) Bioorg Med Chem 17:5520
95. Radi M, Mugnaini C, Petricci E, Corelli F, Botta M (2005) Tetrahedron Lett 46:4361
96. Herdeis L, Bernet B, Augustine A, Kälin RE, Brändli AW, Vasella A (2011) Helv Chim Acta 94:545
97. Moriuchi T, Noguchi S, Sakamoto Y, Hirao T (2011) J Organomet Chem 696:1089

98. Llanes-Pallas A, Matena M, Jung T, Prato M, Stöhr M, Bonifazi D (2008) Angew Chem Int Ed 47:7726
99. Llanes-Pallas A, Palma CA, Piot L, Belbakra A, Listorti A, Prato M, Samori P, Armaroli N, Bonifazi D (2009) J Am Chem Soc 131:509
100. Peifer M, De Giacomo F, Schandl M, Vasella A (2009) Helv Chim Acta 92:1134
101. Zhang F, Kulesza A, Rani S, Bernet B, Vasella A (2008) Helv Chim Acta 91:1201
102. Meng G, Kuang YY, Ji L, Chen FE (2005) Synth Commun 35:1095
103. Loksha YM, Pedersen EB, Bond AD, La Colla P, Secci B, Loddo R (2009) Synthesis 3589
104. Wójtowicz-Rajchel H, Migas M, Koroniak H (2006) J Org Chem 71:8842
105. Aso M, Kaneko T, Nakamura M, Koga N, Suemune H (2003) Chem Commun 1094
106. El Safadi Y, Paillart JC, Lamound G, Aubertin AM, Burger A, Marquet R, Vivet-Boudou V (2010) J Med Chem 53:1534
107. Mieczkowski A, Peltier P, Zevaco T, Agrofoglio LA (2009) Tetrahedron 65:4053
108. Smith K, El-Hili GA, Abdel-Megged MF, Abdo MA (1996) J Org Chem 61:647
109. Busch A, Gautheron-Chapoulaud V, Audoux J, Plé N, Turck A (2004) Tetrahedron 60:5373
110. Le Fur N, Mojovic L, Turck A, Plé N, Quéguiner G, Reboul V, Perrio S, Metzner P (2004) Tetrahedron 60:7983
111. Busch A, Turck A, Nowicka K, Barsella A, Andraud C, Plé N (2007) Heterocycles 71:1723
112. Gautheron-Chapoulaud V, Salliot I, Plé N, Turck A, Quéguiner G (1999) Tetrahedron 55:5389
113. Gautheron-Chapoulaud V, Plé N, Quéguiner G (2000) Tetrahedron 56:5499
114. Le Fur N, Mojovic L, Plé N, Turck A, Marsais F (2005) Tetrahedron 61:8924
115. Gillespie RJ, Adams DR, Bebbington D, Benwell K, Cliffe IA, Dawson CE, Dourish CT, Fletcher A, Gaur S, Giles PR, Jordan AM, Knight AR, Knutsen LJS, Lawrence A, Lerpiniere J, Misra A, Porter RHP, Pratt RM, Shepherd R, Upton R, Ward SE, Weiss SM, Williamson DS (2008) Bioorg Med Chem Lett 18:2916
116. Hernández AR, Kool ET (2011) Org Lett 13:676
117. El-Hiti GA (2004) Monatsh Chem 135:323

Top Heterocycl Chem (2013) 31: 65–92
DOI: 10.1007/7081_2012_91

Published online: 8 November 2012

Metalation Reactions of Pyridines, Quinolines, and Isoquinolines with Ate Bases and Their Alkali Metal Salt-Modified Congeners

Costa Metallinos and Kathryn Stromski

Abstract A survey of the recent uses of metal "ate" type bases for metalation, halogen–metal, and sulfur–metal exchange reactions of the π-deficient aromatics pyridine, quinoline, and isoquinolines is presented, focusing mainly on the period from 2006–2011. "Ate" type bases are defined as reagents where the key metal center bears a formal negative charge or their alkali metal salt-modified (AMSM) counterparts which display very similar reactivity. Selected applications to target molecule synthesis using the preceding processes in combination with cross-coupling reactions, and consideration of aspects of substrate/functional group stability and regioselectivity, are disclosed.

Keywords Alkali metal-modified · Ate bases · Halogen–metal exchange · Metalation · π-Deficient

Contents

C. Metallinos (✉) and K. Stromski
Department of Chemistry, Brock University, 500 Glenridge Avenue, St. Catharines, ON, Canada L2S 3A1,
e-mail: cmetallinos@brocku.ca

Abbreviations

Ac	Acetyl
acac	Acetylacetonate
AMSM	Alkali metal salt-modified
aq	Aqueous
Ar	Aryl
Bn	Benzyl
Bu	Butyl
Bz	Benzoyl
Cbz	Benzyloxycarbonyl
CDI	1,1′-Carbonyldiimidazole
concd	Concentrated
DA	Diisopropylamido
dba	Dibenzylideneacetone
DMA	Dimethylacetamide
DMAE	*N,N*-Dimethylaminoethoxide
DME	Dimethoxyethane
DMF	Dimethylformamide
DMSO	Dimethyl sulfoxide
E^+	Electrophile
equiv	Equivalent(s)
Et	Ethyl
FLP	Frustrated Lewis pair
h	Hour(s)
HMPA	Hexamethylphosphoric triamide
i-Bu	Isobutyl
i-Pr	Isopropyl
KTMP	Potassium 2,2,6,6-tetramethylpiperidide
LDA	Lithium diisopropylamide
LHMDS	Lithium bis(trimethylsilyl)amide
LTMP	Lithium 2,2,6,6-tetramethylpiperidide
M	Molar
Me	Methyl
Mes	Mesityl
MesLi	Mesityllithium
min	Minute(s)
Mol	Mole
n.d.	Not determined

NBS	*N*-Bromosuccinimide
n-Bu	*normal*-Butyl
NFSI	*N*-Fluorobenzenesulfonimide
NMP	*N*-Methyl-2-pyrrolidinone
OAc	Acetoxy
Ph	Phenyl
PMB	*para*-methoxybenzyl
PMDETA	*N,N,N′,N″,N″*-Pentamethyldiethylenetriamine
p-Tol	*para*-Tolyl
Py	Pyridyl
rt	Room temperature
s	Second(s)
satd	Saturated
s-Bu	*sec*-Butyl
TBAF	Tetrabutylammonium fluoride
TBDMS	*tert*-Butyldimethylsilyl
t-Bu	*tert*-Butyl
Tf	Triflate
THF	Tetrahydrofuran
THP	Tetrahydropyran
TIPS	Triisopropylsilyl
TMEDA	*N,N,N′,N′*-Tetramethyl-1,2-ethylenediamine
TMP	2,2,6,6-Tetramethylpiperidyl
TMPH	2,2,6,6-Tetramethylpiperidine
TMS	Trimethylsilyl
μwave	Microwave

1 Introduction

Over the past 5 years, homo- and hetero-leptic “ate” bases have evolved from being considered as exotic reagents [1] to become important alternatives for the metalation of aromatics. Their advantages include greater compatibility with substrates that are normally prone to addition reactions of organolithium reagents [2–7] due either to sensitive functional groups or because of the inherent electrophilicity of the aromatic ring. In general, “ate” reagents allow metalation, halogen–metal and sulfur–metal exchange reactions to proceed at or somewhat below room temperature and without the need for additives, a common requirement when organolithiums are employed for similar transformations [8–11]. In many cases, the metalated intermediates may be exposed directly to conditions of palladium catalyzed cross-coupling, thus obviating the need for additional transmetalation steps.

This chapter will survey the recent applications of metal "ate" type bases for metalation of the π-deficient aromatics pyridine, quinoline, and isoquinoline, focusing mainly on the period from 2006 to 2011. For the sake of brevity, the scope of the chapter will include both ate bases where the key metal center of the reagent bears a formal negative charge, and also their alkali metal salt-modified (AMSM) counterparts which display very similar reactivity. (The acronym AMSM will be used to refer to alkali metal salt-modified reagents in a similar way that AMMM and AMMZ have referred to reactions such as alkali metal-mediated magnesiation and alkali metal-mediated zincation, respectively [12]). From this perspective, the AMSM congeners can be considered to be quasi-ate reagents, although collectively, both formal and quasi "ate" reagents are best viewed as heterobimetallics with highly polarized bonds [13]. Based on the preceding definition, reagents of general structure **A**, **B**, and **C** will be covered here, whereas reagents **D** and **E** in which M is neutral and not associated with alkali metal salts are outside the purview of this chapter (Scheme 1). The discussion has been organized primarily by base type such that pyridine, quinoline, and isoquinoline heterocycles are addressed under each category.

$R_3M^{2+}Li$ **A** $R_2M^{2+}\cdot xLiX$ **B** $RM^{2+}X\cdot xLiX$ **C**

R_2M^{2+} **D** $RM^{2+}X$ **E**

(R = alkyl, aryl or R_2N; X = halide)

Scheme 1 Formulations of ate bases (**A**), quasi-ate bases (**B**, **C**), and classical bases (**D**, **E**)

2 Magnesiates and Their Alkali Metal Salt-Modified Congeners

The considerable growth in the use of organomagnesium reagents in metalation or halogen–metal exchange reactions of aromatics has been achieved by the development of magnesiates, which are produced by the addition of nucleophiles to divalent Grignards. The first magnesiates of this type were described by Wittig and coworkers in 1951 with the compound Ph_3MgLi [14]. While magnesium bases of the form R_2NMgX and $Mg(NR)_2$ reported by Hauser in 1947 [15] are not technically ate bases, the demonstration by Eaton in 1989 [16] that TMP_2Mg may be used in deprotonation of functionalized aromatics created renewed interest in these reagents. The AMSM reagents subsequently developed by Knochel and others were found to increase both the rate and the efficiency of metalation in comparison with the original Eaton system. In general, AMSM reagents are now utilized more often than magnesiates with actual anionic formulations. Within this group, LiCl is the most popular salt used in the formation of modified RMgX bases [17–32]. Bases incorporating TMEDA

[33] or BF_3·LiCl [26] additives are also known. Several types of R groups are utilized when forming these types of pseudo-magnesiates. Examples include *i*-PrMgCl·LiCl [17–20, 22, 29, 30], (TMP)MgCl·LiCl [21, 25–28, 30–32], and $(TMP)_2Mg$·LiCl [21, 24, 27, 28]. Reagents that have formally anionic magnesium include R_3MgLi [33–38], especially for R = Bu.

2.1 *Halogen–Magnesium Exchange Reactions of Pyridines, Quinolines, and Isoquinolines*

Recent halogen–magnesium exchange reactions of π-deficient aromatics have been conducted with one of the two bases: the magnesiate *n*-Bu_3MgLi [35] or the pseudo-magnesiate *i*-PrMgCl·LiCl [17, 39]. The first of these is conveniently prepared by addition of 2 equiv of *n*-BuLi to *n*-BuMgBr [40] (Scheme 2). The *i*-PrMgCl·LiCl reagent can be prepared by standard formation of the Grignard in the presence of LiCl, or by addition of *i*-PrMgCl to a solution of LiCl in THF [17].

2 *n*-BuLi + *n*-BuMgBr (**1**, **2**) —THF, 0 °C→ *n*-Bu_3MgLi (**3**)

i-PrCl + Mg (**4**, **5**) —LiCl, THF→ *i*-PrMgCl·LiCl (**6**)

Scheme 2 Preparation of *n*-Bu_3MgLi and *i*-PrMgCl·LiCl

Halogen–magnesium exchange reactions with R_3MgLi bases have previously been reported for 2-halopyridines [40] and 2,6-, 3,5-, and 3,6-dibromopyridines [41]. Following this earlier work, Mongin and coworkers showed that 2-, 3-, or 4-bromoquinolines also undergo Br/Mg exchange with 0.35 equiv of *n*-Bu_3MgLi in THF at −10°C to give products **8**, **10**, and **12a** in modest to good yields (Scheme 3) [35]. Under these conditions, 3-bromoquinoline gave somewhat better yields with a range of electrophiles. Regardless of the substitution pattern, an intermediate of the form $(quinolinyl)_3MgLi$ may be assumed according to the reaction stoichiometry, as similar intermediates have been invoked with dihalopyridines [41]. Care must be taken in the preparation of *n*-Bu_3MgLi so as to avoid unwanted nucleophilic addition of a butyl group to the heterocycle by adventitious *n*-BuLi. Apart from simple electrophile quench of the intermediate carbanions, Ni or Pd-catalyzed cross-coupling of the magnesiated intermediates is quite facile, especially with other π-deficient haloaromatics (**11** → **12b**) [34, 36].

The development of *i*-PrMgCl·LiCl has provided an alternative method to induce Br/Mg exchange of 2-bromo- and 2,6-dibromo-pyridine. Thus, exposure of **13** or **14** to 1.1 equiv of *i*-PrMgCl·LiCl in THF gives the mono-magnesiated intermediates **15** or **16** that may be trapped with benzaldehydes to afford alcohols **17** or **18** in 72% and 89% yields, respectively (Scheme 4) [17, 42]. For comparison,

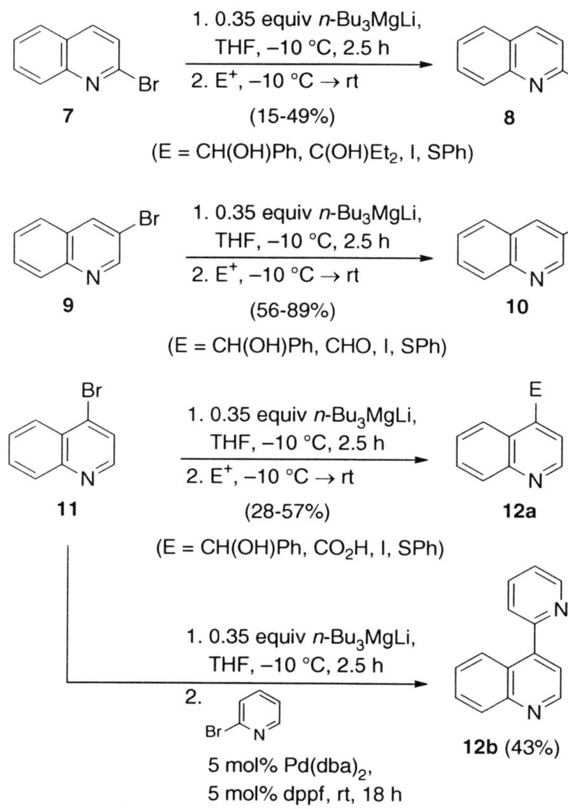

Scheme 3 Bromine–magnesium exchange of 2-, 3- and 4-bromoquinolines

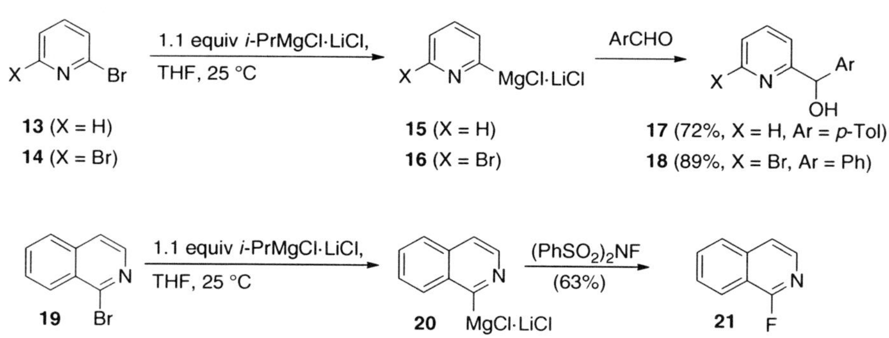

Scheme 4 Bromine–magnesium exchange of 2-bromopyridines and 1-bromoisoquinoline

Br/Mg exchange of **14** with *i*-PrMgCl alone (2 equiv) afforded **18** in only 42% yield [17]. Direct fluorination of magnesio-pyridines and isoquinolines generated from *i*-PrMgCl·LiCl induced Br/Mg exchange has also been demonstrated using *N*-fluorobenzenesulfonimide [$(PhSO_2)_2NF$] (**19** → **21**) [29].

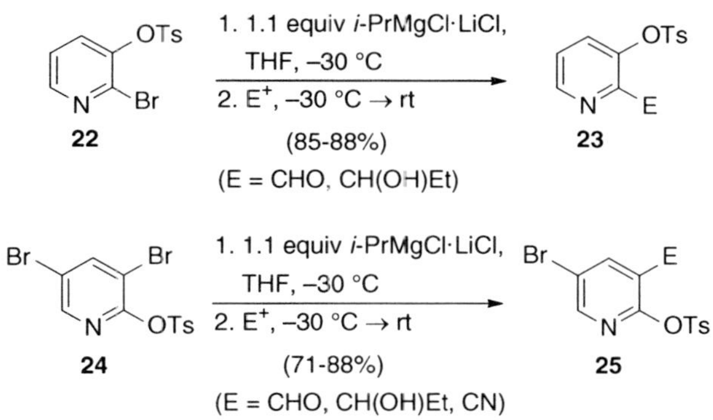

Scheme 5 Bromine–magnesium exchange of pyridinetossylates

Other halopyridines known to react smoothly with *i*-PrMgCl·LiCl include the bromo tosylates **22** and **24** [20], or their corresponding phenolates [43], which have been quenched with a range of electrophiles (Scheme 5). This procedure has been expanded to include oxidative amination of 3,5-dibromopyridine (**26** → **29**) [44] and boronation of should be 3,5-diiodopyridine and 5,7-diiodoquinoline (**30** → **31**) [18] (Scheme 6). In the case of **26**, initial Br/Mg exchange enables sequential transmetalation with zinc chloride and cuprous cyanide. Addition of LiN (TBDMS)Ph and oxidation with $PhI(OAc)_2$ furnishes the silyl-protected aniline **29**. Substrates such as **30** that contain two halogens may undergo a second Br/Mg exchange–electrophile quench, as illustrated by the synthetic sequence **31** → **32**.

2.2 *Deprotonative Magnesiation of Pyridines, Quinolines, and Isoquinolines*

As with halogen–magnesium exchange reactions, direct deprotonative magnesiation of pyridines, quinolines, and isoquinolines has made use of LiCl-modified magnesium amides more often than classical magnesiates. The preference for the former is related to their higher kinetic reactivity, which results in generally better yields of substituted aromatics after electrophile quench. The origin of this enhanced reactivity has been tentatively attributed to de-aggregation of the base by LiCl [19, 45]. This fact allows them to be used in near stoichiometric amounts in contrast to earlier magnesium amides by Eaton [16] and others [46], which required a large excess of both base and electrophiles to achieve adequate yields. Even so, classical magnesiates such as n-Bu_3MgLi have be employed for the metalation of [1,2,3]triazolo[1,5-*a*]pyridines **33**, to give products **34** in 32–75% after quench with 3,4,5-trimethoxybenzaldehyde [38] (Scheme 7).

Among magnesium amides, the most important advancements have been the development of the bases TMPMgCl·LiCl (**35**) and $(TMP)_2Mg$·2LiCl (**36**), which

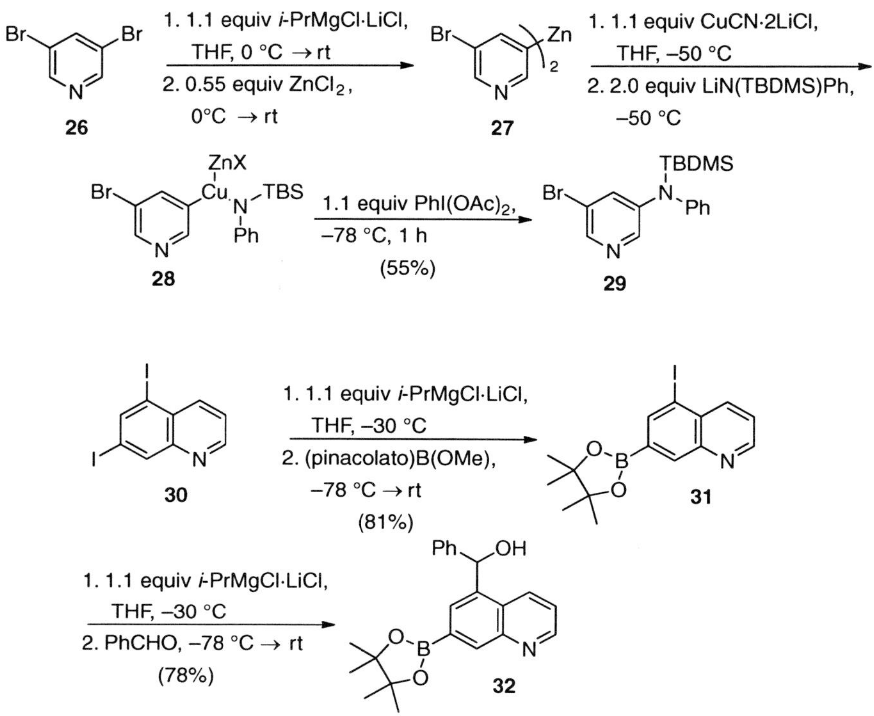

Scheme 6 Halogen–magnesium exchange of 3,5-dibromopyridine and 5,7-diiodoquinoline

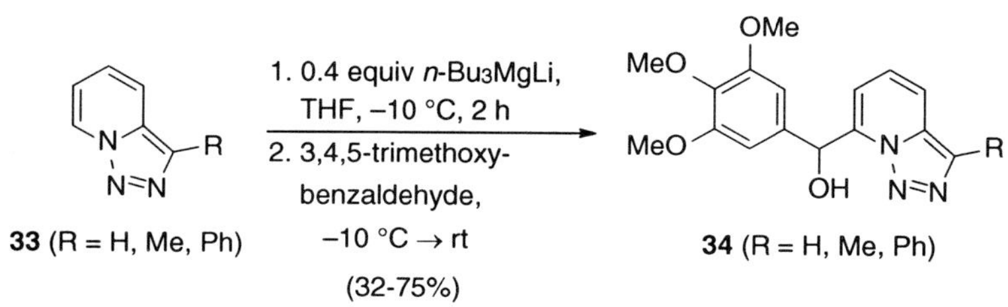

Scheme 7 Magnesiation of [1,2,3]triazolo[1,5-*a*]pyridines

have seen numerous application in metalation of π-deficient aromatics. These reagents are prepared from 2,2,6-6-tetramethylpiperidine (TMPH) according to the procedures in Scheme 8 [19, 21]. TMPMgCl·LiCl is stable for months at room temperature whereas the magnesium bisamide $(TMP)_2Mg{\cdot}2LiCl$ is best prepared for immediate use. Some variations of these reagents have been reported to offset the cost of 2,2,6,6-tetramethylpiperidine for large-scale synthesis [25] or to enhance long-term stability (especially for **36**). The most significant modified reagents are *t*-Bu(*i*-Pr)NMgCl·LiCl and $[t\text{-Bu}(i\text{-Pr})N]_2Mg{\cdot}2LiCl$, which have very similar reactivity profiles to **35** and **36**, respectively [21, 25]. Unlike **36**, $[t\text{-Bu}(i\text{-Pr})N]_2Mg{\cdot}2LiCl$ is stable for three weeks in THF at 4°C.

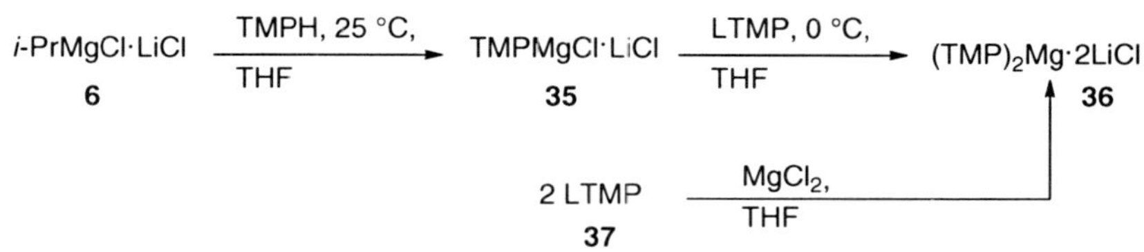

Scheme 8 Synthesis of TMPMgCl·LiCl and $(TMP)_2Mg{\cdot}2LiCl$

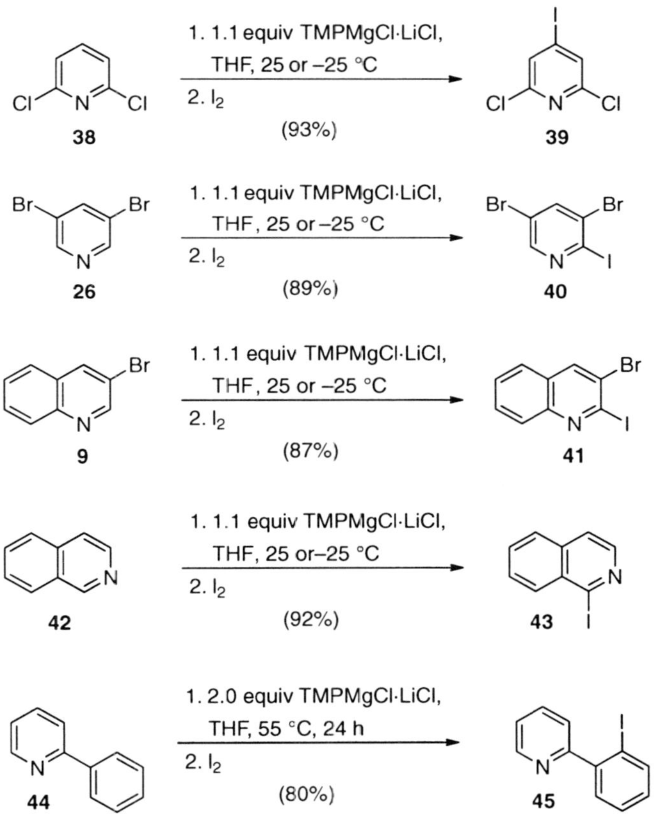

Scheme 9 Metalation of variously substituted pyridines, 3-bromoquinoline and isoquinoline with TMPMgCl·LiCl

An important milestone in development of soluble and kinetically active LiCl-modified magnesium amides for metalation of π-deficient aromatics was attained by the Knochel group in 2006 [19]. Thus, exposure of 2,6-dichloropyridine, 3,5-dibromopyridine, 3-bromoquinoline, or isoquinoline to metalation with 1.1 equiv of TMPMgCl·LiCl in THF at 25 or −25°C gave iodides **39**, **40**, **41**, and **43** in excellent yields after I_2 quench (Scheme 9). Although the metalation of 2-phenylpyridine

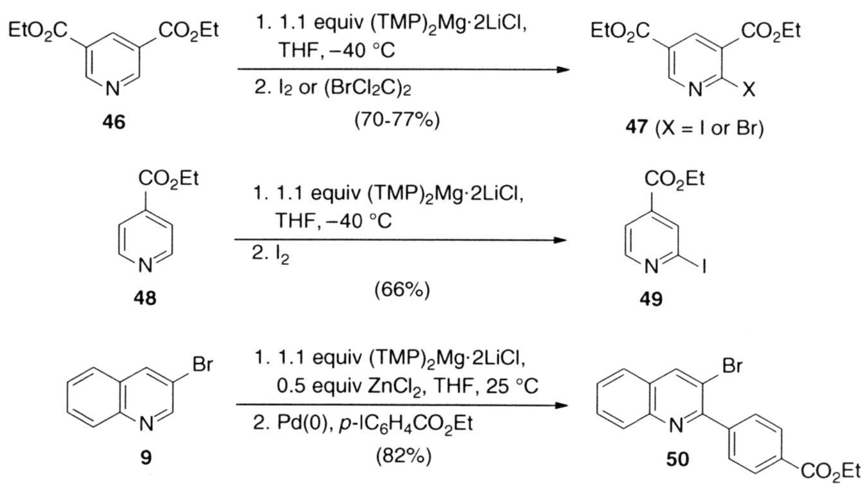

Scheme 10 Metalation of pyridinecarboxylates and 3-bromoquinoline with $TMP_2Mg{\cdot}2LiCl$

44 to give iodide **45** implies complexation of the base to the pyridyl nitrogen in these reactions, the preferential formation of **39** indicates that this effect is not the only one governing regioselectivity.

Despite these promising results, not all aromatics showed good reactivity towards TMPMgCl·LiCl. Benzoate esters in particular gave lower yields with this base [19]. The more reactive and less nucleophilic magnesium bisamide $TMP_2Mg{\cdot}2LiCl$ was developed to address this limitation. Hence, metalation of ethyl pyridinecarboxylates **46** or **48** with **36** proved to be quite facile and afforded the halides **47** and **49** depicted in Scheme 10 [21]. Magnesiation of 3-bromoquinoline (**9**) [24] could be conducted with **36** in the presence of zinc chloride to give the intermediate zinc species for Negishi cross-coupling with *p*-$IC_6H_4CO_2Et$.

The combined utility of reagents *i*-PrMgCl·LiCl, $Mes_2Mg{\cdot}2LiBr$, and MesMgBr·LiCl in Br/Mg exchange, and TMPMgCl·LiCl and $TMP_2Mg{\cdot}2LiCl$ in metalation, has been brought to bear on the synthesis of several quinoline derivatives from bromides **51**–**54** (Scheme 11). This type of multiple regioselective functionalization is exemplified by the rapid preparation of quinoline **57** and the NK3 receptor antagonist talnetant (**60**) [22]. Specifically, 2,4-dibromoquinoline undergoes smooth Br/Mg exchange to afford ester **55** after electrophile quench. Common intermediate **55** may be subjected to successive metalation–electrophile quench sequences using first TMPMgCl·LiCl, and then $TMP_2Mg{\cdot}2LiCl$, to provide biaryl **57** after Negishi cross-coupling. Alternatively, **55** may be used as a precursor to pinacol-protected boronic acid **58**. Three functional group manipulations involving cross-coupling, oxidation, and amide formation furnish the drug talnetant (**60**). It is notable that intermediates with both esters and ketones are tolerated in these synthetic sequences.

A number of the preceding metalations may be performed effectively on 80–100 mmol scale [28] for substrates with halogen atoms, esters, or the aryl nitrogen as directing groups. Recent work has shown that sulfoxides [30], phosphorodiamidates

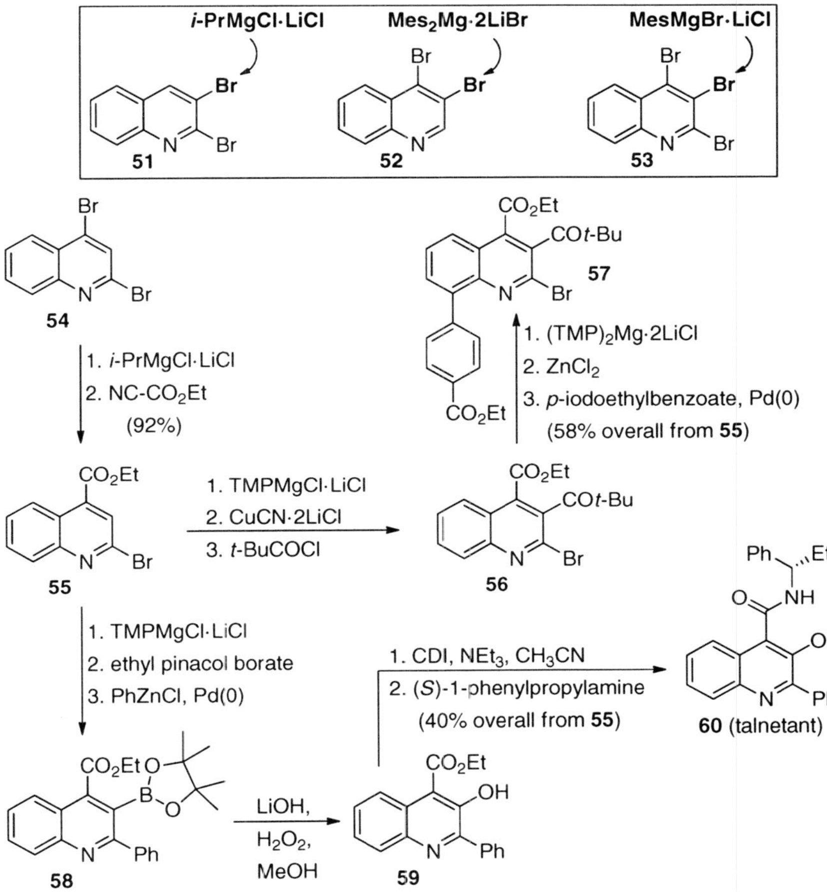

Scheme 11 Regioselective Br/Mg exchange and metalation of quinolines en route to advanced targets

[27], and latent N-BF_3 [26] groups can also serve to direct *ortho* magnesiation in pyridines and quinolines. For example, Br/Mg exchange of **13** followed by sulfinyl chloride quench gives sulfoxide **61** (Scheme 12). This substrate is amenable to magnesiation at the 3-position to give the *p*-cyanophenyl adduct **62** after Negishi coupling. Sulfur–magnesium exchange of **62** with *i*-PrMgCl·LiCl induces Oae-type rearrangement [47] to afford the 2,3-diarylpyridine **63**.

In a similar fashion, synthesis of the COX-2 inhibitor etoricoxib (**67**) has been initiated by magnesiation of pyridine **64** to afford the 3-aryl derivative **66** after cross-coupling and dephosphorylation (Scheme 13). An analogous synthetic protocol has been used in an alternative synthesis of the quinoline derivative talnetant [27].

With respect to boron trifluoride it is perhaps not surprising, based on earlier work of Kessar [48], that exposure of adduct **69** to TMPMgCl·LiCl results in *ortho*-magnesiation. Unlike the LTMP mediated lithiation of BF_3-activated pyridine, the N-

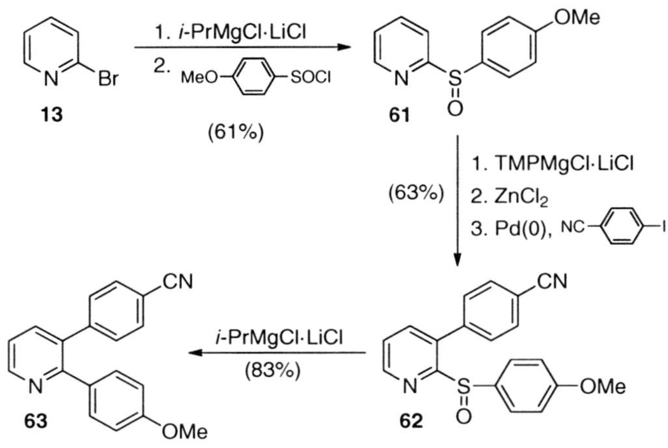

Scheme 12 A 2,3-diarylpyridine via sequential bromine–magnesium exchange, *ortho*-magnesiation, and sulfoxide–magnesium exchange

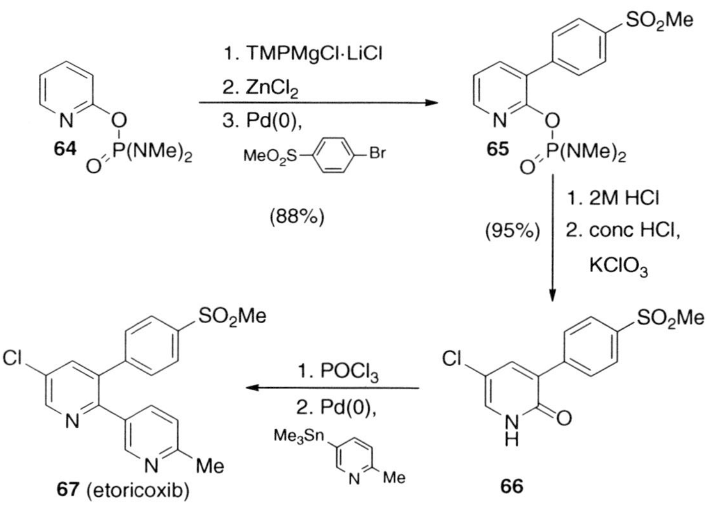

Scheme 13 Synthesis of etoricoxib

BF_3 group is not always a spectator during magnesiation, but may migrate to the 2-position thereby enabling transmetalation with $ZnCl_2$ and cross-coupling (Scheme 14) [26]. An interesting aspect of this transformation is that it takes place just as easily if the reagent TMPMgCl·BF_3·LiCl is preformed and added to uncomplexed pyridine **68** or quinoline, suggesting frustrated Lewis pair (FLP) behavior. In any case, these conditions have allowed for a major expansion of substrate scope leading to metalated DMAP (**72**), nicotine (**73**), and quinine (**74–75**) intermediates for reactions with various electrophiles [32]. Regioselectivity in magnesiation of these kinds of substituted pyridines and quinolines may change according to the presence or absence of BF_3 [26].

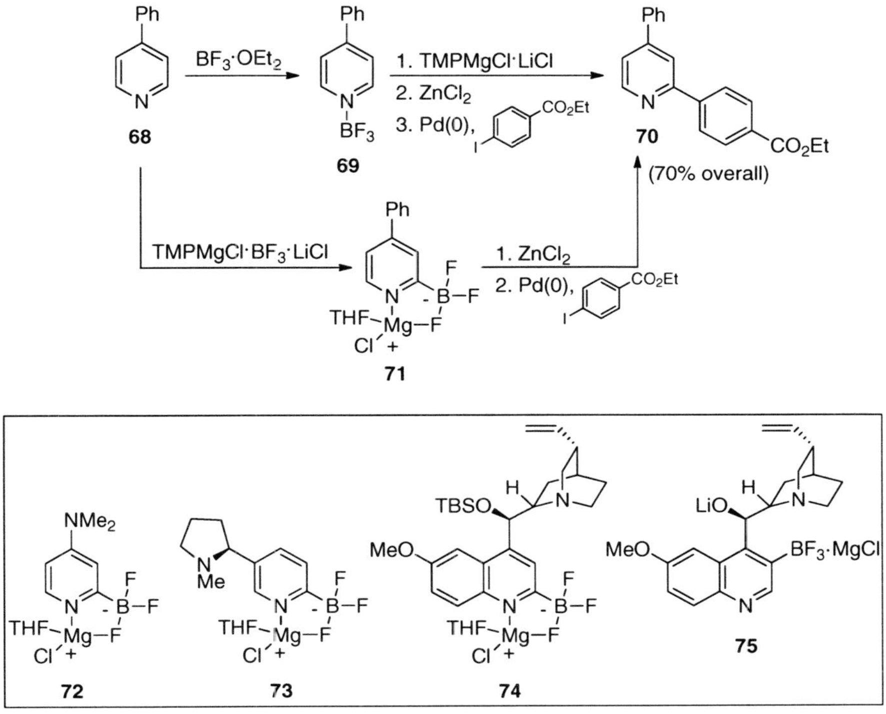

Scheme 14 Boronation of pyridines and quinolines with TMPMgCl·LiCl/BF_3

3 Zincates, Cadmiates, and Their Alkali Metal Salt-Modified Congeners

In terms of popularity, it may be argued that zincates, matched only by magnesiates, are the most widely experimented with metalates currently in use. Zincates have been known for years, dating back to 1858 when Wanklyn prepared Et_3ZnNa and Et_3ZnK [49]. Their reactivity and utility as viable reagents for deprotonation and halogen–metal exchange reactions of pyridines, quinolines, and isoquinolines, has been explored only recently. The metalation of aromatic compounds using zincates was initially demonstrated by Kondo et al. in 1999 with the use of (TMP)Zn(*t*-Bu)$_2$Li [50] but has since expanded in substrate scope. Current popular lithium zincates are mainly found in two forms wherein the zinc atom holds a formal 1^- or 2^- charge. The R groups that are incorporated into zincates vary from simple alkyl groups such as *n*-Bu, *t*-Bu, or Me [51–54] to amides such as TMP [53, 55–58] and DA [55, 59], or combinations of the aforementioned [51, 53]. Lithium is by far the most prevalent counter ion; however, examples of sodium [60] or potassium [56, 57] zincates exist as well. TMP-zinc complexes also readily aggregate with $MgCl_2$ and/or LiCl salts to produce formally neutral Zn complexes such as $(TMP)_2Zn{\cdot}2MgCl_2{\cdot}2LiCl$ [27, 44, 61, 62] or TMPZnCl·LiCl [63, 64], which

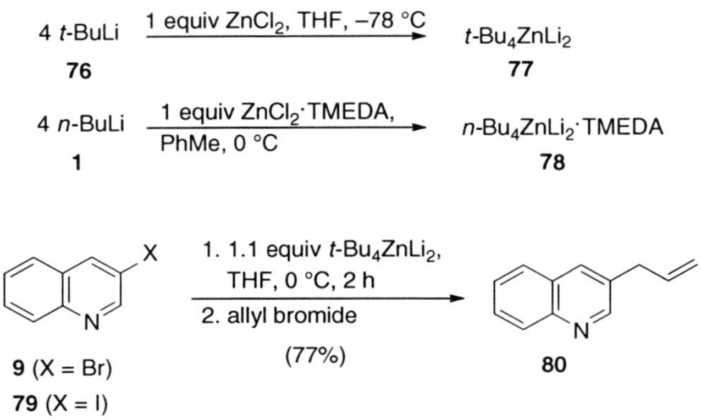

Scheme 15 Preparation of zincates and halogen–zinc exchange of 3-haloquinolines

behave like their anionic and dianionic congeners. The chemical similarities of zinc and cadmium have also led to investigations of $(TMP)_3CdLi$ for deprotonative metalation by Mongin and coworkers beginning in 2008 [53].

3.1 Halogen–Zinc Exchange Reactions of Pyridines, Quinolines, and Isoquinolines

The use of zincates for halogen–metal exchange reactions of π-deficient aromatics has not been as extensive as for the corresponding magnesiates, but there are some important reagents in development for this purpose. Foremost among the new reagents is dianionic $t\text{-}Bu_4ZnLi_2$ (**77**) and $n\text{-}Bu_4ZnLi_2{\cdot}TMEDA$ (**78**), which are prepared by addition of 4 equiv of alkyllithium to $ZnCl_2$ or $ZnLi_2{\cdot}TMEDA$ at low temperature (Scheme 15). In 2006, experiments by Uchiyama and coworkers showed that 3-bromo- or 3-iodo-quinoline undergo clean halogen–zinc exchange with 1.1 equiv of **77** to give the 3-allylquinoline in 77% yield after allyl bromide quench [52, 65]. This reagent is remarkable for the fact that it promotes halogen–zinc exchange even in the presence of acidic hydroxyl groups in some substrates, thereby eliminating the need for their protection.

Unlike the reagent **77**, TMEDA-modified **78** offers the important advantage of requiring as little as 0.25 equiv to affect halogen–magnesium exchange, although 0.33 equiv ensures higher yields [54]. The reaction stoichiometry implies that tri- or tetra-arylzinc intermediates are likely produced. Bromine–zinc exchange using **78** has been applied to a substantial number of variously substituted pyridines to give the products **82–93** as summarized in Scheme 16 after electrophile (E^+) quench. Standard electrophiles such as I_2, allyl bromide, and benzaldehydes are viable, as is Pd(0) catalyzed cross-coupling of the intermediate zincate with other haloaromatics (e.g., **13** → **81**). The use of one or more equivalents of **78** on pyridyl dibromides results in double Br/Zn exchange to give products **91–93** after electrophile quench.

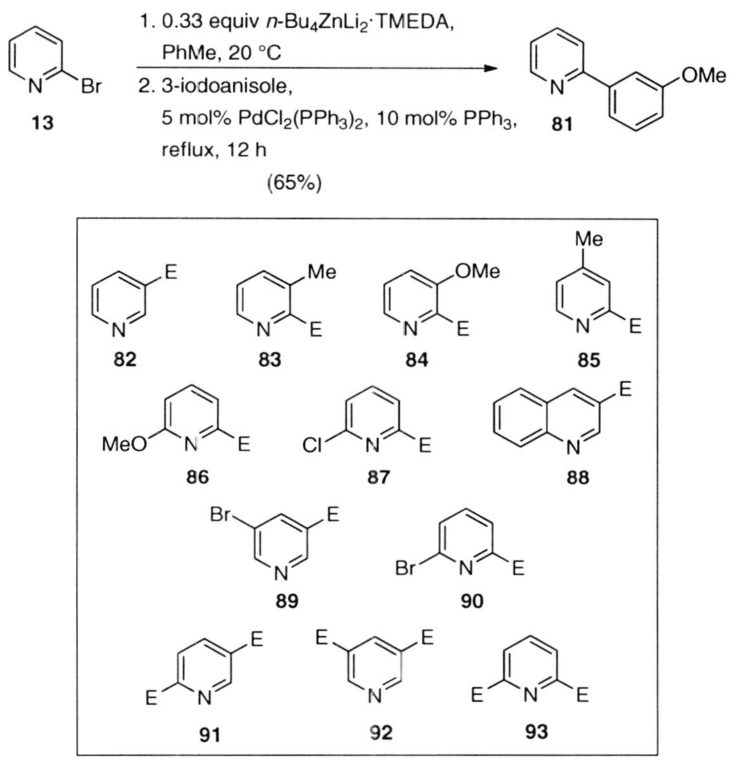

Scheme 16 Bromine–zinc exchange of variously substituted pyridines

Another reason to anticipate further development in halogen–zinc exchange methodology is the recent demonstration by Knochel and coworkers that zincate **95** promotes Cl/Zn exchange in chloropyridines and isoquinolines [66] in the presence of cobalt salts (Scheme 17). The reagent is prepared by I/Mg exchange of aryl ether **94** followed by sequential addition of $ZnCl_2$ and *t*-BuLi. Hence, exposure of chloropyridines **96**, **98**, and **100** or 1-chloroisoquinoline **102** to **95** in the presence of 20 mol% $Co(acac)_2$ and 50 mol% 4-fluorostyrene results in Cl/Zn exchange for each substrate. The corresponding aryl carbanions may be allylated under copper catalysis or trapped with iodine.

3.2 *Deprotonative Zincation of Pyridines, Quinolines, and Isoquinolines*

Current interest in direct zincation of heterocyclic aromatics can be traced to the seminal work of Kondo in 1999 [50] and 2001 [55], which described the use of (TMP)Zn(*t*-Bu)$_2$Li (**104**) and (DA)Zn(*t*-Bu)$_2$Li in regioselective metalation–iodination of

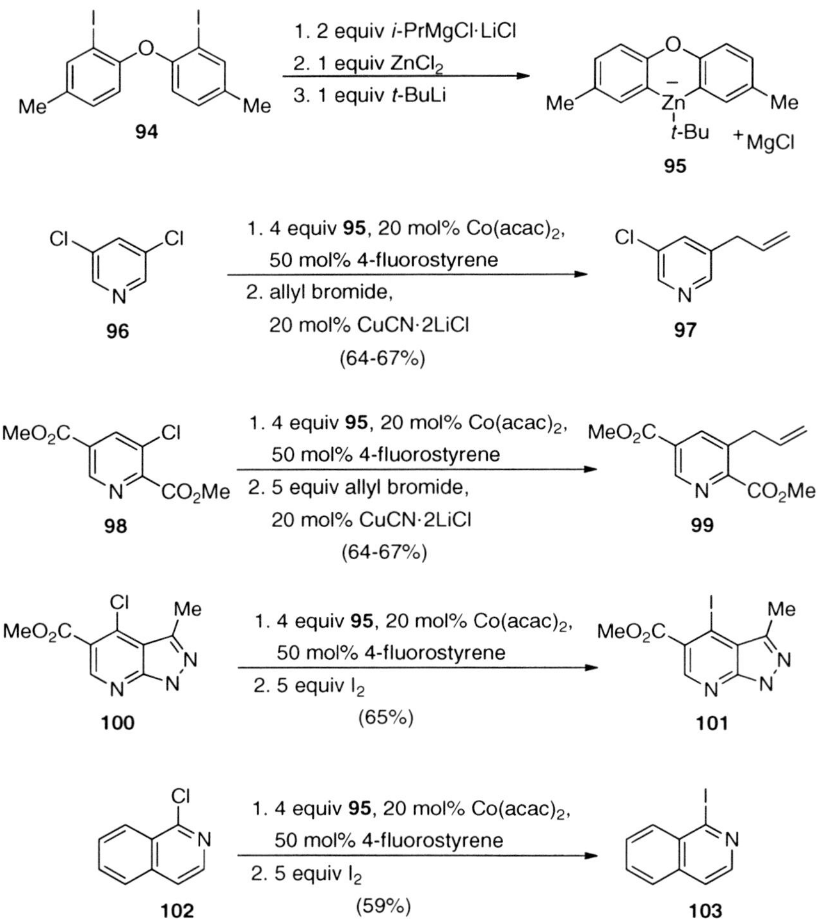

Scheme 17 Chlorine–zinc exchange with zincate **95**

pyridine, quinoline, isoquinoline, and 3-bromopyridine. These amidozincate reagents are readily prepared by addition of t-Bu_2Zn to the appropriate lithium amide in THF at −78°C (e.g., **37** → **104**) (Scheme 18). Metalation with **104** has recently expanded to include 2-, 3-, or 4-methoxypyridines (**105**, **107**, **109**) [67], which afford iodides **106**, **108**, and **110** in 70–92% after I_2 quench.

In related transformations, it has been shown that zincation of quinoline or isoquinoline with (TMP)Zn(t-Bu)$_2$Li in the presence of Lewis acids such as trimethylborate [58] results in generation of the t-Bu adducts **114** and **115** (Scheme 19). The formation of quinoline **114** has been rationalized by N-B (OMe)$_3$ induced migration of the t-Bu group, from zinc in the initial metalated species, to the 2-position of the heterocycle (**111** → **112** → **113** → **114**).

AMSM analogues of **104** have come to the fore in recent years to address the sensitivity of aldehyde and nitro-containing substrates with magnesium-based reagents

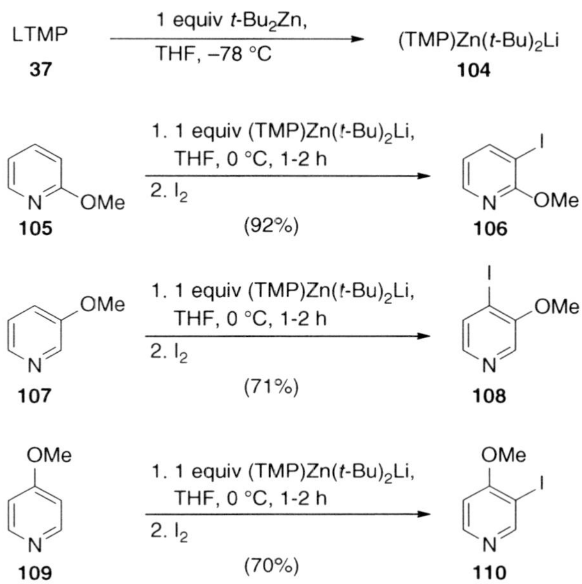

Scheme 18 Preparation of zincate **104** and its use in metalation of methoxypyridines

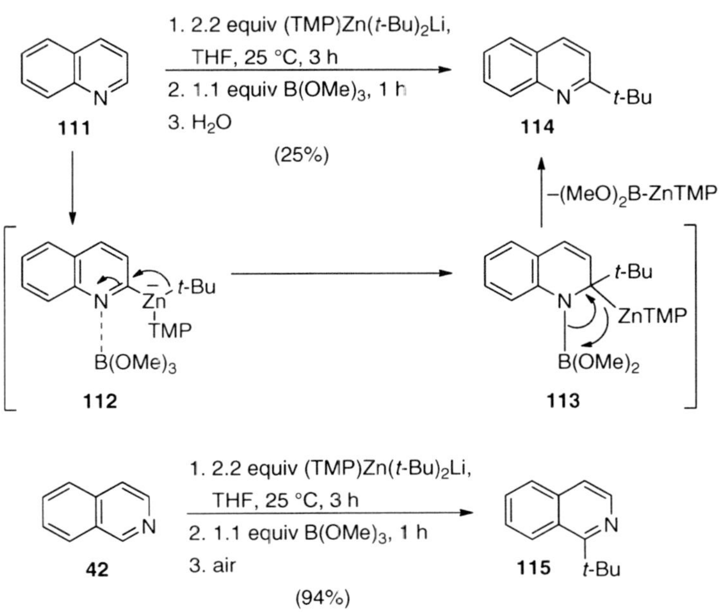

Scheme 19 Zincation of quinoline and $B(OMe)_3$-induced migration of *t*-Bu

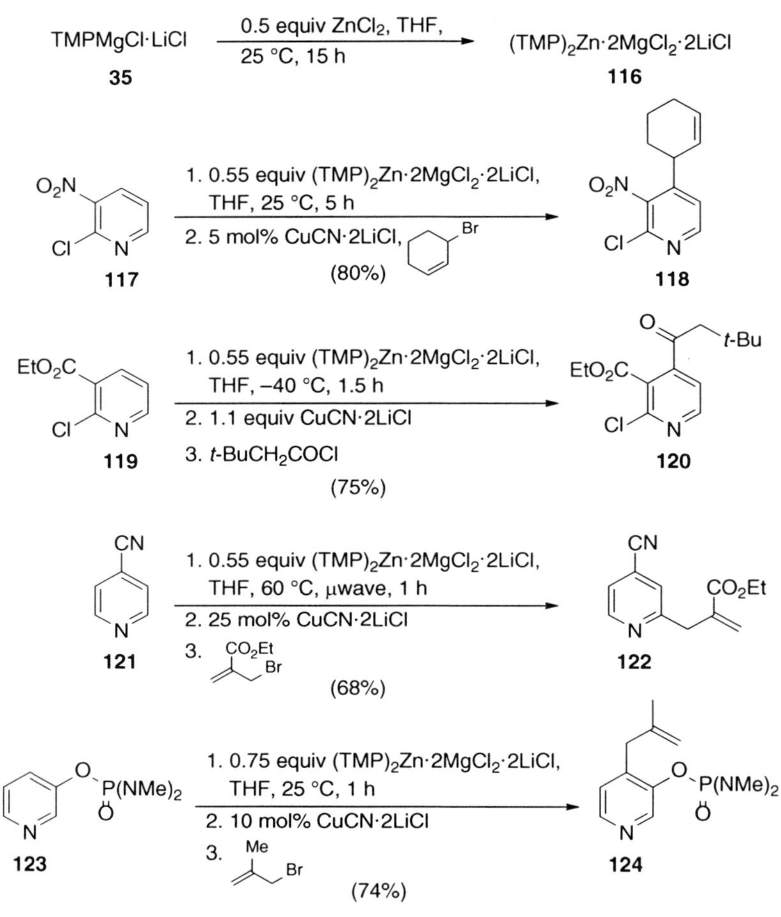

Scheme 20 Preparation and use of $(TMP)_2Zn{\cdot}2MgCl_2{\cdot}2LiCl$ in pyridine zincation

such as TMPMgCl·LiCl. The first within this series, $(TMP)_2Zn{\cdot}2MgCl_2{\cdot}2LiCl$ (**116**) [61], is produced by addition of TMPMgCl·LiCl to 0.5 equiv of $ZnCl_2$ (Scheme 20). Base **116** is capable of deprotonating chloro-, and nitro-pyridines within 5 h at temperatures ranging from −40°C to 25°C to give substituted products **118** and **120** upon CuCN·2LiCl transmetalation and electrophile quench. For some substrates (e.g., **121**) metalation may be completed within 1 h at 60–80°C under microwave irradiation [62]. Similarly, pyridine-3-phosphorodiamidate **123** undergoes clean zincation of the 4-position leading to allylated product **124** [27].

To avoid the need for reaction acceleration under microwave conditions, the more reactive and selective base TMPZnCl·LiCl (**125**), prepared from LTMP and $ZnCl_2$, may supersede **116** for some transformations (Scheme 21). A case-in-point is zincation of 2-chloro-3-nitropyridine **117**, which is conducted at ambient temperature for 45 min to provide the cyclohexenyl adduct **118** or ketone **126** in 73% or 77% yields after copper transmetalation [63, 64]. The latter reaction has been

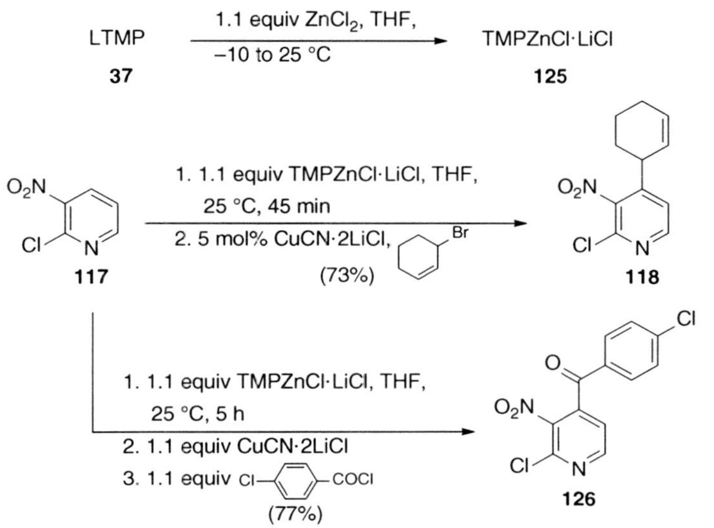

Scheme 21 Preparation and use of TMPZnCl·LiCl in pyridine zincation

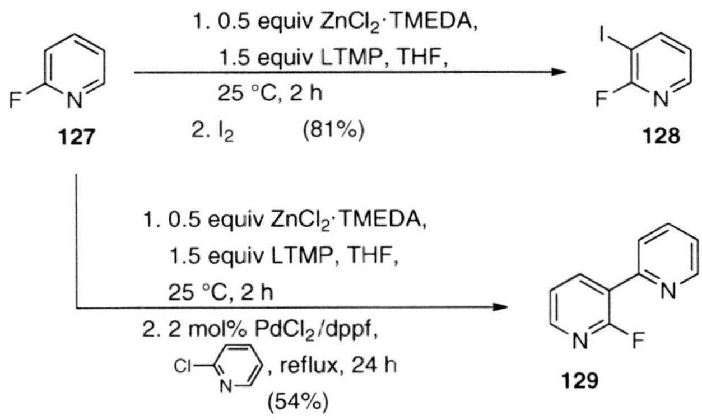

Scheme 22 Zincation of 2-fluoropyridine with LTMP/$ZnCl_2$·TMEDA

performed on 50 mmol scale, indicating the potential suitability of TMPZnCl·LiCl for process chemistry applications.

Although not strictly described as an AMSM zinc reagent, metalation using a synergic combination of LTMP (1.5 equiv) and $ZnCl_2$·TMEDA (0.5 equiv) provides an alternative to TMPZnCl·LiCl [68, 69]. Accordingly, exposure of 2-fluoropyridine to this mixture results in C3 metalation to afford iodide **128** after I_2 quench or bipyridine **129** upon cross-coupling (Scheme 22). The iodide is accompanied by small amounts (6%) of the 3,6-diiodide.

The preceding AMSM zinc reagents make use of lithium and/or magnesium salts to facilitate reactivity and solubility of the base. A potassium-containing ate complex, prepared by combining KTMP, diethylzinc, and the ligand PMDETA (*N,N,N′,N″,N″*-

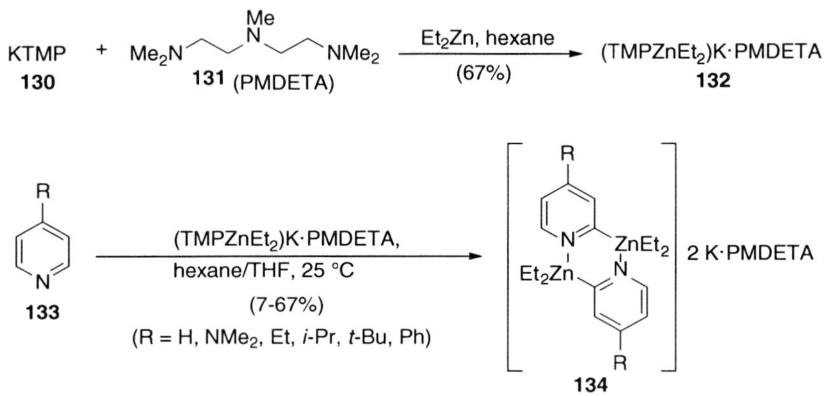

Scheme 23 Isolable pyridyl zincates via metalation with $(TMPZnEt_2)K{\cdot}PMDETA$

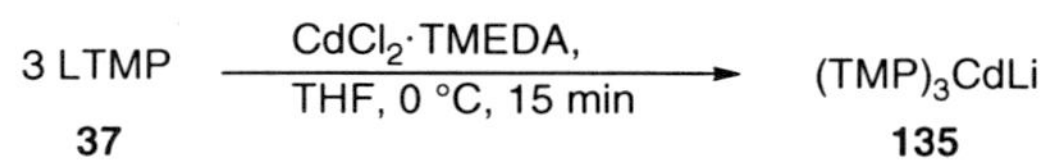

Scheme 24 Preparation of $(TMP)_3CdLi$

pentamethyldiethylenetriamine) to furnish the zincate $(TMPZnEt_2)K{\cdot}PMDETA$ (**132**), has been reported by Mulvey and coworkers [57] (Scheme 23). This reagent generates isolable C2-zincates from a variety of 4-substituted pyridines (**134**), some of which have been characterized by X-ray crystallography. Future synthetic applications of this base, which may be viewed as an updated version of the original Kondo complex, are anticipated.

3.3 Deprotonative Cadmiation of Pyridines

The ease of formation of $(TMP)_3CdLi$ (**135**) from 3 equiv of LTMP and 1 equiv of $CdCl_2{\cdot}TMEDA$ has enabled exploration of this ate complex for metalation of variously substituted pyridines beginning in 2009 (Scheme 24) [38, 70]. Unlike the 3:1 synergic mixture of LTMP and $ZnCl_2{\cdot}TMEDA$ described previously [68, 69], spectroscopic evidence and early metalation studies have shown that all 3 equiv of LTMP are consumed in the presence of $CdCl_2{\cdot}TMEDA$, suggesting that the active base does in fact have the putative formulation of **135**.

Regioselectivity in metalation of pyridyl substrates with $(TMP)_3CdLi$ is very much substrate dependent [71]. Whereas 3-cyanopyridine (**136**), ethyl pyridinecarboxylates (**137**, **138**, **48**), and triazolopyridines (**33**) all undergo Cadmiation–iodination to give the products **139–143** (Scheme 25), the same is not true of 2- or 4-cyanopyridine, 2-, 3-, or 4-chloropyridines, 2,6-dihalopyridines or diethyl pyridine-2,6-dicarboxylate, which give mixtures of regioisomers and/or diiodinated products in varying proportions [72].

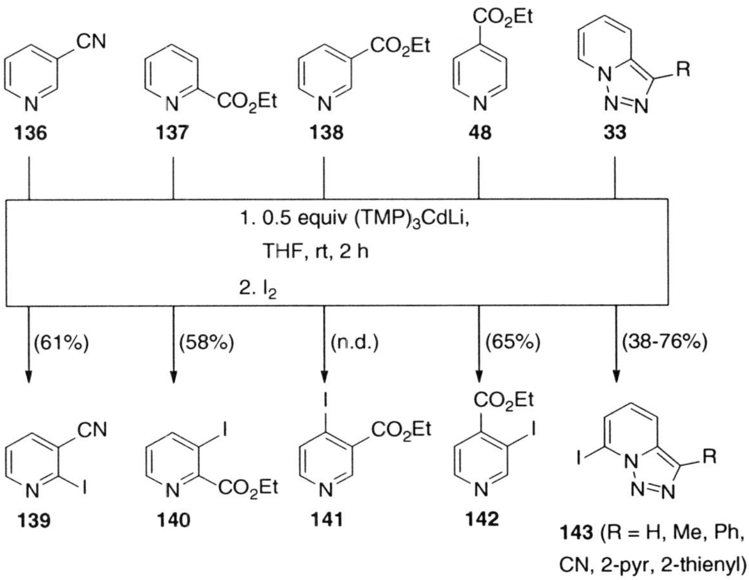

Scheme 25 Cadmiation of various pyridines with $(TMP)_3CdLi$

4 "Ate" Bases of Aluminum, Zirconium, and Lanthanum in Metalation of Pyridines and Quinolines

For this section, ate reagents of aluminum, zirconium, and lanthanum have been grouped together owing to their positions at or near the beginning of their respective p-, d-, or f-blocks in the periodic table. The oxophilicity of metal amide bases derived from these elements gives them similar behavior in metalation of many aromatics [13]. This behavior is typified by i-$Bu_3Al(TMP)Li$ (**144**), described by Uchiyama and coworkers [73], which is an excellent base for metalation of substrates containing methoxy or other oxygen-containing directing groups [74], but displays no propensity for halogen–metal exchange reactions that are common for magnesiates and zincates (Scheme 26). These observations are paralleled by the neutral AMSM trisamide reagents $(TMP)_3Al{\cdot}3LiCl$ (**145**) and $[(t\text{-Bu})((t\text{-Bu})(i\text{-Pr})CH)N]_3Al{\cdot}3LiCl$ (**148**) developed by Knochel and coworkers [75]. Complex **148** promotes C5 alumination of methoxyquinoline **149** to afford derivative **150** after Negishi coupling [26].

The *ortho* alumination of 2-methoxypyridine (**105**) may be performed competently with any of the three reagents **144**, **145**, or **148**, although **148** is the preferred AMSM reagent over **145**. In the case of metalation with aluminate **144**, quench with iodine provides **106** in 82% yield (Scheme 27). Similarly, alumination of **105** followed by sequential transmetalation with zinc chloride, $CuCN{\cdot}2LiCl$, and acid chloride quench affords ketone **151** in 81% yield. 6-Chloro-2-methoxypyridine (**152**) undergoes alumination with the same regioselectivity as **105** to provide aryl ketone **153** following an analogous procedure.

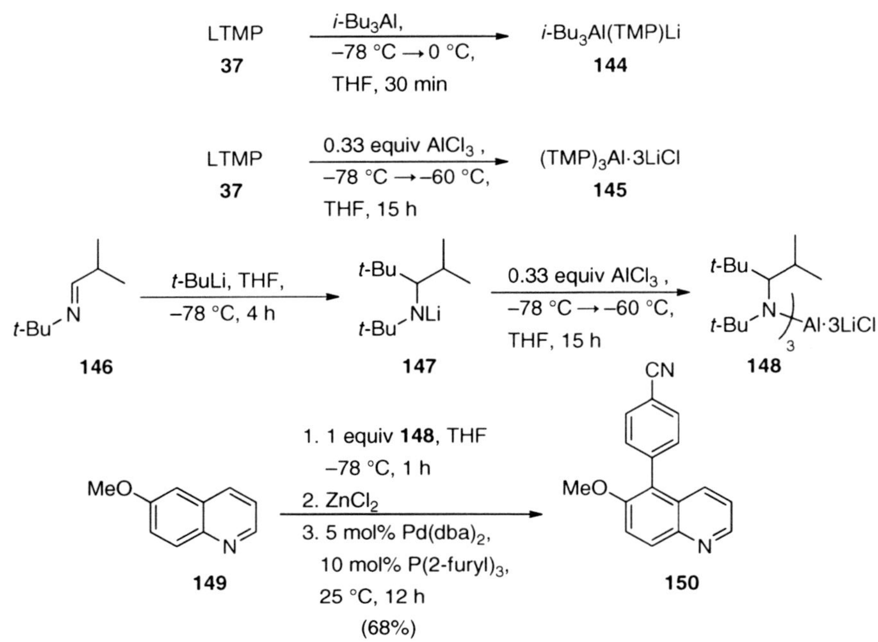

Scheme 26 Synthesis of aluminate bases and *ortho*-alumination of 6-methoxyquinoline

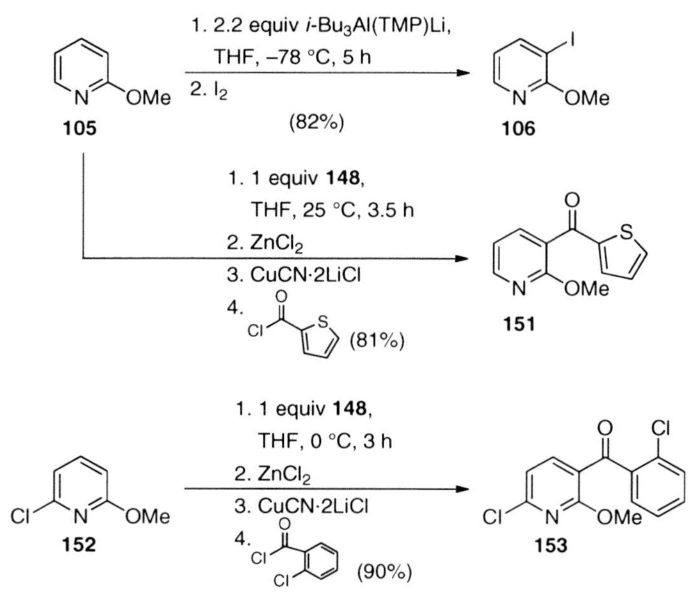

Scheme 27 *Ortho*-alumination of 2-methoxypyridines

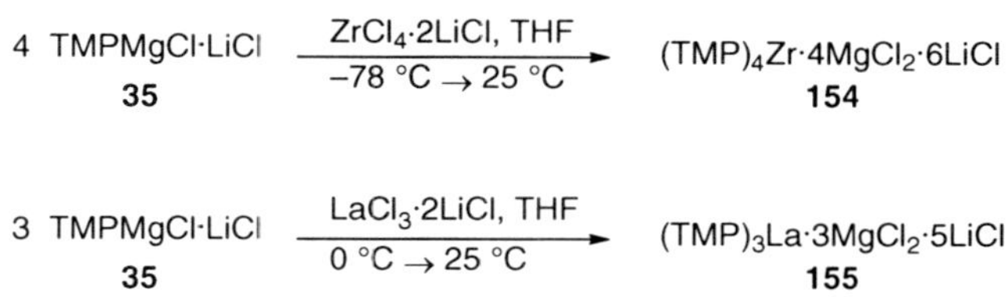

Scheme 28 Preparation of $(TMP)_4Zr{\cdot}4MgCl_2{\cdot}6LiCl$ and $(TMP)_3La{\cdot}3MgCl_2{\cdot}5LiCl$

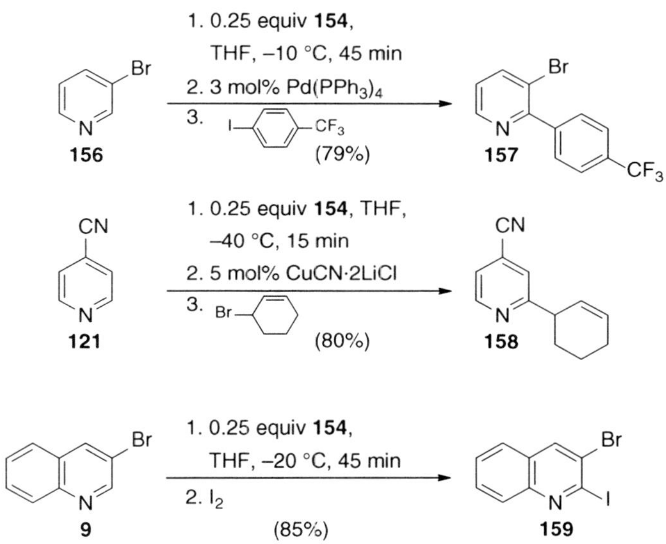

Scheme 29 *Ortho*-zirconation of 3-bromopyridine, 4-cyanopyridine, and 3-bromoquinoline with $(TMP)_4Zr{\cdot}4MgCl_2{\cdot}6LiCl$

On the other hand, metalation of halo- and cyano-pyridines, pyridinecarboxylates, quinoline, and 3-bromoquinoline may also be achieved with the mixed metal species $(TMP)_4Zr{\cdot}4MgCl_2{\cdot}6LiCl$ (**154**) [76] or $(TMP)_3La{\cdot}3MgCl_2{\cdot}5LiCl$ (**155**) [77, 78]. Both reagents are prepared by addition of the requisite equivalents of TMPMgCl·LiCl (**35**) to $ZrCl_4{\cdot}2LiCl$ or $LaCl_3{\cdot}2LiCl$ (Scheme 28).

Ortho-zirconation with **154** proceeds smoothly on 3-bromopyridine (**156**), 4-cyanopyridine (**121**), and 3-bromoquinoline (**9**) to afford the products depicted in Scheme 29. It is notable that zirconation in the presence of boron trifluoride etherate may alter the regioselectivity of the deprotonation for these and other substrates [76].

The advantages of employing lanthanum reagent **155** are related to the low basicity of the derived metalated intermediates, which enables direct electrophile quench with enolizable substrates. Thus, pyridines **119** and **161** afford products **160** or **162** after metalation with **155**, followed by quench with α-tetralone or cycloheptanone (Scheme 30).

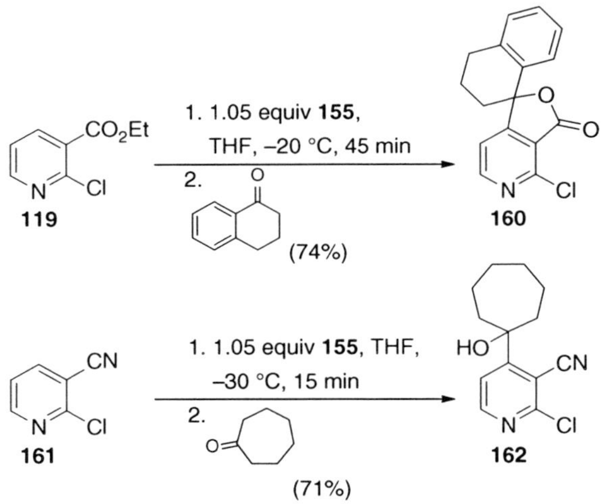

Scheme 30 Metalation of 2-chloropyridines with $(TMP)_3La{\cdot}3MgCl_2{\cdot}5LiCl$

5 "Ate" Bases of Copper, Cobalt, and Manganese in Metalation of Pyridines

The potential utility of copper(I) amides in *N*-heteroaromatic chemistry has been demonstrated by iodine–metal exchange reactions of *N*-phenylsulfonyl-2,3-diiodoindole using $(PhMe_2CCH_2)_2CuLi$, and C2-metalation of *N*-Boc-indole with $MeCu(TMP)(CN)Li_2$ by the groups of Knochel [79] and Uchiyama [80], respectively. These results laid the foundation for the recent extension of deprotonative cupration to variously substituted pyridines by Mongin and coworkers. The key to this advance rested on the development of a reliable synthesis of $(TMP)_2CuLi$ (**164**), which may be considered to be a monoanionic analogue of Uchiyama's dianionic cuprate. Convenient synthesis of this base begins with $CuCl_2{\cdot}TMEDA$, which is reduced to CuCl with 0.5 equiv of *n*-BuLi before in situ addition of 2 equiv of LTMP [81] (Scheme 31). The base prepared in this manner is suitable for C3-cupration of 2-halo- and 2-methoxy-pyridines (**165**) [82]. It has also been employed in C4-cupration of 3,5-dibromopyridine, and C3/C4 cupration of 2,6-dichloropyridine, among other substrates [82, 83].

Mongin and coworkers have explored the possibility of applying amidocobalt ate complexes in metalation of pyridines. For this purpose, a moderately effective base with the putative formulation $(TMP)_3CoLi$ may be prepared by addition of 6 equiv of LTMP to 2 equiv of $CoBr_2$ [84]. Exposure of 2-methoxy- and 2,6-dimethoxypyridines (**105** or **168**) to this mixture afforded the corresponding iodides (**106** and **169**) and homocoupled by-products (**167** and **170**) after iodine quench (Scheme 32).

In contrast, Knochel and coworkers have formed $(TMP)_2Mn{\cdot}2MgCl_2{\cdot}4LiCl$ (**171**) by addition of 2 equiv of TMPMgCl·LiCl to $MnCl_2{\cdot}2LiCl$ in THF at

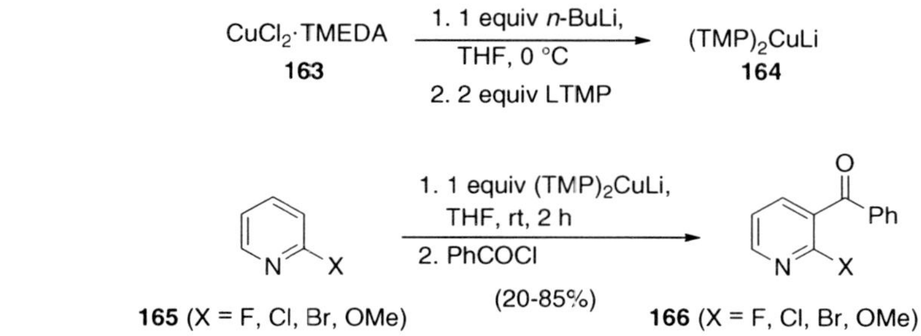

Scheme 31 Preparation of $(TMP)_2CuLi$ and *ortho*-cupration of 2-substituted pyridines

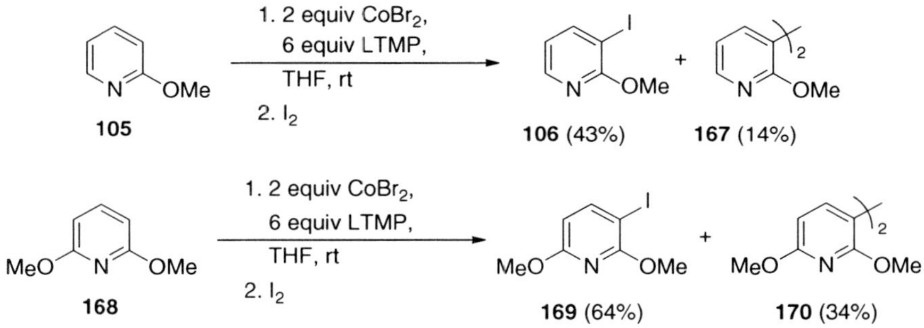

Scheme 32 Metalation of methoxypyridines with "$(TMP)_3CoLi$"

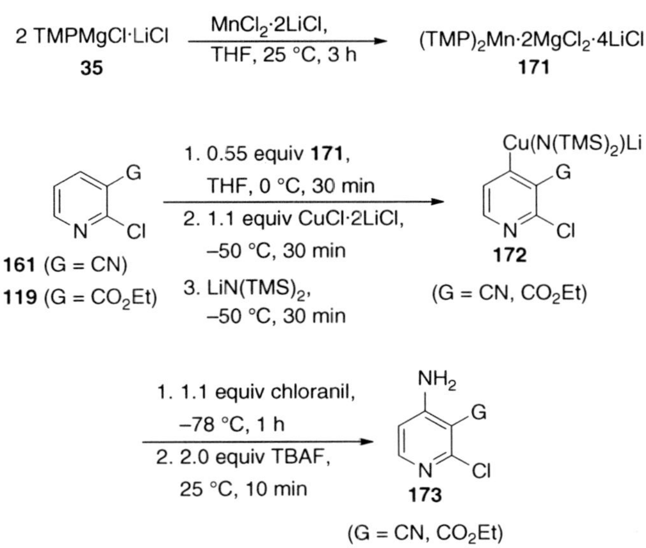

Scheme 33 *Ortho*-manganation of 2-chloropyridines with $(TMP)_2Mn{\cdot}2MgCl_2{\cdot}4LiCl$

room temperature [85]. This reagent promotes C4-manganation of chlorides **119** or **161** to give, after copper-mediated oxidative amination and N-desilylation, 4-aminopyridines **173** in 65–75% yield (Scheme 33).

6 Conclusions

The past several years have been witness to dramatic improvements in the technology to metalate electrophilic π-deficient aromatics such as pyridines, quinolines, and isoquinolines. These advances have been made possible by developments of bimetallic "ate" bases that induce halogen–metal exchange and deprotonation with moderate to high levels of regiocontrol. The new reagents invariably avoid issues of nucleophilic additions to aromatic rings, a problem that is common with alkyllithium reagents. In addition, many functional groups that are otherwise incompatible to metalation conditions can be used without protection or conversion to more robust moieties.

Continued rapid growth in the development of "ate" bases will require that this methodology migrates away from the academic research lab and towards potential large-scale industrial applications. Some steps have already been made in that direction [28, 63, 64]. Practical issues that may need to be overcome include the current requirement to incorporate the TMP group in many bases, which may not be cost effective commercially. This potential limitation has been recognized vis-à-vis alternatives to bases **35** and **36** (e.g., *t*-Bu(*i*-Pr)NMgCl·LiCl and [*t*-Bu(*i*-Pr)N]$_2$Mg·2LiCl) and additional modifications to current reagents may be anticipated in this regard [25]. Finally, the stability of the bases themselves needs to be demonstrated under long-term storage and transport.

Nevertheless, the utility of "ate" bases for metalation of the substrates described in this chapter is undeniable considering their regio- and chemoselectivity, and the generally mild conditions under which they operate. The practicality of being able to use the intermediates directly in subsequent reactions, including cross-coupling, is a potential advantage that may work in favor of future applications both academically and industrially. Despite the flat molecules covered in this chapter, one wonders if chiral analogues of "ate" bases that can induce appreciable asymmetry in prochiral substrates are imminent.

References

1. Mulvey RE (2009) Acc Chem Res 42:743
2. Whisler MC, MacNeil S, Snieckus V, Beak P (2004) Angew Chem Int Ed 43:2206
3. Chinchilla R, Nájera C, Yus M (2004) Chem Rev 104:2667
4. Cottet F, Marull M, Mongin F, Espinosa D, Schlosser M (2004) Synthesis 1619
5. Seyferth D (2009) Organometallics 28:2
6. Rathman T, Bailey WF (2009) Org Process Res Dev 13:144

7. Nakao Y (2011) Synthesis 3209
8. Mulvey RE, Mongin F, Uchiyama M, Kondo Y (2007) Angew Chem Int Ed 46:3802
9. Haag B, Mosrin M, Ila H, Malakhov V, Knochel P (2011) Angew Chem Int Ed 50:9794
10. Mulvey RE (2006) Organometallics 25:1060
11. Armstrong DR, Kennedy AR, Mulvey RE, Robertson SD (2011) Chem Eur J 17:8820
12. Armstrong DR, Clegg W, Dale SH, Hevia E, Hogg LM, Honeyman GW, Mulvey RE (2006) Angew Chem Int Ed 118:3859
13. Mulvey RE, Armstrong DR, Conway B, Crosbie E, Kennedy AR, Robertson SD (2011) Inorg Chem 50:12241
14. Wittig G, Meyer FJ, Lange G (1951) Liebigs Ann 571:167
15. Hauser CR, Walker HG (1947) J Am Chem Soc 69:295
16. Eaton PE, Lee C-H, Xiong Y (1989) J Am Chem Soc 111:8016
17. Krasovskiy A, Knochel P (2004) Angew Chem Int Ed 43:3333
18. Baron O, Knochel P (2005) Angew Chem Int Ed 44:3133
19. Krasovskiy A, Krasovskaya V, Knochel P (2006) Angew Chem Int Ed 45:2958
20. Ren H, Knochel P (2006) Chem Commun 726
21. Clososki GC, Rohbogner CJ, Knochel P (2007) Angew Chem Int Ed 46:7681
22. Boudet N, Lachs JR, Knochel P (2007) Org Lett 9:5525
23. Lin W, Chen L, Knochel P (2007) Tetrahedron 63:2787
24. Dong Z, Clososki CG, Wunderlich SH, Unsinn A, Li J, Knochel P (2009) Chem Eur J 15:457
25. Rohbogner CJ, Wunderlich SH, Clososki GC, Knochel P (2009) Eur J Org Chem 1781
26. Jaric M, Haag BA, Unsinn A, Karaghiosoff K, Knochel P (2010) Angew Chem Int Ed 49:5451
27. Rohbogner CJ, Wirth S, Knochel P (2010) Org Lett 12:1984
28. Wunderlich SH, Rohbogner CJ, Unsinn A, Knochel P (2010) Org Process Res Dev 14:339
29. Yamada S, Knochel P (2010) Synthesis 2490
30. Melzig L, Rauhut CB, Naredi-Ranier N, Knochel P (2011) Chem Eur J 17:5362
31. Knochel P, Diène C (2011) Chimie 14:842
32. Jaric M, Haag BA, Manolikakes SM, Knochel P (2011) Org Lett 13:2306
33. Awad H, Mongin F, Trécourt F, Quéguiner G, Marsais F, Blanco F, Abarca B, Ballesteros R (2004) Tetrahedron Lett 45:6697
34. Dumouchel S, Mongin F, Trécourt F, Quéguiner G (2003) Tetrahedron 59:8629
35. Dumouchel S, Mongin F, Trécourt F, Quéguiner G (2003) Tetrahedron Lett 44:2033
36. Dumouchel S, Mongin F, Trécourt F, Quéguiner G (2003) Tetrahedron Lett 44:3877
37. Awad H, Mongin F, Trécourt F, Quéguiner G, Marsais F (2004) Tetrahedron Lett 45:7873
38. Bentabed-Ababsa G, Blanco F, Derdour A, Mongin F, Trécourt F, Quéguiner G, Ballesteros R, Abarca B (2009) J Org Chem 74:163
39. Ila H, Baron O, Wagner AJ, Knochel P (2006) Chem Lett 35:2
40. Inoue A, Kitagawa K, Shinokubo H, Oshima K (2001) J Org Chem 66:4333
41. Iida T, Wada T, Tomimoto K, Mase T (2001) Tetrahedron Lett 42:4841
42. Brodmann T, Koos P, Metzger A, Knochel P, Ley SV (2012) Org Process Res Dev 16:1102
43. Kopp F, Krasovskiy A, Knochel P (2004) Chem Commun 2288
44. Dunst C, Kienle M, Knochel P (2010) Synthesis 2313
45. Andrews PC, Kennedy AR, Mulvey RE, Raston CL, Roberts BA, Rowlings RB (2000) Angew Chem Int Ed 39:1960
46. Schlecker W, Huth A, Ottow E (1996) J Org Chem 60:8414
47. Oae S, Kawai T, Furukawa N (1984) Tetrahedron Lett 25:69
48. Kessar SV, Singh P, Singh KN, Dutt M (1991) Chem Commun 570
49. Wanklyn JA (1858) Liebigs Ann 107:125
50. Kondo Y, Shilai M, Uchiyama M, Sakamoto T (1999) J Am Chem Soc 121:3539
51. Uchiyama M, Miyoshi T, Kajihara Y, Sakamoto T, Otani Y, Ohwada T, Kondo Y (2002) J Am Chem Soc 124:8514
52. Furuyama T, Yonehara M, Arimoto S, Kobayashi M, Matsumoto Y, Uchiyama M (2008) Chem Eur J 14:10348

53. L'Helgoual'ch J-M, Bentabed-Ababsa G, Chevallier F, Yonehara M, Uchiyama M, Derdourb A, Mongin F (2008) Chem Commun 5375
54. Chau NTT, Meyer M, Komagawa S, Chevallier F, Fort Y, Uchiyama M, Mongin F, Gros PC (2010) Chem Eur J 16:12425
55. Imahori T, Uchiyama M, Sakamoto T, Kondo Y (2001) Chem Commun 2450
56. Conway B, Graham DV, Hevia E, Kennedy AR, Klett J, Mulvey RE (2008) Chem Commun 2638
57. Clegg W, Conway B, Graham DV, Hevia E, Kennedy AR, Mulvey RE, Russo L, Wright DS (2009) Chem Eur J 15:7074
58. Seo HJ, Yoon SJ, Jang SH, Namgoong SK (2011) Tetrahedron Lett 52:3747
59. Clegg W, Garcia-Àlvarez J, Garcia-Àlvarez P, Graham DV, Harrington RW, Hevia E, Kennedy AR, Mulvey RE, Russo L (2008) Organometallics 27:2654
60. Armstrong DR, Clegg W, Dale SH, Hevia E, Hogg LM, Honeyman GW, Mulvey RE (2006) Angew Chem Int Ed 45:3775
61. Wunderlich SH, Knochel P (2007) Angew Chem Int Ed 46:7685
62. Wunderlich S, Knochel P (2008) Org Lett 10:4705
63. Mosrin M, Knochel P (2009) Org Lett 11:1837
64. Bresser T, Monzon G, Mosrin M, Knochel P (2010) Org Process Res Dev 14:1299
65. Uchiyama M, Furuyama T, Kobayashi M, Matsumoto Y, Tanaka K (2006) J Am Chem Soc 128:8404
66. Melzig L, Diène CR, Rohbogner CJ, Knochel P (2011) Org Lett 13:3174
67. Blair VL, Blakemore DC, Hay D, Hevia E, Pryde DC (2011) Tetrahedron Lett 52:4590
68. Snégaroff K, Komagawa S, Chevallier F, Gros PC, Golhen S, Roisnel T, Uchiyama M, Mongin F (2010) Chem Eur J 16:8191
69. L'Helgoual'ch JM, Seggio A, Chevallier F, Yonehara M, Jeanneau E, Uchiyama M, Mongin F (2008) J Org Chem 73:177
70. Snégaroff K, L'Helgoual'ch JM, Bentabed-Ababsa G, Nguyen TT, Chevallier F, Yonehara M, Uchiyama M, Derdour A, Mongin F (2009) Chem Eur J 15:10280
71. Bentabed-Ababsa G, Sid Ely SC, Hesse S, Nassar E, Chevallier F, Nguyen TT, Derdour A, Mongin F (2010) J Org Chem 75:839
72. Mongin F, Uchiyama M (2011) Curr Org Chem 15:2340
73. Uchiyama M, Naka H, Matsumoto Y, Ohwada T (2004) J Am Chem Soc 126:10526
74. Naka H, Uchiyama M, Matsumoto Y, Wheatley AEH, McPartlin M, Morey JV, Kondo Y (2007) J Am Chem Soc 129:1921
75. Wunderlich SH, Knochel P (2009) Angew Chem Int Ed 48:1501
76. Jeganmohan M, Knochel P (2010) Angew Chem Int Ed 49:8520
77. Wunderlich SH, Knochel P (2010) Chem Eur J 16:3304
78. Wunderlich SH, Bresser T, Dunst C, Monzon G, Knochel P (2010) Synthesis 2670
79. Yang X, Althammer A, Knochel P (2004) Org Lett 6:1665
80. Usui S, Hashimoto Y, Morey JV, Wheatley AEH, Uchiyama M (2007) J Am Chem Soc 129:15102
81. Nguyen TT, Chevallier F, Jouikov V, Mongin F (2009) Tetrahedron Lett 50:6787
82. Nguyen TT, Marquise N, Chevallier F, Mongin F (2011) Chem Eur J 17:10405
83. Snégaroff K, Nguyen TT, Marquise N, Halauko YS, Harford PJ, Roisnel T, Matulis VE, Ivashkevich OA, Chevallier F, Wheatley AEH, Gros PC, Mongin F (2011) Chem Eur J 17:13284
84. Dayaker G, Chevallier F, Gros PC, Mongin F (2010) Tetrahedron 66:8904
85. Wunderlich SH, Kienle M, Knochel P (2009) Angew Chem Int Ed 48:7256

Top Heterocycl Chem (2013) 31: 93–130
DOI: 10.1007/7081_2012_92

Published online: 8 November 2012

Lithiations and Magnesiations on Quinoline and Isoquinoline

Floris Chevallier and Florence Mongin

Abstract Recent advances in the formation of lithiated or magnesiated quinolines and isoquinolines by deprotonative metalation and halogen–metal exchange are summarized and discussed. Subsequent trappings with electrophiles show that both methods are efficient to generate functionalized building blocks, which are of foremost importance in supramolecular chemistry, material science, or in the synthesis of biologically active compounds.

Keywords Functionalization · Halogen–metal exchange · Isoquinoline · Lithium · Magnesium · Metalation · Quinoline

Contents

F. Chevallier and F. Mongin (✉)
Institut des Sciences Chimiques de Rennes, UMR 6226 CNRS - Université de Rennes 1, Equipe Chimie et Photonique Moléculaires, Bâtiment 10A, Case 1003, Campus Scientifique de Beaulieu, 35042 Rennes, France
e-mail: florence.mongin@univ-rennes1.fr

Abbreviations

BuLi	*n*-Butyllithium
dba	Dibenzylidene acetone
DEE	Diethyl ether
DMA	Dimethylacetamide
DMAE	*N*,*N*-Dimethylaminoethoxide
DMF	Dimethylformamide
LDA	Lithium diisopropylamide
LTMP	Lithium 2,2,6,6-tetramethylpiperidide
LUMO	Lowest unoccupied molecular orbital
MesLi	Mesityllithium
MOM	Methoxymethyl
NFSI	*N*-Fluorobenzenesulfonimide
Py	Pyridyl
s-BuLi	*sec*-Butyllithium
t-BuLi	*tert*-Butyllithium
Tf	Triflate
THF	Tetrahydrofuran
TMEDA	*N*,*N*,*N*′,*N*′-Tetramethylethylenediamine

1 Introduction

Quinoline and isoquinoline are important heterocyclic cores present in numerous natural products and biologically active agents. Treatment of aromatic compounds such as substituted benzenes by organolithiums and organomagnesiums through hydrogen–metal or halogen–metal exchange in general represents an efficient, chemoselective, and regioselective way to functionalize them. In the quinoline and isoquinoline series, things are less obvious because of the sensitivity of these substrates toward nucleophilic attacks due to their low LUMO energy levels. Thus, the quinoline and isoquinoline functionalization by this way has been delayed until a good knowledge of the different hydrogen–metal or halogen–metal exchange conditions permitted to work at low reaction temperatures.

More recent bases and halogen–metal agents exhibiting reduced nucleophilic behaviors and new protocols performed at higher temperatures which are currently developed will be presented in a separate chapter.

Table 1 Deprotometalation of fluoroquinolines using lithium diisopropylamide

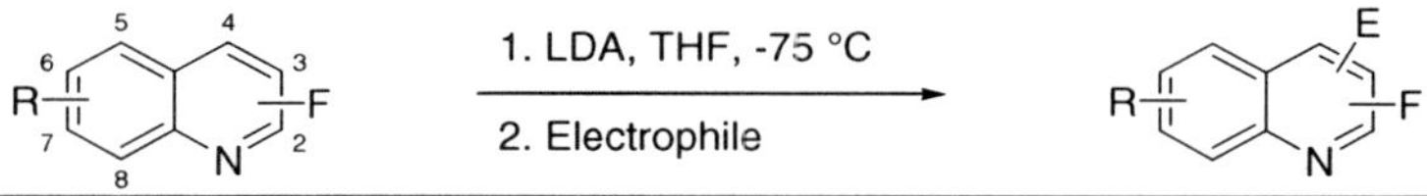

Entry	F position	R	Electrophile	E	Yield (%)
1	2	H	Me_3SiCl	3-$SiMe_3$	58
2	3	H	Me_3SiCl	4-$SiMe_3$	66
3	5	H	Me_3SiCl	6-$SiMe_3$	30
4	6	H	Me_3SiCl	5-$SiMe_3$	65
5	7	H	Me_3SiCl	8-$SiMe_3$	30
6	3	2-OBu	CO_2	4-CO_2H	82
7	3	2-Br	CO_2	4-CO_2H	94
8	3	2-Br, 6-Cl	CO_2	4-CO_2H	87
9	3	2-Br, 7-F	CO_2	4-CO_2H	68
10	3	2-Br, 7-OMe	CO_2	4-CO_2H	83
11	3	2,8-Br	CO_2	4-CO_2H	92
12	3	2-Br, 5,7-Me	CO_2	4-CO_2H	82
13	7	2-Br, 4-CF_3	CO_2	8-CO_2H	87
14	7	2-CF_3, 4-Br	CO_2	8-CO_2H	80
15	7	2-CF_3, 4-Br	Me_3SiCl	8-$SiMe_3$	68
16[a]	3	H	CO_2	4-CO_2H	64

[a]Reaction performed in the presence of *t*-BuOK

2 Deprotonative Metalation of Quinolines and Isoquinolines

2.1 Deprotonative Metalation of Haloquinolines

2.1.1 Fluoroquinolines

Haloquinolines are particularly prone to nucleophilic addition due to their low LUMO energy levels. As a consequence, treatment of fluoroquinolines with BuLi·TMEDA chelate in DEE results in 1,2 addition products. Using the same base in THF, which is a more basic solvent than DEE, allows both deprotonation and nucleophilic addition of the base to the substrate. Chemo- and regioselective metalation occurs at the position adjacent to the fluoro group using LDA in THF at −75°C, as evidenced by subsequent trapping with trimethylsilyl chloride (Table 1, Entries 1–5) [1].

Extension of this reaction to substituted fluoroquinolines such as 2-butoxy-3-fluoroquinoline (Table 1, Entry 6) [2] and various bromofluoroquinolines (Table 1, Entries 7–15) [3–6] has been reported with carbon dioxide and trimethylsilyl chloride as electrophile. It has to be mentioned that while LDA suffices to deprotonate 2-bromo-3-fluoroquinolines at the 4-position owing to the long-range assistance of the bromo group, mixed-metal reagents are required for the regioselective deprotonation-carboxylation of 3-fluoroquinoline (Table 1, Entry 16) [2].

Table 2 Deprotometalation of fluoroquinolines using lithium 2,2,6,6-tetramethylpiperidide

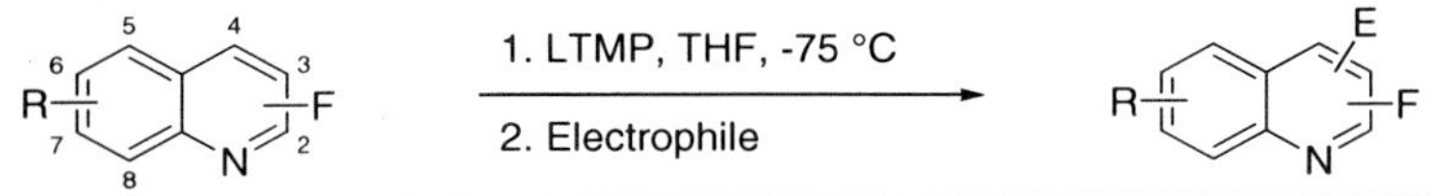

Entry	F position	R	Electrophile	E	Yield (%)
1	6	2-Ph, 3-OMe, 4-carboxamide[a]	I_2	7-I	61
2	7	2-Ph, 3-OMe, 4-carboxamide[a]	I_2	8-I	78
3	8	2-Ph, 3-OMe, 4-carboxamide[a]	I_2	7-I	28
4	8	6-OMOM	Me_3SiCl	7-$SiMe_3$	83

[a](*S*)-*N*-(1-phenylpropyl)carboxamide

Using the same amide deprotonating agent with 3-fluoro-4-iodoquinoline results in a halogen-dance reaction [7, 8] of 3-fluoro-4-iodo-2-lithioquinoline to more stable 3-fluoro-2-iodo-4-lithioquinoline and furnishes the 4-substituted 3-fluoro-2-iodoquinolines after interception with various electrophiles (H_2O, I_2, CCl_3CCl_3, CH_3CHO, PhCHO, HCO_2Et) in 74–98% yield [9–11].

Hydrogen–metal permutations of fluoroquinolines have also been described using the more hindered and basic LTMP in THF at −75 °C (Table 2) [12, 13]. This is currently the only approach reported in the literature allowing deprotonation of 8-fluoroquinoline (Table 2, Entries 3,4) and avoiding strong cation chelation in the vicinity of the fluorine and nitrogen atoms [1].

2.1.2 (Trifluoromethyl)quinolines

(Trifluoromethyl)quinolines attract more and more interest, especially because of their medicinal uses as antimalarial agents (e.g., mefloquine) [14]. Unlike the fluoro group, which only acidifies the adjacent hydrogen, long-range substituent effects of the strong electron-withdrawing trifluoromethyl group appear to be inductive activation counterbalanced by the steric hindrance of this bulky substituent [15].

The deprotonation of trifluoromethyl-substituted quinolines using LDA occurs in THF at −78°C (Table 3). While 2-, 3-, and 4-(trifluoromethyl)quinoline furnish the corresponding 3-, 4-, and 3-quinolinecarboxylic acid in moderate yields after trapping with dry ice (Table 3, Entries 1–3) [16], the presence of a bromo (or chloro) group, which acts as a second activating substituent, is probably at the origin of the better yields obtained but does not change the regioselectivity (Table 3, Entries 4–24) [5, 17, 18]. The low reactivity observed with 5-substituted 4-(trifluoromethyl)quinolines was rationalized by a steric buttressing effect (Table 3, Entry 19) [19].

The use of different bases results in modified selectivities for the (trifluoromethyl)quinolines hydrogen/metal exchange, as evidenced by subsequent interception with CO_2 followed by acidification (Table 4) [16].

In THF, bulkier LTMP attacks both the activated 3- (43% yield after trapping) and the less hindered 4-position (27% yield) of 2-(trifluoromethyl)quinoline

Table 3 Deprotometalation of (trifluoromethyl)quinolines using lithium diisopropylamide

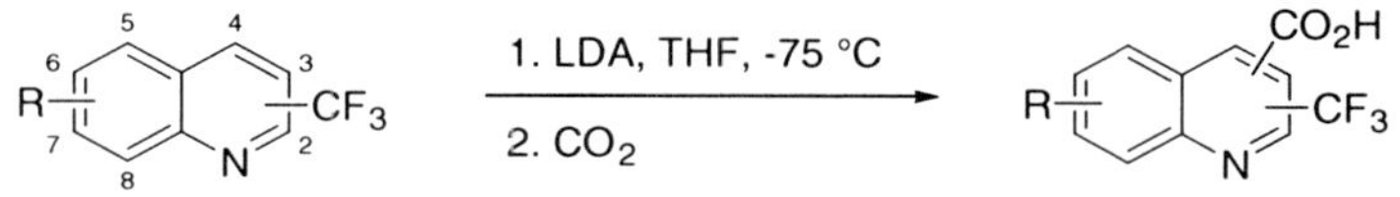

Entry	CF_3 position	R	CO_2H position	Yield (%)
1	2	H	3	31
2	3	H	4	32
3	4	H	3	54
4	2	4-Br	3	88
5	2	4-Cl	3	78
6	2	4-Br, 6-Me	3	74
7	2	4-Br, 6-F	3	81
8	2	4-Br, 6-OMe	3	89
9	2	4-Br, 7-OMe	3	86
10	2	4-Br, 8-F	3	86
11	2	4-Br, 8-Br	3	62
12	2	4-Br, 8-I	3	78
13	4	2-Br	3	81
14	4	2-Br, 6-Me	3	66
15	4	2-Br, 6-F	3	69
16	4	2-Br, 7-Me	3	88
17	4	2-Br, 7-OMe	3	82
18	4	2-Br, 8-Me	3	89
19	4	2-Br, 5,7-Me	3	26
20	4	2-Br, 6-OCF_3	3	81
21	4	2-Br, 8-F	3	83
22	4	2-Br, 8-Cl	3	89
23	4	2-Br, 8-Br	3	78
24	4	2-Br, 8-I	3	83

Table 4 Deprotometalation of (trifluoromethyl)quinolines using either lithium 2,2,6,6-tetramethylpiperidide or butyllithium·lithium *N*,*N*-dimethylaminoethoxide superbase

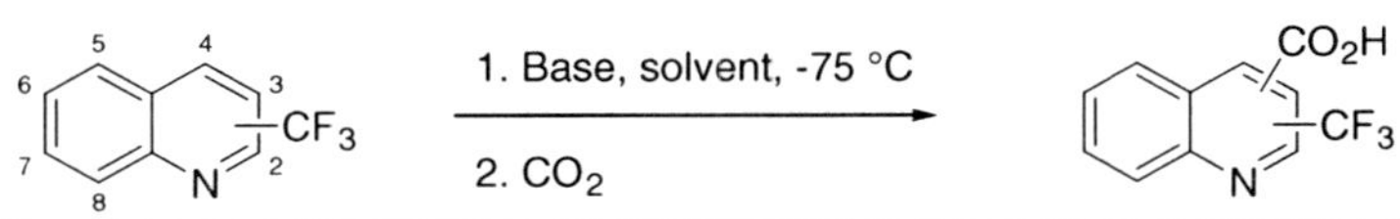

Entry	CF_3 position	Base	Solvent	CO_2H position	Yield (%)
1	2	LTMP	THF	3 + 4	43 + 27
2	2	LTMP	DEE	8	20
3	3	LTMP	THF	2	41
4	3	BuLi·LiDMAE	DEE	2	12
5	4	LTMP	THF	3	76
6	4	BuLi·LiDMAE	DEE	2	36

Table 5 Deprotometalation of chloroquinolines

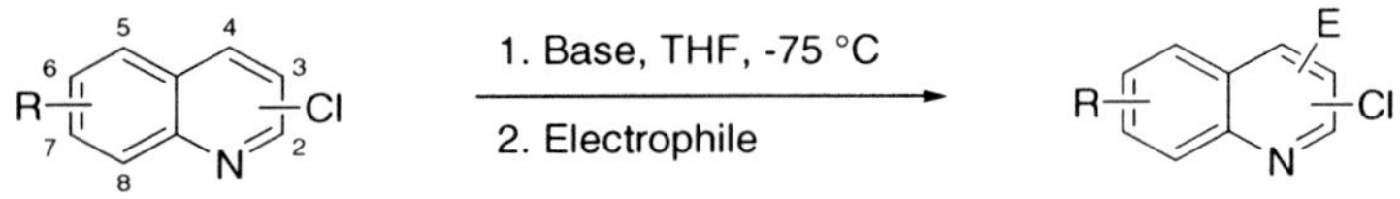

Entry	Cl position	R	Base	Electrophile	E	Yield (%)
1	2	H	LDA	Me_3SiCl	3-$SiMe_3$	50
2	3	H	LDA	Me_3SiCl	4-$SiMe_3$	55
3	4	H	LDA	Me_3SiCl	3-$SiMe_3$	70
4	2	H	LDA	PhCHO	3-CH(OH)Ph	55
5	2	H	LDA	2-(OMe)PhCHO	3-CH(OH) (2-(OMe)Ph)	60
6	2	H	LDA	HCO_2Et	3-CHO	45
7	2	H	LDA	$B(OMe)_3$	3-$B(OH)_2$	85
8	2	H	LDA	I_2	3-I	75
9	2	8-Br	LDA	I_2	3-I	79
10	2	H	LDA	CO_2	3-CO_2H	67
11	2	H	BuLi·TMEDA	CO_2	3-CO_2H	65
12	2	8-OMe	LTMP	$B(OMe)_3$	3-$B(OH)_2$	82

(Table 4, Entry 1). When DEE is used instead of THF, lithiation is regioselectively observed at the 8-position (20% yield), probably due to coordination of the reagent by the nitrogen lone pair (Table 4, Entry 2).

Unlike LDA, LTMP in THF and BuLi·LiDMAE in DEE metalate 3-(trifluoromethyl)quinoline at its less congested 2-position (absence of *peri*-hydrogen atom), and the corresponding carboxylic acid can be obtained after electrophilic trapping in 41% and 12% yield, respectively (Table 4, Entries 3–4).

In the case of 4-(trifluoromethyl)quinoline, while LTMP in THF exclusively abstracts the proton from the 3-position (76% yield), BuLi·LiDMAE in DEE orients the reaction to the nitrogen-adjacent 2-position but to provide the carboxylic acid in a low 36% yield (Table 4, Entries 5–6).

2.1.3 Chloroquinolines

The deprotonation of 2-, 3-, and 4-chloroquinoline has been reported mainly using LDA as a base. As observed with 2- and 3-fluoroquinoline [1], lithiation in THF at −75°C occurs at the 3- and 4-position (adjacent to the chloro group), respectively (Table 5). A wide variety of substituents could be introduced in yields ranging from 45% to 85% (Table 5, Entries 1–10) [20, 21]. BuLi·TMEDA [22] and hindered LTMP [23] can also be used to perform deprotometalation of 2-chloroquinolines without regioselectivity change (Table 5, Entries 11–12). The use of these bases was extended in 1997 to the regioselective deprotonation of 1-aryl-3-chloroisoquinolines at their 4-position [24].

Table 6 Deprotometalation of 2-bromoquinolines using lithium diisopropylamide

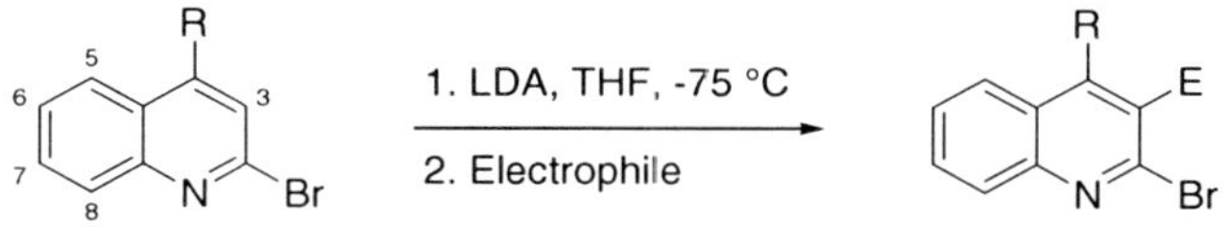

Entry	R	Electrophile	E	Yield (%)
1	H	CH_2O	CH_2OH	58
2	H	DMF	CH_2OH[a]	78[a]
3	Br	DMF	CHO	78
4	H	RCHO[b]	CH_2R[b,c]	55[c]
5	H	R′CHO[d]	CH_2R'[c,d]	65[c]

[a]Over two steps, after reduction with $NaBH_4$
[b]R = methyl 1-methyl-1*H*-3-indolyl-2-carboxylate
[c]Over two steps, after reduction with Et_3SiH
[d]R′ = methyl 1-(*tert*-butoxycarbonyl)-1*H*-3-indolyl-2-carboxylate

2.1.4 Bromoquinolines

The metalation of bromoquinolines has been less studied than other halogenoquinolines because of lower availability of these substrates. The deprotonation of 2-bromo- and 2,4-dibromoquinoline has been reported using LDA in THF at low temperature with subsequent trapping of the 3-lithiated quinolines with carbonyl electrophiles (Table 6) [25–27]. The products thus obtained were used as key intermediates in the synthesis of Camptothecin and Calothrixin alkaloids.

2.2 Deprotonative Metalation of Oxygen-Bearing Quinolines

2.2.1 Hydroxyquinolines and Quinolones

Quinolones and their tautomeric hydroxy forms are a group of synthetic broad-spectrum antibiotics whose first generation is derived from Nalidixic acid [28].

Low temperature lithiation of *N*-methyl-4-quinolone and -quinolinethione using an excess of LDA in THF takes place at the 2-position after reverse addition. The resulting lithio species react with a large range of electrophiles to give 2-substituted 4-quinolones and -quinolinethiones (Table 7) [29].

In contrast to *N*-methyl-2-quinolone, for which a complex mixture of unidentified products is obtained, 2-quinolone is cleanly deprotonated at the 3-position using BuLi·TMEDA chelate in THF (Table 8, Entries 1–12) [30, 31]. Metalation at the same position is also observed when a methoxy group is present at the 5- or 6-position (Table 8, Entries 13–16), whereas a single methoxy group at the 8-position promotes the addition of the butylated base to the quinoline ring [32]. However, the

Table 7 Deprotometalation of 4-quinolones and 4-quinolinethiones using lithium diisopropylamide

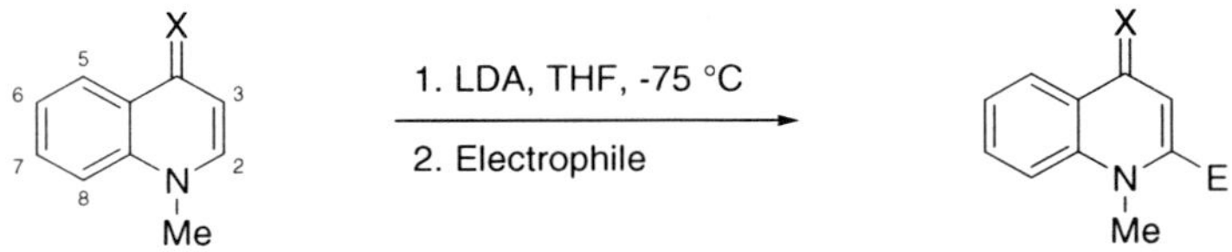

Entry	X	Electrophile	E	Yield (%)
1	O	CD_3OD	D	15
2	O	EtI	Et	64
3	O	PhCHO	CH(OH)Ph	71
4	O	4-(NO_2)PhCHO	CH(OH)(4-(NO_2)Ph)	36
5	O	4-(OMe)PhCHO	CH(OH)(4-(OMe)Ph)	14
6	O	3-PyCHO	CH(OH)(3-Py)	60
7	O	PhC(O)Cl	C(O)Ph	17
8	O	3-PyC(O)Cl	C(O)(3-Py)	25
9	O	HCHO	CH_2OH	40
10	O	PhC(O)Me	C(OH)PhMe	31[a]
11	O	$Ph_2C(O)$	$C(OH)Ph_2$	35
12	S	PhCHO	CH(OH)Ph	40
13	S	3-PyCHO	CH(OH)(3-Py)	82
14	S	4-(OMe)PhCHO	CH(OH)(4-(OMe)Ph)	46

[a]20% of the 3-substituted regioisomer were isolated

presence of an additional methoxy group at the 4-position contributes to stabilize the 3-lithio derivative with the phenoxide anion (Table 8, Entries 17–26) [33].

2.2.2 Alkoxyquinolines

First attempts to metalate 2-ethoxyquinoline with BuLi in DEE at room temperature led to the major 1,2-addition product and the expected 2-ethoxy-3-substituded quinolines in only low yields after electrophilic trapping (Table 9, Entry 1) [34–37]. Deprotonation of 2,4-dimethoxyquinolines with BuLi in DEE has also been studied and proved successful at 0°C (Table 9, Entry 2) [38–41]. After interception of the lithio compounds, the key intermediate 3-functionalized derivatives were used in the synthesis of quinoline alkaloids Dictamnine [38, 42], Kokusaginine [39], Skimmianine [40], Atanine [41], γ-Fagarine [42], Evolitrine [42], and Pteliene [42].

To functionalize 2-alkoxyquinolines, a trick consists in adding LDA to the substrate in the presence of an electrophile. Higher yields of the corresponding 3-quinolineboronic acids are obtained by this way in THF using triisopropylborate (Table 10) [43].

Table 8 Deprotometalation of 2-quinolones using butyllithium·*N*,*N*,*N*′,*N*′-tetramethylethylenediamine chelate

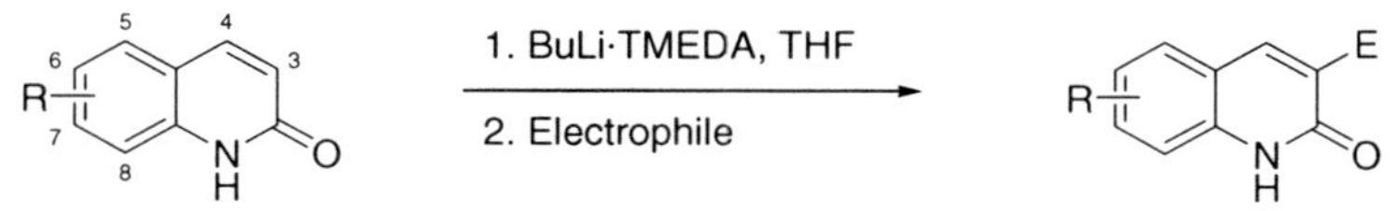

Entry	R	Electrophile	E	Yield (%)
1	H	CH_3I	CH_3	73
2	H	PhCHO	CH(OH)Ph	26
3	H	4-(OMe)PhCHO	CH(OH)(4-(OMe)Ph)	23
4	H	4-FPhCHO	CH(OH)(4-FPh)	25
5	H	2-FPhCHO	CH(OH)(2-FPh)	27
6	H	4-ClPhCHO	CH(OH)(4-ClPh)	25
7	H	2-ClPhCHO	CH(OH)(2-ClPh)	26
8	H	$Ph_2C(O)$	$C(OH)Ph_2$	18
9	H	Me_3SiCl	$SiMe_3$	56
10	H	$(SMe)_2$	SMe	90
11	H	DMF	CHO	60
12	H	CO_2	CO_2H	96
13	5-OMe	Me_3SiCl	$SiMe_3$	25
14	5-OMe	CH_3I	CH_3	25
15	6-OMe	Me_3SiCl	$SiMe_3$	38
16	6-OMe	CH_3I	CH_3	47
17	4,8-OMe	DMF	CHO	60
18	4,8-OMe	$(SMe)_2$	SMe	93
19	4,8-OMe	PhCHO	CH(OH)Ph	62
20	4,8-OMe	Me_3SiCl	$SiMe_3$	77
21	4,8-OMe	CH_3I	CH_3	77
22	4,8-OMe	$Ph_2C(O)$	$C(OH)Ph_2$	44
23	4,8-OMe	I_2	I	27
24	4,8-OMe	allylBr	allyl	10
25	4,8-OMe	$B(OMe)_3$	$B(OH)_2$	90
26	4,7,8-OMe	DMF	CHO	54

Table 9 Deprotometalation of alkoxyquinolines using butyllithium

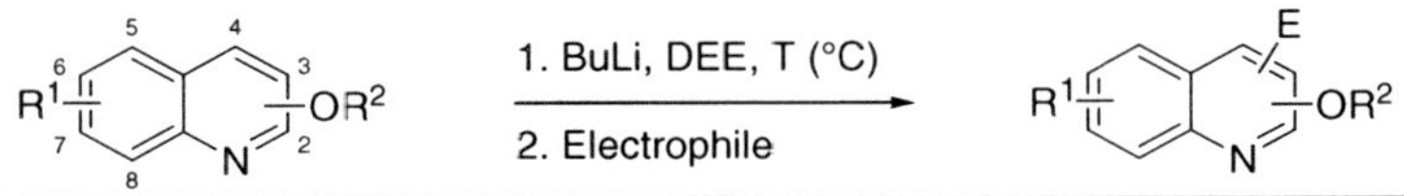

Entry	R^1	OR^2	T (°C)	Electrophile	E	Yield (%)
1	H	2-OEt	rt	CO_2	3-CO_2H	7[a]
2	6,7,8 = H or OMe	2,4-OMe	0	DMF or $Me_2C{=}CHCH_2Br$	CHO or $CH_2CH{=}CMe_2$	48–76

[a]2-Butylquinoline is isolated in 60% yield

Table 10 Deprotometalation of alkoxyquinolines using lithium diisopropylamide

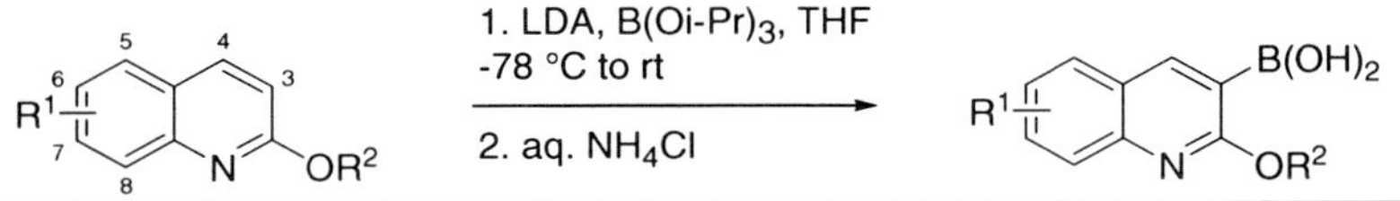

Entry	R^1	OR^2	Yield (%)
1	H	2-OMe	73
2	H	2-OEt	85
3	7-(2-OMe-3-quinolyl)	2-OMe	77
4	7-(2-OEt-3-quinolyl)	2-OEt	83

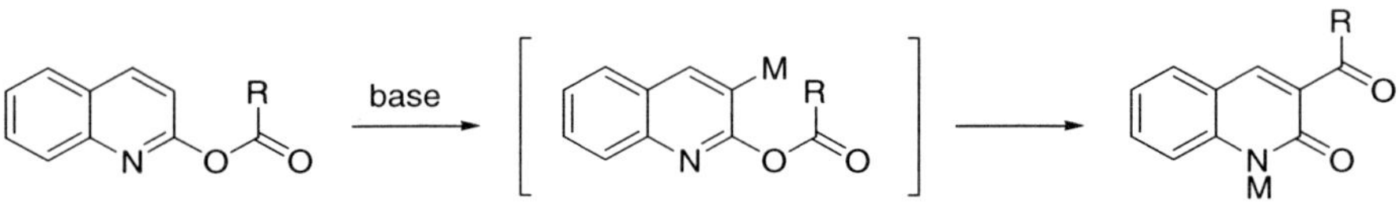

Scheme 1 Anionic Fries rearrangement

2.2.3 *O*-(Quinolyl)carbamates

Deprotonative metalation of all isomeric *O*-(quinolyl)carbamates has been studied. The results obtained with this kind of substrates show that the lithio derivatives can either survive until the trapping step or stabilize themselves by anionic Fries rearrangement (Scheme 1) [44, 45].

Hydrogen–metal permutations of *N*,*N*-diethyl-2-quinolinecarbamate have been described using LDA and *s*-BuLi·TMEDA chelate in THF at −105°C (Table 11, Entries 1–4) [45]. Using the latter reduces the Fries rearrangement product formation (Table 11, Entries 1–2) but the electrophilic interception with aldehydes induces the competitive formation of isomeric compounds resulting from the nucleophilic attack of the intermediate lithium alkoxides to the carbamate function (Table 11, Entries 3–4).

The treatment of *N*,*N*-dimethyl- and *N*,*N*-diethyl-3-quinolinecarbamate with LDA in THF results in a regioselective deprotonation at the 4-position without Fries rearrangement before trapping with methanol-d_4 or trimethylsilyl chloride (Table 11, Entries 5–7) [45]. Electrophilic interception with aldehydes leads to a mixture including the expected secondary alcohols but also the isomeric and decarboxylated isomeric compounds in unreproducible yields [44, 45].

Using LDA with *N*,*N*-dialkyl-4-quinolinecarbamate allows anionic Fries rearrangement of the *N*,*N*-dimethyl lithio compound, whereas the corresponding *N*,*N*-diethyl lithio compound remains intact until trapping to furnish the 3-functionalized quinolinecarbamates (Table 11, Entries 8–11) [45].

N,*N*-dimethyl-5-, 7- and 8-quinolinecarbamate can be regioselectively converted into the corresponding 6-, 8-, and 7-functionalized quinolines by simultaneous

Table 11 Deprotometalation of *O*-(quinolyl)carbamates using either lithium diisopropylamide or chelate *sec*-butyllithium·*N*,*N*,*N*′,*N*′-tetramethylethylenediamine

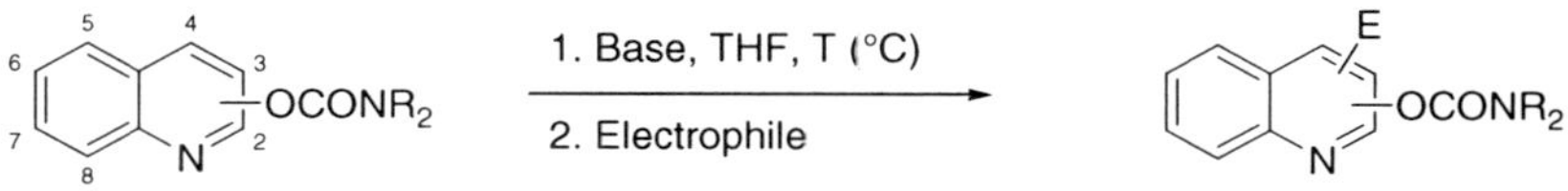

Entry	$OCONR_2$	Base	Electrophile	E	T (°C)	Yield (%)
1	2-$OCONEt_2$	LDA	CD_3OD	3-D	−105	40
2	2-$OCONEt_2$	*s*-BuLi·TMEDA	CD_3OD	3-D	−105	80
3	2-$OCONEt_2$	*s*-BuLi·TMEDA	EtCHO	3-CH(OH)Et	−105	30
4	2-$OCONEt_2$	*s*-BuLi·TMEDA	PhCHO	3-CH(OH)Ph	−105	28
5	3-$OCONMe_2$	LDA	CD_3OD	4-D	−78	80
6	3-$OCONMe_2$	LDA	Me_3SiCl	4-$SiMe_3$	−78	90
7	3-$OCONEt_2$	LDA	CD_3OD	4-D	−78	80
8	4-$OCONEt_2$	LDA	CD_3OD	3-D	−78	85
9	4-$OCONEt_2$	LDA	CH_3I	3-CH_3	−78	75
10	4-$OCONEt_2$	LDA	Me_3SiCl	3-$SiMe_3$	−78	95
11	4-$OCONEt_2$	LDA	EtCHO	3-CH(OH)Et	−78	43
12	5-$OCONMe_2$	LDA	Me_3SiCl[a]	6-$SiMe_3$	−78	70
13	6-$OCONMe_2$	LDA	Me_3SiCl[a]	–[b]	−78	–[b]
14	7-$OCONMe_2$	LDA	Me_3SiCl[a]	8-$SiMe_3$	−78	90
15	8-$OCONMe_2$	LDA	Me_3SiCl[a]	7-$SiMe_3$	−78	40
16	7-$OCONMe_2$	LDA	I_2	8-I	−78	65

[a]Under in situ trapping conditions
[b]Mixture of 5-$SiMe_3$, 7-$SiMe_3$, and 5,7-$SiMe_3$ in a 2:2:1 ratio

treatment with LDA and trimethylsilyl chloride (Table 11, Entries 12, 14–15). Under these conditions, *N*,*N*-dimethyl-6-quinolinecarbamate shows poor selectivity and gives a mixture of mono- and disilylated compounds (Table 11, Entry 13). It has to be noted that, in the absence of electrophile, the lithio derivatives decompose by Fries rearrangement at a specific temperature for each isomer [46].

It is also possible to functionalize *N*,*N*-dimethyl-7-quinolinecarbamate at the 8-position by lithiation using LDA followed by addition of the resulting metalated intermediate to a solution of iodine in THF (Table 11, Entry 16). The 8-iodo product was used in the synthesis of 7,7′-bis(dimethylaminocarbonyloxy)-8,8′-biquinolyl by Ullmann coupling. Quinoline dimers were also obtained after deprotometalation by oxidation using ferric chloride [47].

2.3 *Deprotonative Metalation of Nitrogen-Bearing Quinolines*

Resonance effects induce a low availability of the nitrogen lone pair of the nitrogen-based substituents for coordination with the metalating agents [48]. Moreover, compared with non-coordinating functions such as heavy halogens, the acidifying

Table 12 Deprotometalation of protected aminoquinoline

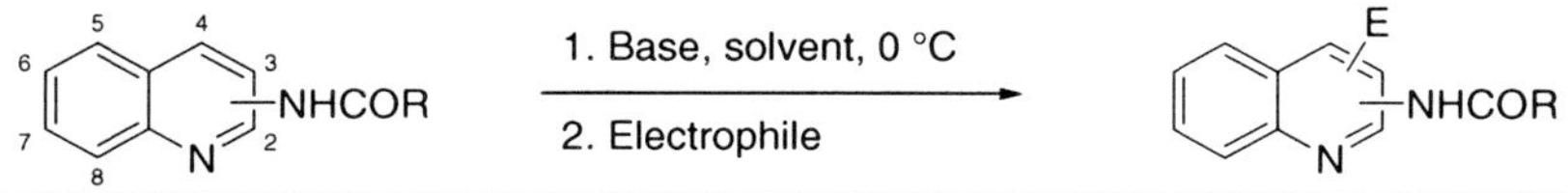

Entry	NHCOR	Base	Solvent	Electrophile	E	Yield (%)
1	2-NHCO*t*-Bu	BuLi	DEE	DCl/D_2O	3-D	80
2	2-NHCO*t*-Bu	BuLi·TMEDA	DEE	DCl/D_2O	3-D	65
3	2-NHCO*t*-Bu	BuLi	DEE	I_2	3-I	90
4	2-NHCO*t*-Bu	BuLi·TMEDA	DEE	I_2	3-I	45
5	2-NHCO*t*-Bu	BuLi	DEE	PhCHO	3-CH(OH)Ph	70
6	2-NHCO*t*-Bu	BuLi·TMEDA	DEE	PhCHO	3-CH(OH)Ph	31
7	2-NHCO*t*-Bu	BuLi·TMEDA	DEE	MeCHO	3-CH(OH)Me	11
8	2-NHCO*t*-Bu	BuLi	DEE	Me_3SiCl	3-$SiMe_3$	95
9	2-NHCO*t*-Bu	BuLi·TMEDA	DEE	$(SMe)_2$	3-SMe	28
10	2-NHCO*t*-Bu	BuLi	DEE	$CH_2(O)CHCH_3$	3-$CH_2CH(OH)CH_3$	42
11	2-NHCO*t*-Bu	BuLi	DEE	$Et_2C(O)$	3-$CH(OH)Et_2$	12
12	2-NHCO*t*-Bu	BuLi	DEE	$ClCO_2Et$	CO_2Et	17
13[a]	3-$NHCONMe_2$	LDA	THF	Me_3SiCl	2-$SiMe_3$	93
14[b]	4-NHCO*t*-Bu	*s*-BuLi	THF	Me_3SiCl	8-$SiMe_3$	45

[a]Reaction performed at −78°C and Me_3SiCl introduced at the same time as the base
[b]Reaction performed at −90°C

effect of the amino group on adjacent hydrogens is weak. Therefore, nucleophilic bases such as alkyllithiums tend to add to the ring at the 2- or, when occupied, 4-position.

2-Pivaloyl-aminoquinoline is selectively metalated at its 3-position using BuLi in DEE at 0°C; subsequent trapping reactions with various electrophiles give the corresponding 3-substituted 2-pivaloyl-aminoquinolines in 12–95% yield (Table 12, Entries 1–12) [49]. Using BuLi·TMEDA chelate as the metalating agent affords the functionalized products in lower yields because of competitive 1,4-addition of the alkyl base.

Deprotometalation attempts of 3-pivaloyl-aminoquinoline using alkyllithiums predominantly furnish the 1,2-addition products, which are isolated after oxidation [50]. Only the corresponding *N,N*-dimethyl urea can be functionalized at its 2-position in THF under in situ trapping conditions, that is using LDA in the presence of trimethylsilyl chloride (Table 12, Entry 13) [50].

1,2-Addition is similarly observed when 4-pivaloyl-aminoquinoline is treated with BuLi·TMEDA complex at −50°C. Nucleophilic addition is limited by using more basic *s*-BuLi in THF at −90°C and subsequent trapping with trimethylsilyl chloride. Under these conditions, the 8-silylated compound is obtained in 45% yield; this unexpected regioselectivity can be rationalized by coordinative interactions between the nitrogen ring and the base (Table 12, Entry 14) [50].

Table 13 Deprotometalation of quinolinecarboxylic acids using lithium 2,2,6,6-tetramethylpiperidide

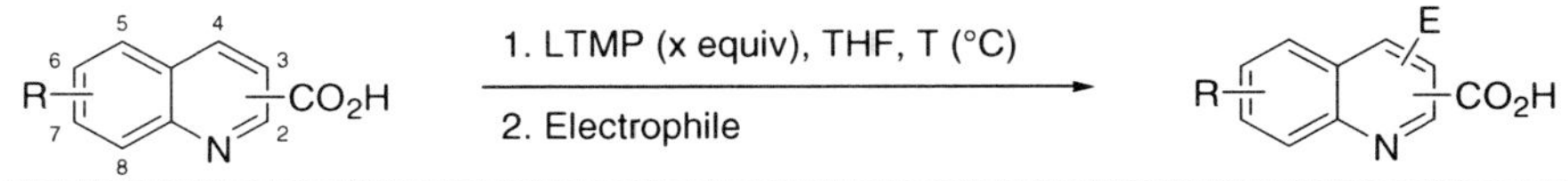

Entry	CO_2H position	R	x	T (°C)	Electrophile	E	Yield (%)
1	2	H	3	−25	D_2O	3-D	57
2	2	H	3	−25	CO_2	3-CO_2H	75
3	2	H	3	−25	PhCHO	3-CH(OH)Ph[a]	48
4	2	4-OMe	3	−25	D_2O	3-D	60
5	2	4-OMe	3	−25	CO_2	3-CO_2H	60
6	2	4-OMe	3	−25	I_2	3-I	45
7	3	H	3	−50	D_2O	4-D	59
8	3	H	3	−50	CO_2	4-CO_2H	80
9	4	H	6	−50	D_2O	3-D	33

[a]Product isolated after lactonization under acidic conditions

2.4 Deprotonative Metalation of Carbon-Bearing Quinolines

2.4.1 Quinolinecarboxylic Acids

To avoid protection and deprotection steps, the metalation was attempted on quinolinecarboxylic acids lithium salts. As established in the pyridine series [51], LTMP was preferred over BuLi in order to avoid competitive nucleophilic addition.

On 2-, 3-, and 4-quinolinecarboxylic acid, deprotonation can be performed using LTMP at the 3-, 4-, and 3-position, respectively (Table 13) [52]. It has to be mentioned that to avoid competitive 1,2- and 1,4-addition of the lithio intermediates to the substrates, the reaction of the 3- and 4-isomers was carried out at −50°C (Table 13, Entries 7–9). Furthermore, 5 equiv of LTMP are required to deprotonate the 4-quinolinecarboxylic acid. The non-coplanarity between the C=O and C–Li bonds, due to steric hindrance of the *peri* hydrogen at the 5-position, could be responsible for the decreased efficiency in this case of the directing power of the lithium carboxylate.

2.4.2 Quinolinecarboxamides

Metalation of *N*,*N*-diethyl-4-quinolinecarboxamide requires an excess of LTMP in THF (Table 14, Entries 1–2); *s*-BuLi, *t*-BuLi, HMDSLi, and LDA attempts fail to furnish 3-functionalized derivatives in satisfactory yields [53]. This reaction was used as a key step in the synthesis of Calothrixin A and B [53].

Lithiation of (*S*)-*N*-(1-phenylpropyl)-2-phenylquinoline-4-carboxamide using a large excess of BuLi·TMEDA complex in THF occurs regioselectively at the

Table 14 Deprotometalation of quinolinecarboxamides using either lithium 2,2,6,6-tetramethylpiperidide or the chelate butyllithium·*N*,*N*,*N*′,*N*′-tetramethylethylenediamine

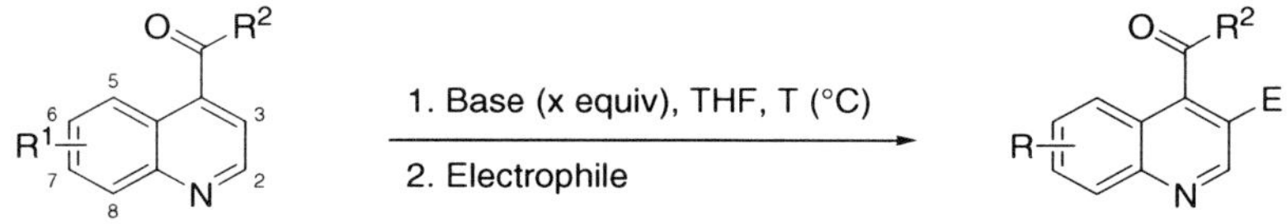

Entry	R^1	R^2	Base	x	T (°C)	Electrophile	E	Yield (%)
1	H	NEt_2	LTMP	2	−78	Me_3SiCl	$SiMe_3$	50[a]
2	H	NEt_2	LTMP	4	−90	*N*-MOM-3-indoleCHO	–[b]	12[b]
3	2-Ph	NHCH(Ph)Et	BuLi·TMEDA	10	−60	CD_3OD	D	74
4	2-Ph	NHCH(Ph)Et	BuLi·TMEDA	10	−60	I_2	I	59
5	2-Ph	NHCH(Ph)Et	BuLi·TMEDA	10	−60	CBr_4	Br	63
6	2-Ph	NHCH(Ph)Et	BuLi·TMEDA	10	−60	C_2Cl_6	Cl	69
7	2-Ph	NHCH(Ph)Et	BuLi·TMEDA	10	−60	Me_3SnCl	$SnMe_3$	71
8	2-Ph	NHCH(Ph)Et	BuLi·TMEDA	10	−60	Me_3SiCl	$SiMe_3$	70

[a]Conversion determined by ^{1}H-NMR
[b]Product isolated after trapping, cyclization, and oxidation

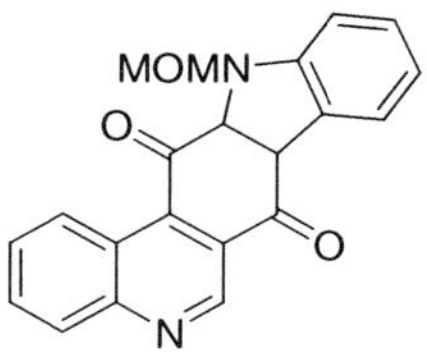

3-position without racemization of the stereogenic center after interception with electrophiles (Table 14, Entries 3–8) [54]. Using hindered lithium amides LDA or LTMP, as well as with *t*-BuLi, no reaction is detected.

2.4.3 Other Carbon-Bearing Quinolines

The selective deprotonation–cyclization reactions of methyl 2-(3-quinolylcarbonyl) benzoate and 2-(3-quinolylcarbonyl)benzoic acid affording benzo[*j*] phenanthridine-7,12-dione can be performed in THF by treatment with 3 equiv of LTMP (Table 15, Entries 1–2) [55]. A better yield is obtained from the ester, the metalation step being carried out at a lower temperature.

A regioselective lithiation of a quinazolinoylquinoline was used as a key step for the synthesis of human DNA topoisomerase I poisons Luotonin A, B, and E. While the reaction of the 2-quinolylquinazolinone using BuLi, *s*-BuLi, and *t*-BuLi, combined or not with TMEDA, or even with LDA, leads to the formation of complex mixtures. The use of non-nucleophilic mesityllithium furnishes the expected

Table 15 Deprotometalation of carbon-bearing quinolines using either lithium 2,2,6,6-tetramethylpiperidide or mesityllithium

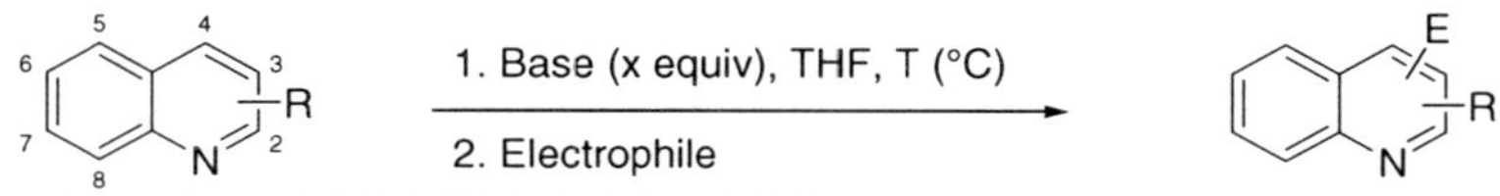

Entry	R	Base	x	T (°C)	Electrophile	E	Yield (%)
1	3-COPh(2′-COR′)[a]	LTMP	3	−75	Methyl ester[b]	–[b]	34
2	3-COPh(2′-COR′)[c]	LTMP	3	rt	Lithium carboxylate[b]	–[b]	10
3	2-(4′-oxo-3′*H*-2′-quinazolinyl)	MesLi	2.2	−78	HCHO	CH_2OH	86
4	2-(4′-oxo-3′*H*-2′-quinazolinyl)	MesLi	2.2	−78	DMF	CHO	81[d]

[a]R′ = OMe (starting material) or R′ = linked to the 4-position (product)
[b]Intramolecular cyclization
[c]R′ = OH (starting material) or R′ = linked to the 4-position (product)
[d]Product isolated after trapping and intramolecular cyclization

Table 16 Deprotometalation of *N*-activated quinolines and isoquinolines using lithium 2,2,6,6-tetramethylpiperidide

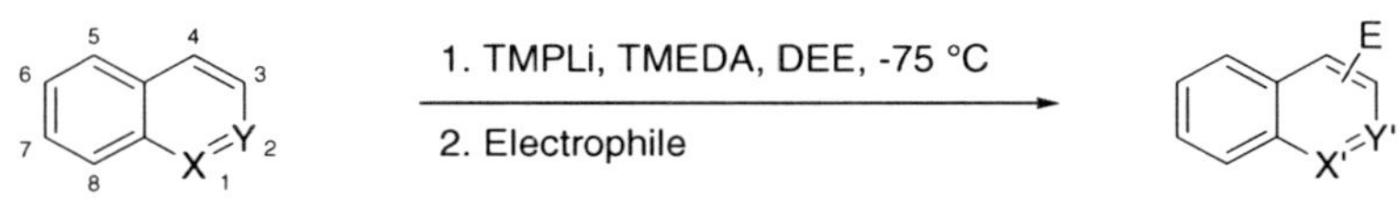

Entry	X	Y	Electrophile	X′	Y′	E	Yield (%)
1	$N^+OB^-F_3$	CH	PhCHO	N^+O^-	C-E	2-CH(OH)Ph	12
2	$N^+OB^-F_3$	CH	cyclohexanone	N^+O^-	C-E	$2\text{-}C(OH)(CH_2)_5$	11
3	CH	$N^+OB^-F_3$	PhCHO	C-E	N^+O^-	1-CH(OH)Ph	19[a]
4	CH	N^+O^-	PhCHO	C-E	N^+O^-	1-CH(OH)Ph	28[a]

[a]The 1,3-difunctionalized isoquinoline is also isolated in 17% yield

dilithiated intermediate. Indeed, subsequent quenching with formaldehyde or DMF provides the expected quinazolinones in 81–86% yield (Table 15, Entries 3–4) [56].

2.5 *Deprotonative Metalation of* N*-Activated Quinolines and Isoquinolines*

The acidity of the hydrogen at the 2-position of quinoline *N*-oxide is enhanced by the inductive effect of the oxide and by the complexing ability of the lone pair on oxygen with lithium. The deprotonation of quinoline *N*-oxide leads to 2,2′-dimer, and starting from the corresponding BF_3 complex only mixtures including the expected compounds are obtained (Table 16, Entries 1–2) [57, 58].

Table 17 Deprotometalation of triazoloquinolines

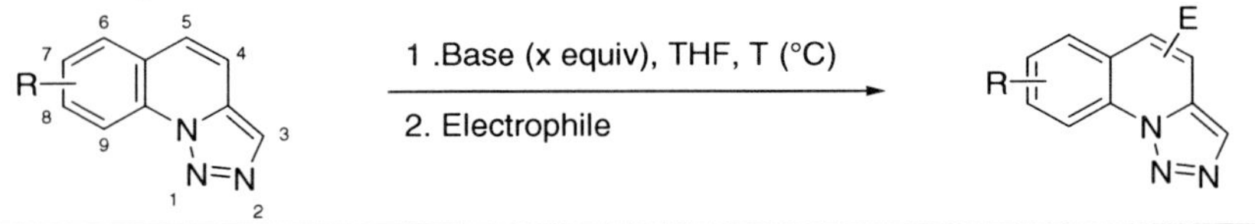

Entry	R	Base	x	T (°C)	Electrophile	E	Yield (%)
1	H	LDA	1	−40	$R^1C(O)R^{2a}$	$3\text{-}R^1C(OH)R^{2a}$	74–84
2	H	LDA	1	−40	CO_2	$3\text{-}CO_2H$	66
3	$3\text{-}CONEt_2$	LDA	1	−40	$R^1C(O)R^{2b}$	$4\text{-}R^1C(OH)R^{2b}$	54–65
4	H	LTMP	1	−78	CH_3I	$3\text{-}CH_3$	86
5	H	LTMP	1	−78	CD_3OD	3-D	96
6	H	LTMP	1	−78	Me_3SiCl	$3\text{-}SiMe_3$	99
7	H	LTMP	1	−78	I_2	3-I	72
8	H	LTMP	1	−78	Menthyl toluenesulfinate	3-S(O)-*p*-Tol	82
9	$3\text{-}SiMe_3$	BuLi	1.7	−78	CH_3I	$9\text{-}CH_3$	98
10	H	BuLi	3	−78	CH_3I	$3,9\text{-}CH_3$	82
11	H	BuLi	3	−78	I_2	3,9-I	79
12	H	BuLi	3	−78	$CBrF_2\text{-}CBrF_2$	3,9-Br	74
13	H	BuLi	3	−78	$CClF_2\text{-}CCl_2F$	3,9-Cl	71
14	H	BuLi	3	−78	NFSI	9-F	60

[a]R^1 = H, Me; R^2 = Me, Et, i-Pr
[b]R^1 = H, Me; R^2 = Me, 4-(OMe)Ph

Similar results were obtained starting from isoquinoline *N*-oxide and its BF_3 complex (Table 16, Entries 3–4) [59]. In this case, mono- and difunctionalized compounds are obtained because the oxide activates the 1- and 2-position of the ring.

2.6 Deprotonative Metalation of Other Substituted Quinolines and Isoquinolines

2.6.1 Triazoloquinolines

Metalation of 1,2,3-triazolo[1,5-*a*]quinolines was first reported using LDA in THF at −40°C with aldehydes, ketones, and carbon dioxide as electrophiles (Table 17, Entries 1–2). The resulting 3-carboxylic acid was converted into the corresponding *N,N*-diethylcarboxamide, a substrate for which the lithiation is observed at the 4-position under the same reaction conditions and after interception with carbonyl electrophiles (Table 17, Entry 3) [60].

More recently, LTMP was used in THF at low temperature for the selective metalation of [1,2,3]triazolo[1,5-*a*]quinoline at its 3-position (Table 17, Entries 4–8). When BuLi is employed instead of LTMP, 3,9-difunctionalized quinolines are obtained except if the 3-position is already occupied (Table 17, Entry 9) or when

Table 18 Deprotometalation of triazoloisoquinolines using either lithium diisopropylamide or butyllithium

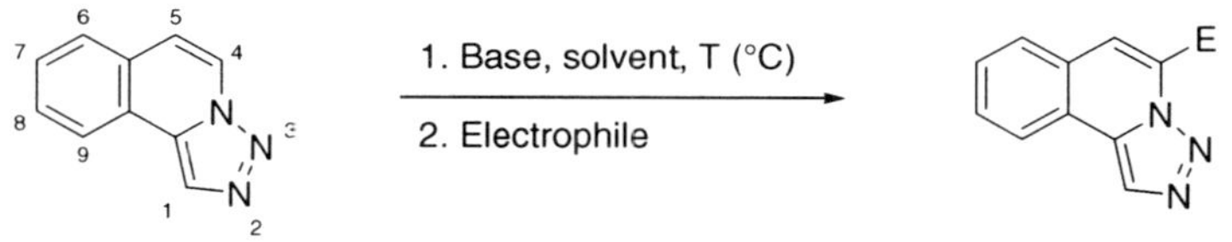

Entry	Base	Solvent	T (°C)	Electrophile	E	Yield (%)
1	LDA	DEE	−40	ClC(O)NEt$_2$	CONEt$_2$	45–70
				Me$_3$SiCl	SiMe$_3$[a]	
				MeCHO	CH(OH)Me	
				4-(OMe)PhCHO	CH(OH)(4-(OMe)Ph)	
2	BuLi	PhMe	−70	Me$_3$SnCl	SnMe$_3$	43
3	LDA	THF	−40	B(Oi-Pr)$_3$	B(OH)$_2$	78

[a]The 1,4-disilylated derivative is also obtained in 23% yield

N-fluorobenzenesulfonimide is used as the electrophile (Table 17, Entries 10–14); in these cases, no difluorinated product can be detected [61].

2.6.2 Triazoloisoquinolines and Pyrazoloisoquinolines

Unlike 1,2,3-triazolo[1,5-*a*]quinoline, 1,2,3-triazolo[5,1-*a*]isoquinoline is lithiated with LDA or BuLi at the 4-position, as seen from the products resulting from electrophilic quenching (Table 18) [62, 63].

Lithiation at both the 2- and 7-position of 1-(4,5-dihydro-4,4-dimethyl-2-oxazolyl)pyrazolo[5,1-*a*]isoquinoline using 2 equiv of BuLi in THF at −75°C followed by reaction with benzaldehyde gives the 3-substituted isoquinoline-2-acetonitrile in 90% yield after ring cleavage of the pyrazole ring [64].

2.7 Deprotonative Metalation of Unsubstituted Quinoline and Isoquinoline

Unlike alkyllithiums, hindered lithium amides show only little tendency to add to the quinoline and isoquinoline ring. Nevertheless, when treated with LDA, quinoline and isoquinoline undergo complex reactions due to the formation of 2,2′- and 1,1′-dimers, respectively [65].

The unimetallic complex base BuLi·LiDMAE can react with unsubstituted quinoline to give 2-metalated species, which lead to 2-substituted heterocycles after trapping by electrophiles (Table 19, Entries 1–7) [66].

Table 19 Deprotometalation of unsubstituted quinolines using butyllithium·lithium *N*,*N*-dimethylaminoethoxide superbase

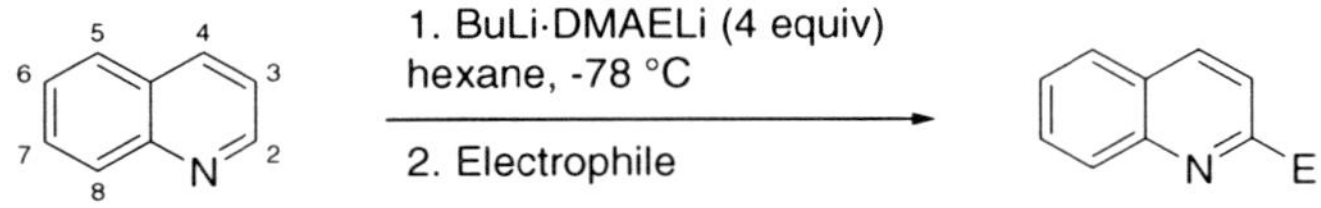

Entry	Electrophile	E	Yield (%)
1	Me_3SiCl	$SiMe_3$	65
2	CH_3I	Et	30
3	MeSSMe	SMe	55
4	*t*-BuCHO	CH(OH)*t*-Bu	25
5	PhCHO	CH(OH)Ph	45
6	MeC(O)Et	C(OH)MeEt	25
7	$Ph_2C(O)$	$C(OH)Ph_2$	40

3 Halogen–Metal Exchange and Related Reactions of Quinolines and Isoquinolines

Halogen–lithium exchange is particularly sensitive to reaction temperature in the quinoline and isoquinoline series where derivatives are more prone to nucleophilic attacks than the corresponding pyridines due to their lower LUMO levels. This limits the conditions under which metal-halogen exchange can occur.

The conversion of bromoquinolines to the corresponding Grignard reagents has been considered due to possible subsequent electrophilic trapping or coupling reactions. Sachs and Sachs studied at the beginning of the twentieth century the direct access to quinoline Grignard compounds using metallic magnesium, and concluded that it was not possible [67]. Their findings were later confirmed by Howitz and Köpke [68] and by Wibaut and coworkers [69, 70].

3.1 Halogen–Metal Exchange and Related Reactions of Quinolines

3.1.1 2-Substituted Quinolines

Gilman and Spatz documented in 1941 the halogen–metal exchange of 2-iodo-4-methylquinoline using BuLi. 4-Methyl-2-quinolinecarboxylic acid was isolated in an optimized 53% yield after a metalation time of 15 min at −5°C followed by carboxylation and acidification [71].

Gilman and Soddy reported in 1957 the first halogen–lithium exchange of 2-bromoquinoline. The optimization of the reaction conditions showed that THF is a more appropriate solvent than DEE (less basic and solvating) when the reaction

Table 20 Halogen–lithium exchange of 2-bromoquinoline using butyllithium

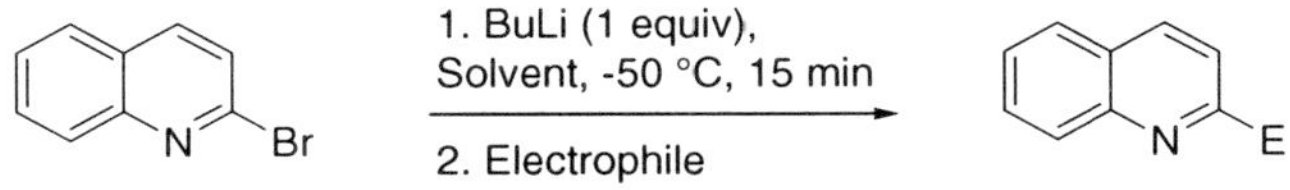

Entry	Solvent	Electrophile	E	Yield (%)
1	DEE	PhC(O)Ph	$C(OH)Ph_2$	70[a]
2	DEE	CO_2[b]	CO_2H	25[c]
3	THF	CO_2	CO_2H	50

[a]Yield before crystallization
[b]Trapping at −100°C (if the trapping is performed at −50°C, only traces of the expected acid and 34% of 2,2′-biquinolyl ketone are obtained)
[c]The 2,2′-biquinolyl ketone is also isolated in 21% yield

is carried out using BuLi at −50°C, affording the corresponding carboxylic acid in 50% yield (Table 20) [72].

Starting from 2-bromoquinoline and 1-bromoisoquinoline, Tanji and coworkers performed halogen-tellurium exchanges using BuTeLi (prepared from BuLi and Te) in THF; after subsequent treatment with BuLi at −78°C to generate the corresponding lithio compounds, trapping with pivalaldehyde, and hydrolysis, they obtained the expected alcohols in 75% and 36% yield, respectively [73].

Schlosser and coworkers have performed since 2003 different halogen–metal exchange reactions from trifluoromethyl-substituted 2-bromoquinolines using BuLi at low temperatures (Table 21, Entries 1–15) [4, 18, 21]. Fluoro-substituted 2-bromoquinolines were involved in 2005 in similar reactions using either BuLi (Table 21, Entries 16–23) or Bu_3MgLi (Table 21, Entry 24) [3].

In the course of the synthesis of indolizinones, Sotomayor and coworkers prepared a 2-quinolyllithium derivative, from the corresponding bromide or iodide and BuLi, and studied its intramolecular reaction with remote *N*-methoxy-*N*-methyl and morpholine amides (Table 22) [74].

From 2-chloroquinolines, the naphthalene-catalyzed reductive lithiation in the presence of aldehydes or ketones developed by Yus and coworkers furnished, after hydrolysis, the functionalized heterocycles in modest yields (Table 23) [75].

3.1.2 3-Substituted Quinolines

In their study reported in 1941, Gilman and Spatz optimized the halogen–metal interconversion of 3-bromoquinoline using butyllithium in DEE. They obtained their best result, 3-quinolinecarboxylic acid in 52% yield, using a 15 min contact time between the bromide and BuLi at −35°C (Table 24, Entry 1) [71]. Gilman and Soddy in 1957 similarly intercepted 3-quinolyllithium, generated in DEE after treatment of 3-bromoquinoline with BuLi for 10 min at −50°C, by CO_2 to afford the expected carboxylic acid after subsequent acidification in 50% yield (Table 24, Entry 2) [72]. Wibaut and Heeringa performed the same reaction in DEE, but

Table 21 Halogen–lithium exchange of trifluoromethyl-substituted 2-bromoquinolines using butyllithium

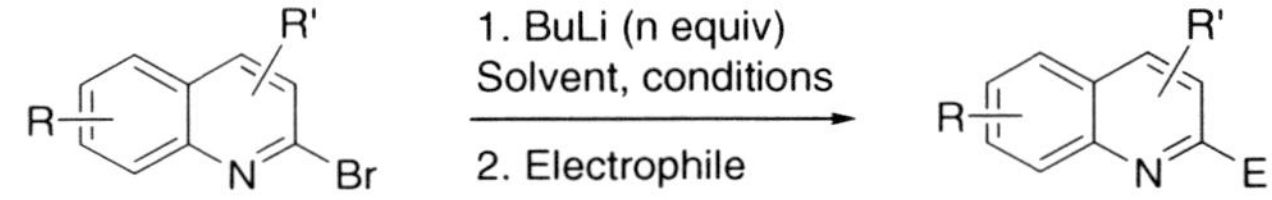

Entry	R	R′	n	Solvent, conditions	Electrophile	E	Yield (%)
1	H	3-CF_3	1	PhMe, −75°C, 0.75 h	CO_2	CO_2H	82
2	H	4-CF_3	1	THF, −75°C, 0.75 h	CO_2	CO_2H	72
3	H	3-CO_2H-4-CF_3	2	DEE, −100°C, 2 h	MeOH	H	64
4	6-Me	4-CF_3	1	THF, −75°C, 0.25 h	CO_2	CO_2H	72
5	6-F	4-CF_3	1	THF, −75°C, 0.25 h	CO_2	CO_2H	71
6	6-F	3-CO_2H-4-CF_3	2	DEE, −100°C, 2 h	MeOH	H	66
7		6-$OCF_3$4-CF_3	1	THF, −75°C, 0.75 h	CO_2	CO_2H	69
8	7-Me	4-CF_3	1	THF, −75°C, 0.25 h	CO_2	CO_2H	73
9	7-OMe	4-CF_3	1	THF, −75°C, 0.25 h	CO_2	CO_2H	66
10	7-F	4-CF_3	1	THF, −75°C, 0.25 h	CO_2	CO_2H	52
11	8-Me	4-CF_3	1	THF, −75°C, 0.25 h	CO_2	CO_2H	75
12	8-Me	3-CO_2H-4-CF_3	2	DEE, −100°C, 2 h	MeOH	H	67
13	8-Me	3-CO_2H-4-CF_3	2	DEE, −100°C, 2 h	DMF	CHO	57
14	8-F	4-CF_3	1	THF, −75°C, 0.75 h	CO_2	CO_2H	73
15	8-Cl	4-CF_3	1	THF, −75°C, 0.75 h	CO_2	CO_2H	61
16	H	3-F	1	DEE, −75°C, 0.75 h	CO_2	CO_2H	72
17	6-Cl	3-F	1	PhMe, −75°C, 0.75 h	CO_2	CO_2H	82
18	7-F	3-F	1	PhMe, −75°C, 0.75 h	CO_2	CO_2H	67
19	7-OMe	3-F	1	DEE, −75°C, 0.75 h	CO_2	CO_2H	82
20	8-Br	3-F	1	PhMe, −100°C, 0.75 h	MeOH	H	90
21	8-Br	3-F	1	PhMe, −75°C, 0.75 h	CO_2	CO_2H	79
22	5,7-Me	3-F	1	PhMe, −75°C, 0.75 h	CO_2	CO_2H	67
23	H	3-F-4-($CH(OCH_2)_2$)	1	THF-DEE, −100°C, 6 h	CO_2	CO_2H	94
24[a]	5,7-Me	3-F	1	THF, 0°C, 2 h	DMF[b]	CHO	78

[a]Using Bu_3MgLi instead of BuLi

[b]Followed by addition of thiosemicarbazide in order to obtain the 2-[(thiocarbamoyl) hydrazonomethyl] derivative

Table 22 Halogen–lithium exchange of *N*-(2-halogenoquinolylmethyl)pyrrole-2-carboxamides using butyllithium

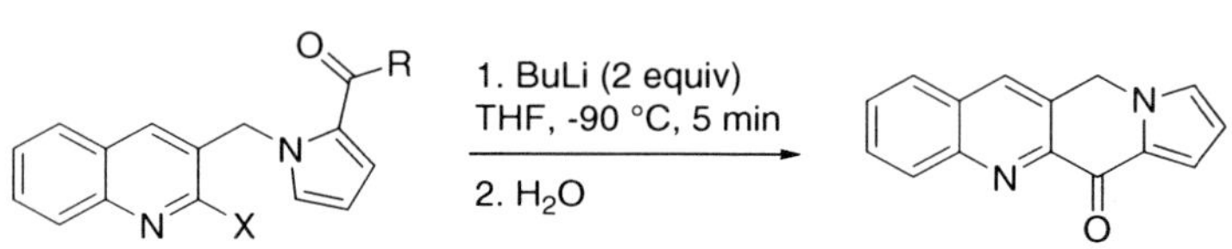

Entry	X	R	Yield (%)
1	Br	NMe(OMe)	61
2	Br	NEt_2	28
3	I	NMe(OMe)	85
4	I	NEt_2	83

Table 23 Reductive lithiation of 2-chloroquinolines

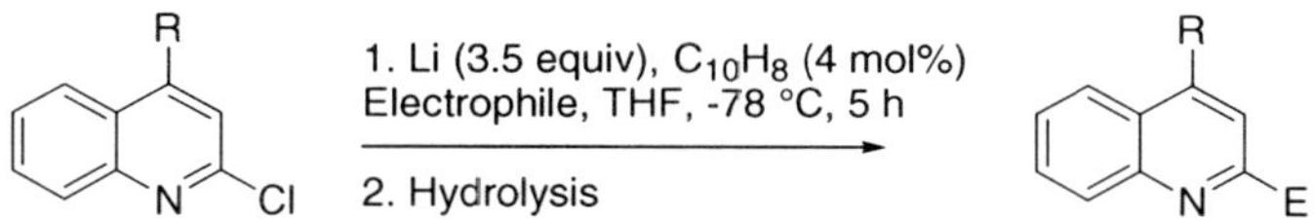

Entry	R	Electrophile	E	Yield (%)
1	H	*t*-BuCHO	C(OH)*t*-Bu	29
2	H	PhCHO	CH(OH)Ph	–[a]
3	H	$Et_2C(O)$	$C(OH)Et_2$	21
4	H	PhC(O)Me	C(OH)PhMe	20
5	Me	*t*-BuCHO	C(OH)*t*-Bu	25
6	Me	$Et_2C(O)$	$C(OH)Et_2$	25

[a]The corresponding phenyl ketone is obtained in 56% yield instead of the expected alcohol

Table 24 Halogen–lithium exchange of 3-bromoquinoline using butyllithium

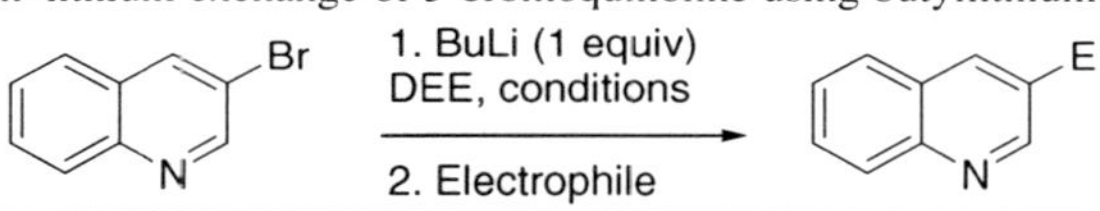

Entry	Conditions	Electrophile	E	Yield (%)
1	−35°C, 15 min	CO_2	CO_2H	70
2	−50°C, 10 min	CO_2	CO_2H	50
3	−100°C, 2 h	2-BrPhCHO	CH(OH)(2-BrPh)	77
4	−100°C, 2 h	DMF	CHO	61

at −75°C [70], and Harrowven and coworkers at −100°C but for 2 h before trapping with 2-bromobenzaldehyde (Table 24, Entry 3) [76] or dimethylformamide (Table 24, Entry 4) [77]. The use of *t*-BuLi (2 equiv) in DEE at −78°C was also reported in 1994 by Maguire and coworkers in the course of the synthesis of the corresponding trimethylstannane (50% yield) [78].

While 2- and 4-lithioquinoline only gives 2- and 4-ethylquinoline, respectively, by treatment with diethylmethoxyborane (1,2-migration), it is possible to replace the metal of 3-lithioquinoline with 9-borabicyclo[3.3.1]nonane [79]. Ota and Terashima intercepted 3-quinolyllithium with tributylborane; such generated lithium tributyl(3-quinolyl)borate was then trapped with allylic bromides in the presence of CuCN (60–69% yield using 3-cyclohexyl bromide, allyl bromide, and 2-bromoallyl bromide) [80]. In the course of the synthesis of phosphodiesterase-4 inhibitors, 3-lithioquinoline was intercepted with 1-[3-methoxy-4-(2-pyridyloxy) phenyl]-1-butanone, affording the expected alcohol in 53% yield [81].

Atkins and coworkers synthesized in 1997 3-butyryl-8-methoxy-4-[(2-methylphenyl)amino]quinoline in good yields via halogen–metal exchange of the 3-bromo substrate using BuLi in THF at −70°C followed by quenching with *N*-methoxy-*N*-methylbutyramide or *N*,*N*-dimethylbutyramide (Scheme 2) [82].

In 2011 Knochel and coworkers remedied the difficult insertion of magnesium into the carbon–halogen bond of 3-bromoquinoline by using a LiCl-promoted

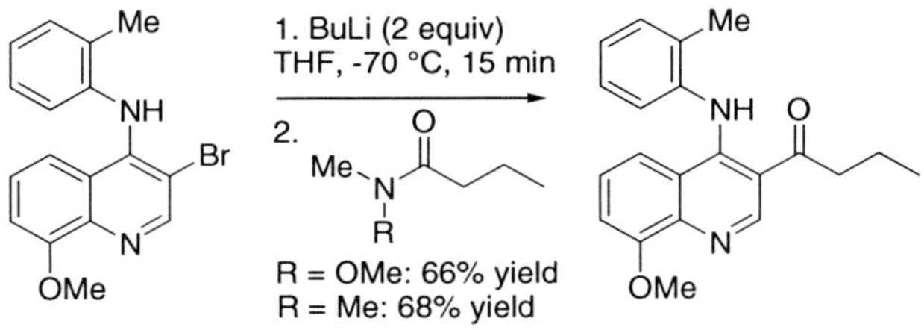

Scheme 2 Halogen–lithium exchange of a 3-bromoquinoline derivative using butyllithium

Table 25 Halogen–lithium exchange of 4-bromoquinoline using butyllithium

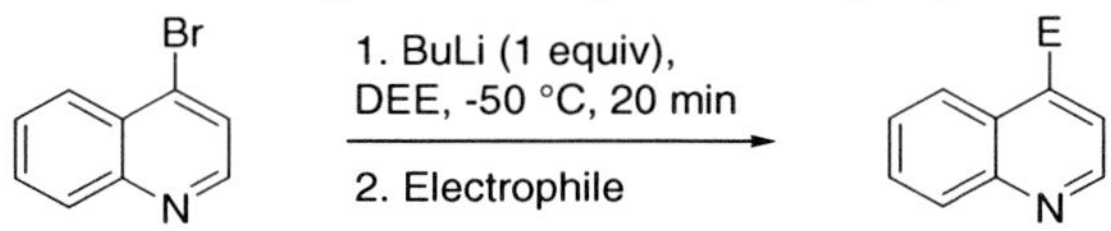

Entry	Electrophile	E	Yield (%)
1	$Ph_2C(O)$	$C(OH)Ph_2$	80[a]
2	CO_2	CO_2H	39

[a]Yield before crystallization

reaction [83]. i-PrMgCl was also employed to functionalize 3-iodo-4-methoxyquinolin-2(1*H*)-ones [84].

3.1.3 4-Substituted Quinolines

Gilman and Soddy published in 1958 the halogen–lithium interconversion of 4-bromoquinoline using BuLi in DEE at −50°C for 20 min. The corresponding lithio derivative was evidenced using benzophenone and CO_2, with the formation of the expected alcohol and carboxylic acid in 80% and 39% yield, respectively (Table 25) [85].

In the course of the synthesis of mefloquine and quinine analogues, Biot and coworkers reported in 2000 the halogen–metal exchange of 4-bromo-2,8-bis (trifluoromethyl)quinoline using BuLi in DEE at −65°C; the lithio derivative was intercepted by different 2-(dialkylaminomethyl)ferrocenecarboxaldehydes (Scheme 3) [86].

Schlosser and coworkers documented in 2003 several halogen–metal exchange reactions from trifluoromethyl-substituted 4-bromoquinolines using BuLi at low temperatures (Table 26) [5, 21]. It is interesting to note that the reaction is compatible with the presence of other substituents (methoxy, fluoro, trifluoromethoxy, and, above all, methyl and carboxylic acid). In addition, it takes place regioselectively at C4 when a second bromo group is present at C8.

Comins and coworkers studied the halogen–lithium exchange of 2,4-dibromoquinoline (including 3-substituted derivatives), and observed a

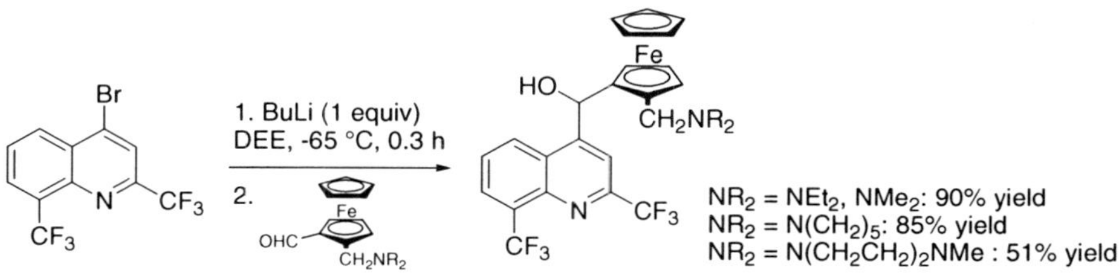

Scheme 3 Halogen–lithium exchange of 4-bromo-2,8-bis(trifluoromethyl)quinoline using butyllithium

Table 26 Halogen–lithium exchange of trifluoromethyl-substituted 4-bromoquinolines using butyllithium

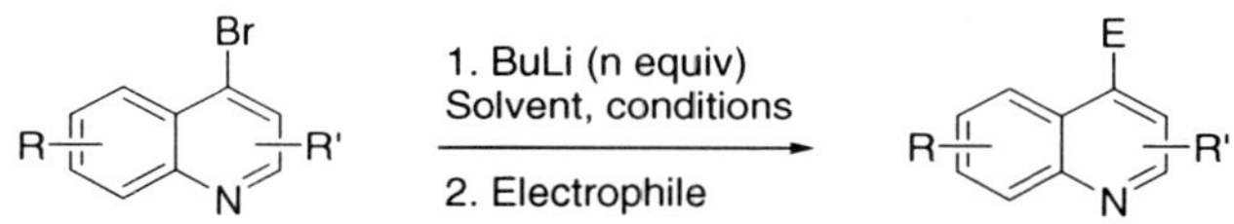

Entry	R	R′	n	Solvent, conditions	Electrophile	E	Yield (%)
1	H	2-CF_3	1	THF, −75°C, 0.25 h	MeOH	H	81
2	H	2-CF_3	1	THF, −75°C, 0.75 h	CO_2	CO_2H	74
3	H	2-CF_3-3-CO_2H	2	DEE, −75°C, 6 h	MeOH	H	67
4	5-OMe	2-CF_3	1	THF, −75°C, 0.75 h	CO_2	CO_2H	89
5	5-F	2-CF_3	1	THF, −75°C, 0.75 h	CO_2	CO_2H	82
6	6-Me	2-CF_3	1	THF, −75°C, 0.25 h	MeOH	H	74
7	6-Me	2-CF_3	1	THF, −75°C, 0.75 h	CO_2	CO_2H	77
8	6-Me	2-CF_3-3-CO_2H	2	HexH, −75°C, 6 h	MeOH	H	68
9	6-F	2-CF_3	1	THF, −75°C, 0.25 h	MeOH	H	89
10	6-F	2-CF_3	1	THF, −75°C, 0.75 h	CO_2	CO_2H	83
11	6-OCF_3	2-CF_3	1	THF, −75°C, 0.75 h	CO_2	CO_2H	74
12	6-OCF_3	2-CF_3-3-CO_2H	2	HexH, −75°C, 6 h	MeOH	H	57
13	7-OMe	2-CF_3	1	THF, −75°C, 0.25 h	MeOH	H	79
14	7-OMe	2-CF_3	1	THF, −75°C, 0.75 h	CO_2	CO_2H	74
15	7-OMe	2-CF_3-3-CO_2H	2	HexH, −75°C, 6 h	MeOH	H	84
16	7-F	2-CF_3	1	THF, −75°C, 0.25 h	MeOH	H	80
17	7-F	2-CF_3	1	THF, −75°C, 0.75 h	CO_2	CO_2H	84
18	7-F-2-CO_2H	2-CF_3	2	DEE, −100°C, 0.25 h	MeOH	H	58
19	8-Br	2-CF_3	1	THF, −75°C, 0.25 h	MeOH	H	76
20	8-Br	2-CF_3	1	THF, −75°C, 0.75 h	CO_2	CO_2H	57
21	8-F	2-CF_3	1	THF, −75°C, 0.75 h	CO_2	CO_2H	79
22	8-F	2-CF_3-3-CO_2H	2	HexH, −75°C, 6 h	MeOH	H	58
23	H	3-CF_3	1	THF, −100°C, 0.25 h	CO_2	CO_2H	81

regioselective reaction in favor of the 4-position using BuLi in a DEE-THF mixture at −78°C, as demonstrated by subsequent quenching with different electrophiles (Table 27) [87].

Table 27 Halogen–lithium exchange of 2,4-dibromoquinolines using butyllithium

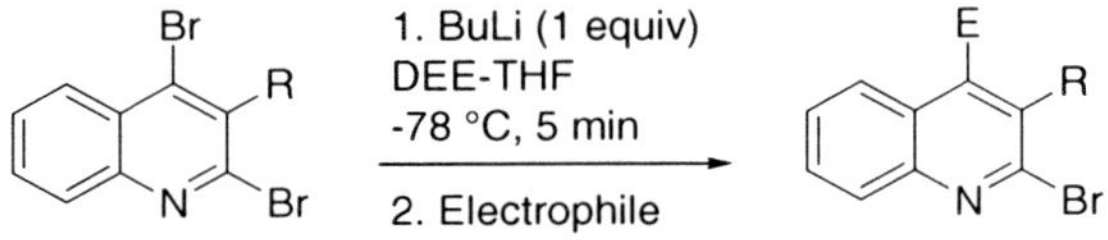

Entry	R	Electrophile	E	Yield (%)
1	H	H_2O	H	76
2	H	I_2	I	87
3	H	MeI	Me	85
4	H	Me_3SiCl	$SiMe_3$	70
5	H	DMF	CHO	74
6	H	PhCHO	CH(OH)Ph	82
7	$CH(OLi)N(CH_2CH_2)_2NMe$	I_2	I	90
8	$CH(OLi)N(CH_2CH_2)_2NMe$	MeSSMe	SMe	50
9	CH_2OLi	I_2	I	89

Table 28 Halogen–lithium exchange of 4,6,8-tribromoquinoline using butyllithium

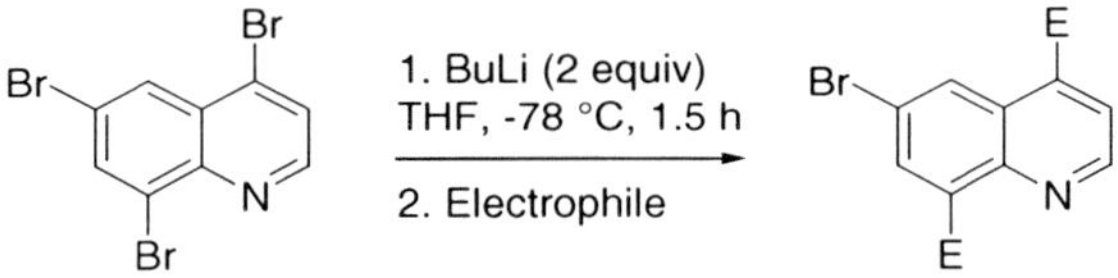

Entry	Electrophile	E	Yield (%)
1	H_2O	H	65
2	I_2	I	10[a]
3	Me_3SiCl	$SiMe_3$	85
4	MeSSMe	SMe	85

[a]6-Bromo-8-iodoquinoline was obtained in 50% yield

Using 2 equiv of BuLi in THF at −78°C, Cakmak and coworkers achieved the regioselective double halogen–metal exchange of 4,6,8-tribromoquinoline at the 4-and 8-position, as evidenced by subsequent trapping with different electrophiles (Table 28) [88].

In the course of the synthesis of 1,8-bis(4,4′-diquinolyl)naphthalenes, Wolf and coworkers prepared 2-substituted 4-(tributylstannyl)quinolines by iodine–lithium exchange from the 4-iodo substrates using BuLi in DEE at −78°C (Table 29, Entries 1–3) [89, 90]. Starting from 4,8-diiodo-2-(trifluoromethyl)quinoline, prepared from 4,8-dibromo-2-(trifluoromethyl)quinoline by double halogen–metal exchange with BuLi (2 equiv, THF, −75°C, 0.75 h) followed by diiodination (69% yield), the reaction proved to be regioselective at the 4-position using i-PrMgCl in THF at 0°C (Table 29, Entry 4) [5].

Table 29 Halogen–lithium exchange of 2-substituted 4-iodoquinolines using either butyllithium or isopropylmagnesium chloride

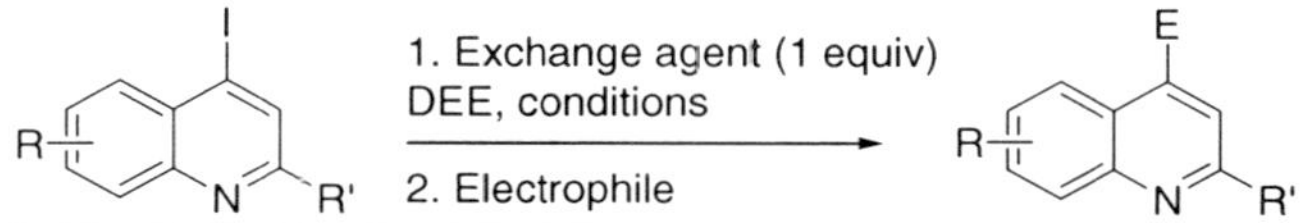

Entry	R	R′	Exchange agent	Solvent	Conditions	Electrophile	E	Yield (%)
1	H	Me	BuLi	DEE	−78°C, 0.5 h	Me_3SnCl	$SnMe_3$	80
2	H	i-Pr	BuLi	DEE	−78°C, 0.5 h	Me_3SnCl	$SnMe_3$	72
3	H	Ph	BuLi	DEE	−78°C, 0.5 h	Me_3SnCl	$SnMe_3$	57
4	8-I	CF_3	i-PrMgCl	THF	0°C, 0.75 h	CO_2	CO_2H	45

Table 30 Halogen–lithium exchange of 5,7-dibromo-8-hydroxyquinoline (sodium salt) using butyllithium

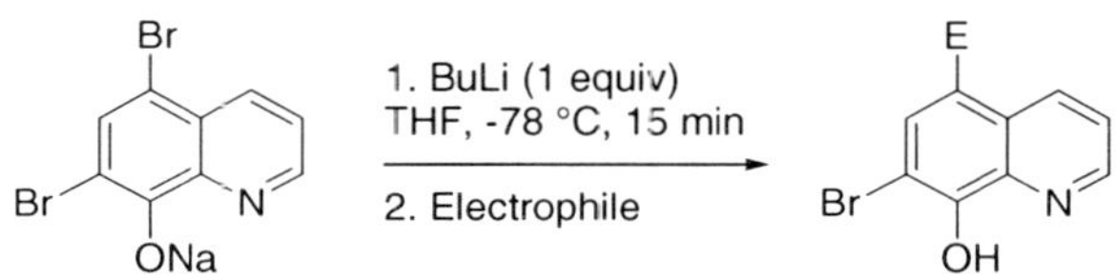

Entry	Electrophile	E	Yield (%)
1	EtOD	D	90
2	PhCHO	CH(OH)Ph	60
3	$(CH_2)_5NCHO$	CHO	70
4	$I(CH_2)_2I$	I	65
5	CCl_3CCl_3	Cl	63
6	Me_3SiCl	$SiMe_3$	58
7	MeI	Me	48
8	$(CH_2)_2O$	$(CH_2)_2OH$	68
9	HeptBr	Hept	47
10	4-(OMe)PhCHO	CH(OH)(4-(OMe)Ph)	50

3.1.4 5-Substituted Quinolines

The sodium salt of 5,7-dibromo-8-hydroxyquinoline was found to be resistant to nucleophilic additions of lithium compounds and its halogen–metal exchange can thus be carried out using BuLi in THF at −78°C for 15 min. Subsequent interception with a large range of electrophiles followed by hydrolysis showed a regioselective functionalization at the 5-position (Table 30) [91].

Upon treatment with PhMgCl in THF at −40°C for 5 min and subsequent benzaldehyde addition, sensitive 5-iodo-6-nitroquinoline was converted to the expected alcohol in 78% yield by Knochel and coworkers [92].

Table 31 Halogen–lithium exchange of 7-bromo-8-methoxyquinoline and 5,7-dibromo-8-methoxyquinoline using phenyllithium

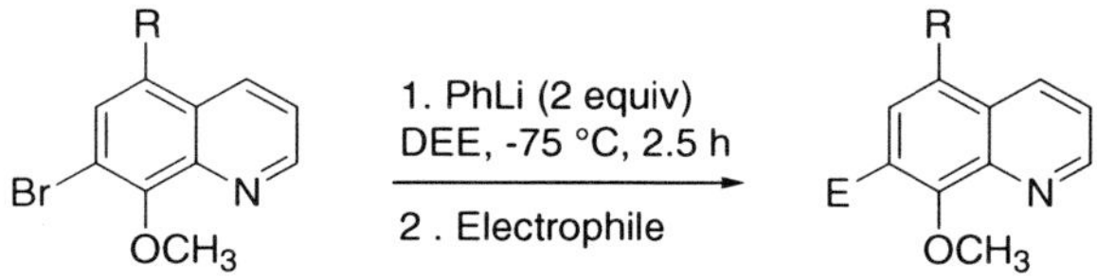

Entry	R	Electrophile	E	Yield (%)
1	H	HCl-H_2O	H	78
2	H	DCl-H_2O	D	76
3	H	PhCHO	CH(OH)Ph	69
4	H	I_2	I	70
5	H	$B(OMe)_3$ then $MeCO_3H$	OH	63
6	Br	HCl-H_2O	H	56
7	Br	DCl-H_2O	D	55
8	Br	PhCHO	CH(OH)Ph	54
9	Br	$B(OMe)_3$ then $MeCO_3H$	OH	55
10	Br	$(CH_2)_5NCHO$	CHO	43
11	Br	2-F-3-PyCHO	CH(OH)(2-F-3-Py)	46

3.1.5 7-Substituted Quinolines

Quéguiner and coworkers reported in 1995 their study about the halogen–lithium exchange of 7-bromo-8-methoxyquinoline. While the use of alkyllithiums in THF was not successful, even at −75°C due to competitive 1,2- nucleophilic additions, recourse to PhLi in DEE allowed for a chemoselective reaction, affording after trapping the expected derivatives in yields ranging from 63% to 78% (Table 31, Entries 1–5). The reaction conditions were extended to 5,7-dibromo-8-methoxyquinoline, and regioselective functionalizations (still at the 7-position) were achieved (Table 31, Entries 6–11). It is interesting to note that starting from 5,7-diiodo-8-methoxyquinoline and using PhLi (2 equiv) in THF at −75°C for 2.5 h led to a mixture of 5-iodo-8-methoxyquinoline (27% yield) and 8-methoxyquinoline (52% yield), after hydrolysis [93]. More recently, Baron and Knochel reported the regioselective halogen–metal exchange of 5,7-diiodo-8-tosyloxyquinoline at its 7-position using i-PrMgCl, a result evidenced by subsequent boronation using a dioxaborolane [94]. If the tosyl group is replaced by $4\text{-ClC}_6H_4SO_2$, the quinolylmagnesium halide is less stable, and elimination is evidenced at temperatures above −78°C by cycloaddition in the presence of furan (Scheme 4) [95].

Starting from 7-bromo-8-hydroxyquinoline, halogen-metal exchange is possible using BuLi in THF at −78°C for 1 h provided that the substrate is preliminary converted to its sodium salt; otherwise, the exchange is taking place more rapidly than the removal of the OH proton, and 8-hydroxyquinoline is the only product obtained after work-up. Subsequent interception with trimethylsilyl chloride or

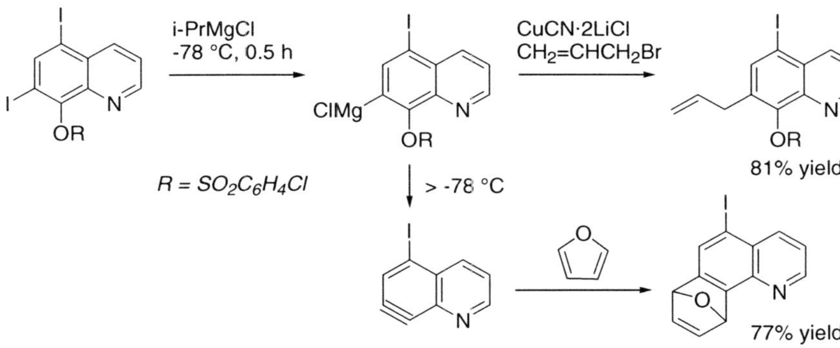

Scheme 4 Halogen–lithium exchange of *O*-substituted (4-$ClC_6H_4SO_2$) 5,7-diiodo-8-hydroxyquinoline using isopropylmagnesium chloride

Table 32 Halogen–lithium exchange of 7-bromo-8-hydroxyquinoline (sodium salt) using butyllithium

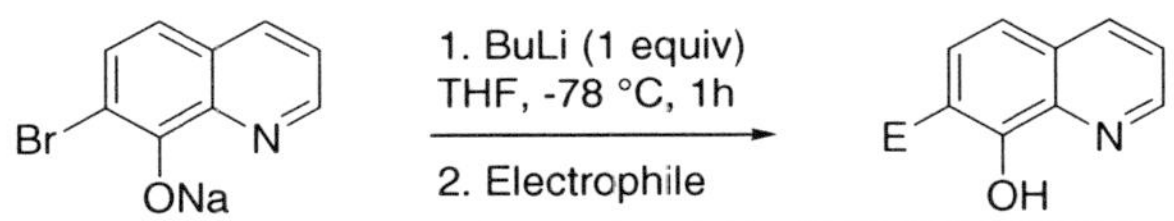

Entry	Electrophile	E	Yield (%)
1	Me_3SiCl	$SiMe_3$	63
2	4-(OMe)PhCHO	CH(OH)(4-(OMe)Ph)	57

anisaldehyde afforded after hydrolysis the expected derivatives in medium yields along with 8-hydroxyquinoline (Table 32) [91].

3.1.6 8-Substituted Quinolines

Pearson and coworkers described in 1969 the conversion of 8-bromoquinoline to the corresponding lithio compound using BuLi (2 equiv) in THF at −70°C [96]. Issleib and Haftendorn trapped in 1970 8-lithioquinoline by R_2PCl, $R(Me_2N)PCl$, and $(Me_2N)_2PCl$ in the course of syntheses of 8-quinolyl-phosphinite, -phosphonite, and -phosphines [97]. Using iodomethane as trapping agent, Suggs and Pearson later optimized the halogen–metal reaction of 8-bromoquinoline using different alkyllithiums (BuLi, *s*-BuLi, *t*-BuLi), solvents (THF, DEE), and conditions (−78 and −100°C). They achieved their best conversion using *s*-BuLi in THF at −78°C for 5 min, conditions for which the 8-methylated derivative was obtained in 87% yield. The study was then extended to various other electrophiles (Table 33). The authors observed that the reaction cannot be extended to 8-chloroquinoline for which only products resulting from addition to C=N are

Table 33 Halogen–lithium exchange of 8-bromoquinoline using *sec*-butyllithium

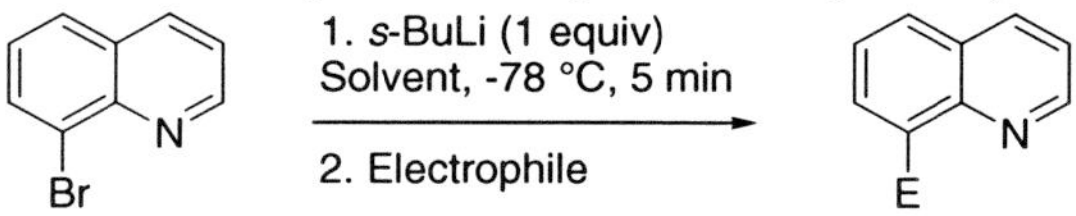

Entry	Solvent	Electrophile	E	Yield (%)
1	THF	MeI	Me	87[a]
2	THF	DMF	CHO	42[a]
3	THF	d^7-DMF	CDO	43
4	THF	MeCHO	CH(OH)Me	64
5	THF	Ph_2PCl	PPh_2	32
6	THF	Me_3SnCl	$SnMe_3$	55
7	DEE	allylBr[b]	allyl	65
8	DEE	$(CH_2)_2O$[b]	$(CH_2)_2OH$	50

[a]Determined by NMR integration using acetophenone as a standard
[b]After transmetalation using CuI

Table 34 Halogen–lithium exchange of trifluoromethyl-substituted 8-halogenoquinolines using either butyllithium or isopropylmagnesium chloride

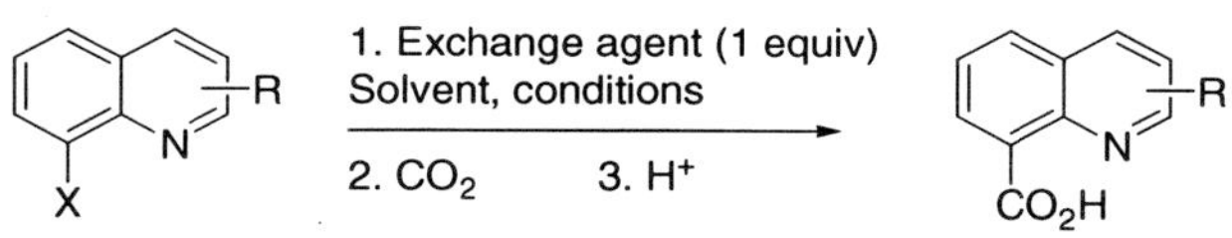

Entry	X	R	Exchange agent	Solvent, conditions	Yield (%)
1	Br	2-CF_3	BuLi	THF, −75°C, 0.75 h	73
2	Br	2-Cl-3-CF_3	BuLi	PhMe, −75°C, 0.25 h	68
3	I	2-Br-4-CF_3	BuLi	PhMe, −75°C, 0.25 h	81
4	I	4-Br-2-CF_3	i-PrMgCl	DEE, 0°C, 0.75 h	84

observed under the same reaction conditions [98]. The method was employed in 2001 by Don Tilley and coworkers, who used methyldichlorosilane as the electrophile to synthesize bis(8-quinolyl)methylsilane (37% yield) [99]. Kozlowski and coworkers employed in 2003 BuLi (2 equiv) in DEE at −78°C for 30 min to afford, after trapping with chlorotributylstannane the expected 8-substituted derivative in 83% yield [100].

Schlosser and coworkers reported in 2003 halogen–metal exchanges from trifluoromethyl-substituted 8-halogenoquinolines using BuLi [21] or i-PrMgCl [5] at low temperatures (Table 34). From 4-bromo-8-iodo-2-(trifluoromethyl)quinoline, the reaction first takes place at the 8-position using BuLi in THF at −75°C, but initially formed 8-lithioquinoline then isomerizes to a 4-lithio derivative through bromine–lithium interconversion affording after trapping with CO_2 8-bromo-2-(trifluoromethyl)-4-quinolinecarboxylic acid in 86% yield [5].

Quinoline-8-carboxamides were prepared in 2009 by Threadgill and coworkers by halogen–metal exchange of the corresponding bromides using BuLi in THF at −78°C (Table 35) [101].

Table 35 Halogen–lithium exchange of 2-substituted 8-bromoquinolines using butyllithium

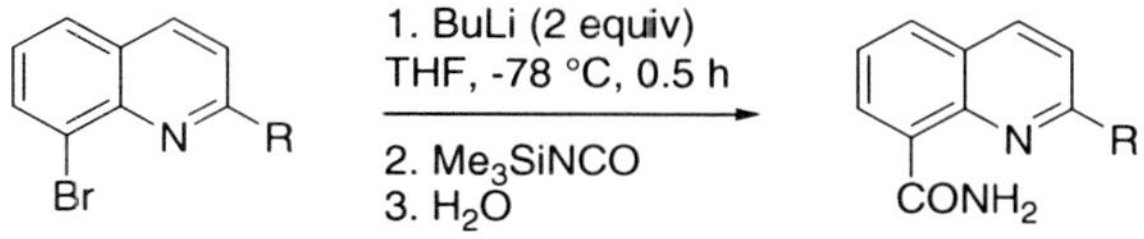

Entry	R	Yield (%)
1	H	97
2	Me	52
3	Et	43
4	Ph	34
5	4-(OMe)Ph	43

Scheme 5 Halogen–lithium exchange of 5,8-dibromoquinoline using butyllithium

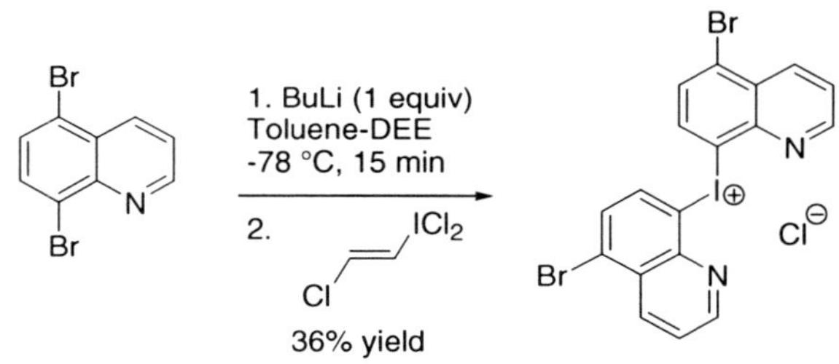

In the course of the synthesis of bis(quinolyl)iodonium salts, Stang and coworkers in 1995 observed in the case of 5,8-dibromoquinoline a regioselective halogen–metal exchange at the 8-position (next to nitrogen) upon treatment with BuLi in toluene-DEE at −78°C for 15 min (Scheme 5) [102].

Knochel and coworkers showed in 2003 that i-PrMgCl is a suitable reagent, when used in THF at −30°C, to perform the halogen–metal exchange of 8-iodoquinolines bearing functional groups; the quinoline Grignard reagents were trapped by electrophiles either directly or after a transmetalation step (Table 36) [103].

3.2 Halogen–Metal Exchange and Related Reactions of Isoquinolines

3.2.1 1-Substituted Isoquinolines

Gilman and Soddy reported in 1957 the first halogen–lithium exchange of 1-bromoisoquinoline. Treatment by BuLi in DEE at −50°C for 15 min afforded after quenching with benzophenone the expected alcohol in 68% yield. The method cannot be extended to the synthesis of 1-isoquinolinecarboxylic acid since the

Table 36 Halogen-magnesium exchange of functionalized 8-iodoquinolines using isopropylmagnesium chloride

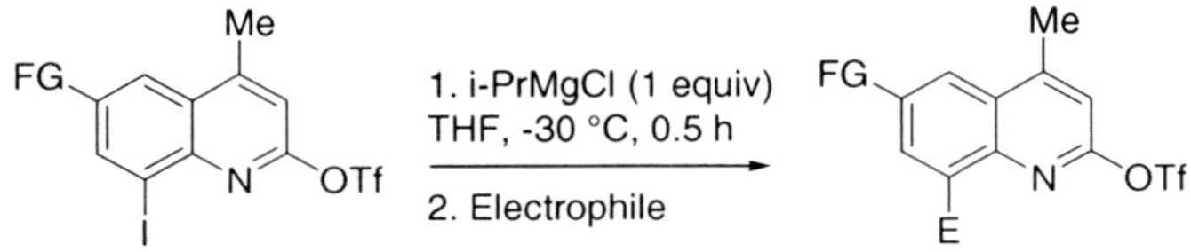

Entry	FG	Electrophile	E	Yield (%)
1	CO_2Et	4-(OMe)PhCHO	CH(OH)(4-(OMe)Ph)	52
2	CO_2Et	CCl_2BrCCl_2Br	Br	78
3	CO_2Et	allylBr[a]	allyl	77
4	CO_2Et	$HC{\equiv}CCO_2t$-Bu[a]	(E)-CH=$CHCO_2t$-Bu	41
5	CO_2Et	$HC{\equiv}CCH_2Br$[a]	$CH_2C{\equiv}CH$	70[b]
6	CO_2Et	4-(I)$PhCO_2Et$[c]	4-$PhCO_2Et$	74
7	CO_2Et	2-(I)pyrimidine[c]	2-pyrimidyl	48
8	CF_3	allylBr[a]	allyl	70
9	CF_3	4-(I)$PhCO_2Et$[c]	4-$PhCO_2Et$	63

[a]After transmetalation using CuCN·2LiCl
[b]The product contains 10% of the corresponding allene
[c]After transmetalation using $ZnCl_2$ and in the presence of catalytic $Pd(dba)_2$ and tri(2-furyl) phosphine

Table 37 Halogen–lithium exchange of 1-bromoisoquinoline using butyllithium

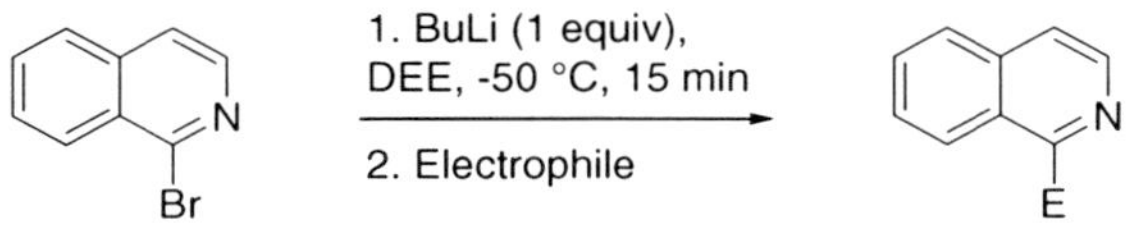

Entry	Electrophile	E	Yield (%)
1	$Ph_2C(O)$	$C(OH)Ph_2$	68[a]
2	CO_2	CO_2H	–[b]

[a]Yield before crystallization
[b]1,1′-Biisoquinolyl ketone is isolated instead in 44% yield

carboxylate is again attacked by the organolithium reagent in the course of the trapping step furnishing instead 1,1′-biisoquinolyl ketone (Table 37) [72].

3.2.2 3-Substituted Isoquinolines

Muchowski studied the access to 3-lithio isoquinolines. He observed in 2005 a 20 to 1 regioselective halogen–metal exchange from 1,3-dibromoisoquinoline in favor of the 1-position using BuLi in THF at temperatures below −70°C (3-bromoisoquinoline, a compound very sensitive to nucleophilic attacks, was isolated in 70% yield). Starting from 3-bromo-1-(methylthio)isoquinoline, the halogen–metal exchange was readily

Table 38 Halogen–lithium exchange of 3-bromo-1-(methylthio)isoquinoline using butyllithium

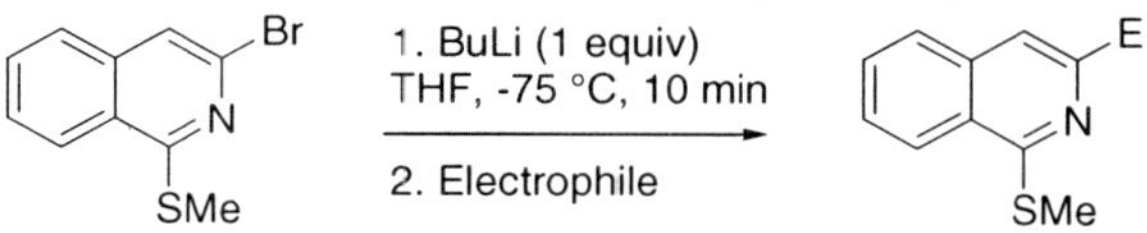

Entry	Electrophile	E	Yield (%)
1	EtOH	H	85
2	MeI	Me	74
3	BuI	Bu	19
4	DMF	CHO	89
5	DMA	COMe	60
6	$ClCO_2Me$	CO_2Me	45

performed under the same reaction conditions at the desired position, affording different derivatives in moderate to high yields (Table 38) [104].

3.2.3 4-Substituted Isoquinolines

Gilman and Soddy reported in 1957 the synthesis of 4-isoquinolinecarboxylic acid (46% yield) starting from the corresponding bromide by treatment with BuLi in DEE at −50°C for 10 min followed by addition of CO_2 and acidification [72]. A 58% yield was noted using isobutyraldehyde to intercept the lithium derivative [105].

4-Isoquinolyllithium was quenched by tributylborane, and the formed lithium tributyl(4-isoquinolyl)borate then trapped with allylic bromides in the presence of CuCN (61–62% yield using 3-cyclohexyl bromide and allyl bromide) [80].

In the course of the synthesis of an inhibitor of B-raf kinase, Bänziger, Yusuff, and coworkers documented in 2006 the halogen–metal exchange of 1-substituted 4-bromoisoquinolines using BuLi in THF at low temperatures (Scheme 6) [106].

3.2.4 5-Substituted Isoquinolines

Threadgill and coworkers attempted the halogen–metal exchange of 5-bromoquinoline using BuLi (1 equiv) in THF at −116°C; after subsequent trapping with benzaldehyde and hydrolysis, the expected alcohol was obtained, but in a moderate 34% yield [107].

3.2.5 6-Substituted Isoquinolines

Tsou and coworkers performed the halogen–metal exchange of a 1,3-disiloxy 6-bromoisoquinoline using *t*-BuLi in THF at −78°C; the 6-substituted isoquinoline-1,3(2*H*,4*H*)-diones were isolated in yields ranging from 56% to 57% after trapping with iodine and *N*,*N*-dimethylcarbamoyl chloride (Table 39) [108].

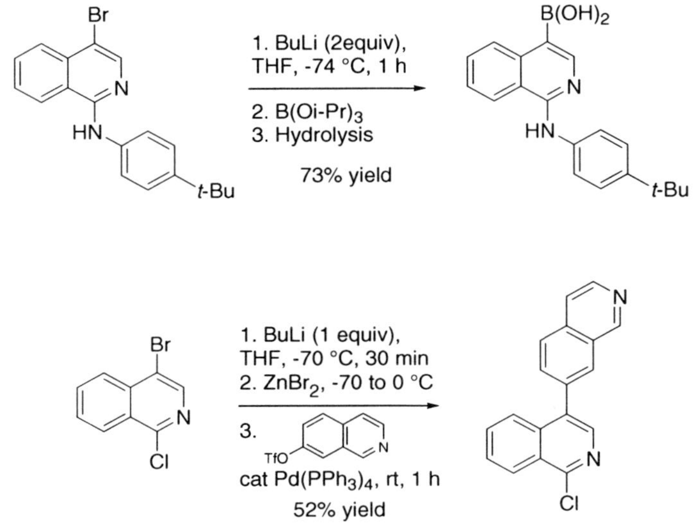

Scheme 6 Halogen–lithium exchange of 1-substituted 4-bromoisoquinolines using butyllithium

Table 39 Halogen–lithium exchange of a 1,3-disiloxy 6-bromoisoquinoline using *tert*-butyllithium

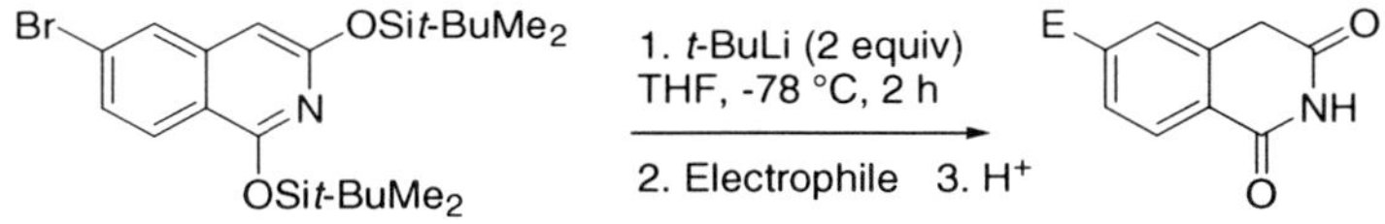

Entry	Electrophile	E	Yield (%)
1	I_2	I	56
2	$Me_2NC(O)Cl$	$C(O)NMe_2$	57

4 Conclusion

This chapter has underlined the importance of lithiation and magnesiation processes in the functionalization of quinoline and isoquinoline substrates. In spite of the sensitivity of these aromatic heterocycles toward nucleophile species (in relation with their low LUMO energy levels), numerous scaffolds have been elaborated through lithium or magnesium intermediates at low reaction temperatures.

One way to free oneself from the use of such cryogenic conditions has already been developed. It is based on the use of polymetal compounds, either acting as bases or as halogen–metal exchange agents. When properly chosen, these reagents can behave synergically, allowing both efficient and chemoselective reactions. They are thus particularly suitable for the functionalization of sensitive substrates such as quinolines and isoquinolines, notably those for which monometal reagents failed, and their use will be presented in a separate chapter.

Acknowledgment The Institut Universitaire de France is gratefully acknowledged.

References

1. Marsais F, Bouley E, Quéguiner G (1979) New reaction of quinoline: regioselective metalation of fluoroquinolines. J Organomet Chem 171:273–282
2. Shi GQ, Takagishi S, Schlosser M (1994) Metalated fluoropyridines and fluoroquinolines as reactive intermediates: new ways for their regioselective generation. Tetrahedron 50:1129–1134
3. Ondi L, Volle JN, Schlosser M (2005) Metalation/functionalization sequences applied to 2-bromo-3-fluoroquinolines. Tetrahedron 61:717–725
4. Lefebvre O, Marull M, Schlosser M (2003) 4-(Trifluoromethyl)quinoline derivatives. Eur J Org Chem 2115–2121
5. Marull M, Schlosser M (2003) Selective and efficient structural elaboration of 2-(trifluoromethyl)quinolinones. Eur J Org Chem 1576–1588
6. Marull M, Schlosser M (2004) Regioflexibility in the functionalization of multiply halogenated quinolines. Eur J Org Chem 1008–1013
7. Schnürch M, Spina M, Khan AF et al (2007) Halogen dance reactions – a review. Chem Soc Rev 36:1046–1057
8. Schnürch M (2011) Recent progress on the halogen dance reaction on heterocycles. Top Heterocycl Chem 27:185–218
9. Arzel E, Rocca P, Marsais F et al (1998) First halogen-dance reaction in quinoline series: application to a new synthesis of quindoline. Tetrahedron Lett 39:6465–6466
10. Arzel E, Rocca P, Marsais F et al (1999) First total synthesis of cryptomisrine. Tetrahedron 55:12149–12156
11. Arzel E, Rocca P, Grellier P et al (2001) New synthesis of benzo-δ-carbolines, cryptolepines, and their salts: in vitro cytotoxic, antiplasmodial, and antitrypanosomal activities of δ-carbolines, benzo-δ-carbolines, and cryptolepines. J Med Chem 44:949–960
12. Bennacef I, Tymciu S, Dhilly M et al (2004) Lithiation of functionalized fluoroquinolines: synthesis of dihalo-2-phenylquinoline-4-carboxamides and in vitro evaluation as NK-3 receptor ligands for medical imaging studies. J Org Chem 69:2622–2625
13. Stadlwieser J, Barbier P, Taylor S (1998) 8-Fluoro-6-(methoxymethoxy)quinoline. Synthesis and regioselective functionalization via reaction with organolithium compounds. Helv Chim Acta 81:1088–1094
14. Wiesner J, Ortmann R, Jomaa H et al (2003) New antimalarial drugs. Angew Chem Int Ed 42:5274–5293
15. Schlosser M, Mongin F, Porwisiak J et al (1998) Bis- and oligo(trifluoromethyl)benzenes: hydrogen/metal exchange rates and gas-phase acidities. Chem Eur J 4:1281–1286
16. Schlosser M, Marull M (2003) The direct metalation and subsequent functionalization of trifluoromethyl-substituted pyridines and quinolines. Eur J Org Chem 1569–1575
17. Isobe A, Takagi J, Katagiri T et al (2008) Palladium-catalyzed chloroimination of imidoyl chlorides to a triple bond: an intramolecular reaction leading to 4-chloroquinolines. Org Lett 10:2657–2659
18. Marull M, Lefebvre O, Schlosser M (2004) An improved access to 4-trifluoromethyl-2(1H)-quinolinones: the "watering protocol". Eur J Org Chem 54–63

19. Schlosser M, Cottet F, Heiss C et al (2006) Relay propagation of crowding: the trifluoromethyl group as both an emitter and transmitter of steric pressure. Eur J Org Chem 729–734
20. Marsais F, Godard A, Quéguiner G (1989) Directed ortho-lithiation of chloroquinolines. Application to synthesis of 2,3-disubstituted quinolines. J Heterocycl Chem 26:1589–1594
21. Cottet F, Marull M, Lefebvre O et al (2003) Recommendable routes to trifluoromethyl-substituted pyridine- and quinolinecarboxylic acids. Eur J Org Chem 1559–1568
22. Leleu S, Papamicael C, Marsais F et al (2004) Preparation of axially chiral quinolinium salts related to NAD+ models: new investigations of these biomimetic models as 'chiral amide-transferring agents'. Tetrahedron Asymmetry 15:3919–3928
23. Tagawa Y, Yamashita H, Nomura M et al (1998) Preparation of new 3-hydroxyquinoline alkaloid, jineol and its ether derivatives using directed ortho-lithiation of chloroquinoline as the key step. Heterocycles 48:2379–2387
24. Ford A, Sinn E, Woodward S (1997) Exploitation of differential reactivity of the carbon–chlorine bonds in 1,3-dichloroisoquinoline. Routes to new N,N-chelate ligands and 1,3-disubstituted isoquinolines. J Chem Soc, Perkin Trans 1 927–934
25. Comins DL, Hong H, Saha JK et al (1994) A six-step synthesis of (±)-camptothecin. J Org Chem 59:5120–5121
26. Nolan JM, Comins DL (2003) Regioselective Sonogashira couplings of 2,4-dibromoquinolines. A correction. J Org Chem 68:3736–3738
27. Bennasar ML, Roca T, Ferrando F (2006) A new radical-based route to calothrixin B. Org Lett 8:561–564
28. Hooper DC, Rubinstein E (eds) (2002) Quinolone antimicrobial agents, 3rd edn. ASM Press, Washington
29. Alvarez M, Salas M, Rigat L et al (1992) Hetero-ring lithiation of N-methyl-4-quinolone and N-methylquinoline-4-thione. J Chem Soc, Perkin Trans 1 351–356
30. Gonzalez R, Ramos MT, De la Cuesta E et al (1993) Base-catalyzed electrophilic substitution in 2(1H)-quinolinones. Heterocycles 36:315–322
31. Fernandez M, de la Cuesta E, Avendano C (1995) Metalation of 2(1H)-quinolinone: synthesis of 3-substituted compounds. Synthesis 1362–1364
32. Moreno T, Fernandez M, de la Cuesta E et al (1996) Metalation of methoxy-2(1H)-quinolinones. Heterocycles 43:817–828
33. Tagawa Y, Kawaoka T, Goto Y (1997) A convenient preparation of 4,8-dimethoxy-3-substituted-2(1H)-quinolones by an electrophilic reaction through base-induced deprotonation and its synthetic application for the synthesis of new alkaloids, 3,4,8-trimethoxy-2(1H)-quinolone and 3-formyl-4,7,8-trimethoxy-2(1H)-quinolone (glycocitridine). J Heterocycl Chem 34:1677–1683
34. Gilman H, Beel JA (1951) The metalation of 2-ethoxyquinoline. J Am Chem Soc 73:32
35. Narasimhan NS, Paradkar MV (1967) A new synthesis of the furo[2,3-b]quinolines: a synthesis of dictamnine. Chem Ind (London) 831–832
36. Narasimhan NS, Alurkar RH (1968) Synthesis of dihydro-γ-fagarine and edulitine. Chem Ind (London) 515
37. Narasimhan NS, Paradkar MV, Alurkar RH (1971) Synthetic application of lithiation reactions. IV. Novel synthesis of linear furoquinoline alkaloids and a synthesis of edulitine. Tetrahedron 27:1351–1356
38. Narasimhan NS, Mali RS (1974) Synthetic application of lithiation reactions. VI. New synthesis of linear furoquinoline alkaloids. Tetrahedron 30:4153–4157
39. Narasimhan NS, Mali RS, Gokhale AM (1979) Synthetic application of lithiation reaction. Part XI. Synthesis of kokusaginine, a furoquinoline alkaloid. Indian J Chem, Sect B 18B:115–116
40. Collins JF, Donnelly WJ, Grundon MF et al (1972) Aromatic hydroxylation in quinoline alkaloids. Biosynthesis of skimmianine from dictamnine, and a convenient synthesis of furanoquinoline alkaloids. J Chem Soc, Chem Commun 1029–1030

41. Jones K, Roset X, Rossiter S et al (2003) Demethylation of 2,4-dimethoxyquinolines: the synthesis of atanine. Org Biomol Chem 1:4380–4383
42. Bhoga U, Mali RS, Adapa SR (2004) New synthesis of linear furoquinoline alkaloids. Tetrahedron Lett 45:9483–9485
43. Broch S, Aboab B, Anizon F et al (2010) Synthesis and in vitro antiproliferative activities of quinoline derivatives. Eur J Med Chem 45:1657–1662
44. Robin Y, Godard A, Quéguiner G (1987) Unexpected course of the reaction of aromatic aldehydes on lithiated N, N-dimethyl O-(3-quinolyl)carbamate. J Heterocycl Chem 24:1487–1488
45. Jacquelin JM, Robin Y, Godard A et al (1988) Metalation of carbamates of 2-, 3-, and 4-hydroxyquinolines. Can J Chem 66:1135–1140
46. Godard A, Robin Y, Quéguiner G (1987) Metalation of 5-, 6-, 7-, and 8-hydroxyquinolyl carbamates. J Organomet Chem 336:1–12
47. Blakemore PR, Kilner C, Milicevic SD (2005) Harnessing anionic rearrangements on the benzenoid ring of quinoline for the synthesis of 6,6'-disubstituted 7,7'-dihydroxy-8,8'-biquinolyls. J Org Chem 70:373–376
48. Gschwend HW, Rodriguez HR (1979) Heteroatom-facilitated lithiations. Org React 26:1–360
49. Jacquelin JM, Marsais F, Godard A et al (1986) Lithiation of 2-aminoquinoline derivatives. Functionalization at the 3-position. Synthesis 670–672
50. Godard A, Jacquelin JM, Quéguiner G (1988) Metalation of 2-, 3- and 4-pivalamidoquinolines. J Organomet Chem 354:273–285
51. Mongin F, Trecourt F, Quéguiner G (1999) Directed lithiation of unprotected pyridine-carboxylic acids. Tetrahedron Lett 40:5483–5486
52. Rebstock A-S, Mongin F, Trecourt F et al (2002) Directed lithiation of unprotected quinoline-carboxylic acids. Tetrahedron Lett 43:767–769
53. Kelly TR, Zhao Y, Cavero M et al (2000) Synthesis of the potent antimalarials calothrixin A and B. Org Lett 2:3735–3737
54. Bennacef I, Perrio C, Lasne M-C et al (2007) Functionalization through lithiation of (S)-N-(1-phenylpropyl)-2-phenylquinoline-4-carboxamide. Application to the labeling with carbon-11 of NK-3 receptor antagonist SB 222200. J Org Chem 72:2161–2165
55. Rebstock A-S, Mongin F, Trécourt F et al (2004) Metallation of pyridines and quinolines in the presence of a remote carboxylate group. New syntheses of heterocyclic quinones. Org Biomol Chem 2:291–295
56. Mhaske SB, Argade NP (2004) Regioselective quinazolinone-directed ortho lithiation of quinazolinoylquinoline: practical synthesis of naturally occurring human DNA topoisomerase I poison luotonin A and luotonins B and E. J Org Chem 69:4563–4566
57. Tagawa Y, Hama K, Goto Y et al (1992) Electrophilic reaction of quinoline 1-oxide through base-induced deprotonation. Heterocycles 34:2243–2246
58. Mongin O, Rocca P, Thomas-dit-Dumont L et al (1995) Metalation of pyridine N-oxides and application to synthesis. J Chem Soc, Perkin Trans 1 2503–2508
59. Tagawa Y, Hama K, Goto Y et al (1995) Electrophilic reaction of pyridine, quinoline, isoquinoline, their N-oxides and their boron trifluoride complexes through base-induced deprotonation. Heterocycles 40:809–816
60. Abarca B, Gomez-Aldaravi E, Jones G (1984) Triazolopyridines. Part 4. Directed lithiation using 1,2,3-triazolo[1,5-a]quinoline. J Chem Res, Synop 140–141
61. Ballesteros-Garrido R, Leroux FR, Ballesteros R et al (2009) The deprotonative metalation of [1,2,3]triazolo[1,5-a]quinoline. Synthesis of 8-haloquinoline-2-carboxaldehydes. Tetrahedron 65:4410–4417
62. Abarca B, Ballesteros R, Gómez-Aldaraví E et al (1985) Triazolopyridines. Part 5. The reactions of 1,2,3-triazolo[5,1-a]isoquinoline: a new route to 1,3-disubstituted isoquinolines. J Chem Soc, Perkin Trans 1 1897–1901

63. Abarca B, Ballesteros R, Gay B et al (2004) Triazolopyridines 23. Synthesis of 5,5'-bi[1,2,3] triazolo[5,1-α]isoquinoline. ARKIVOC 42–47
64. Miyashita A, Sato Y, Watanabe S et al (1995) Synthesis of 2,6-disubstituted pyridines by lithiation of pyrazolo[1,5-a]pyridines. Chem Pharm Bull 43:174–176
65. Clarke AJ, McNamara S, Meth-Cohn O (1974) Novel aspects of the metalation of heterocycles. Side-chain metalation of thiophene and ring metalation of six-membered nitrogen heterocycles. Tetrahedron Lett 15:2373–2376
66. Gros P, Fort Y, Caubère P (1997) Aggregative activation in heterocyclic chemistry. Part 5. Lithiation of pyridine and quinoline with the complex base BuLi·Me2N(CH2)2OLi (BuLi·LiDMAE). J Chem Soc, Perkin Trans 1 3597–3600
67. Sachs F, Sachs L (1904) Das Verhalten tertiärer Amine gegen magnesium-organische Verbindungen. Ber Dtsch Chem Ges 37:3088–3092
68. Howitz J, Köpke O (1913) Über o-Chinolylketone und ihre Derivate. Justus Liebigs Ann Chem 396:38–52
69. Jansen HE, Wibaut JP (1937) Some reactions of 2-bromo- and 3-bromo-quinolines. Recl Trav Chim Pays-Bas Belg 56:709–713
70. Wibaut JP, Heeringa LG (1955) Pyridine derivatives. XCIX. Syntheses, with the aid of 4-pyridylmagnesium chloride, of 4-lithiopyridine and of 3-lithioquinoline. Recl Trav Chim Pays-Bas Belg 74:1003–1020
71. Gilman H, Spatz SM (1941) Organometallic derivatives of carbazole and quinoline. Amides of 3-quinolinecarboxylic acid. J Am Chem Soc 63:1553–1557
72. Gilman H, Soddy TS (1957) Carbonation of lithium derivatives of quinolines and isoquinolines. J Org Chem 22:565–566
73. Sugimoto O, Sudo M, Tanji K (2001) The tellurium-lithium exchange reaction: selective functionalization of electron-deficient heteroaromatics. Tetrahedron 57:2133–2138
74. Ruiz J, Lete E, Sotomayor N (2006) Intramolecular cyclization of functionalized heteroaryllithiums. Synthesis of novel indolizinone-based compounds. Tetrahedron 62:6182–6189
75. Gómez I, Alonso E, Ramón DJ et al (2000) Naphthalene-catalyzed lithiation of chlorinated nitrogenated aromatic heterocycles and reaction with electrophiles. Tetrahedron 56:4043–4052
76. Harrowven DC, Sutton BJ, Coulton S (2001) Intramolecular radical additions to quinolines. Tetrahedron Lett 42:2907–2910
77. Harrowven DC, Sutton BJ, Coulton S (2002) Intramolecular radical additions to quinolines. Tetrahedron 58:3387–3400
78. Maguire MP, Sheets KR, McVety K et al (1994) A new series of PDGF receptor tyrosine kinase inhibitors: 3-substituted quinoline derivatives. J Med Chem 37:2129–2137
79. Ishikura M, Mano T, Oda I et al (1984) An alternative synthesis of dialkylpyridylboranes. Heterocycles 22:2471–2474
80. Ota T, Terashima M (1987) Reactions of lithium tributyl heteroaryl borates with allylic bromides in the presence of copper(I) cyanide. J Heterocycl Chem 24:377–386
81. Zheng S, Kaur G, Wang H et al (2008) Design, synthesis, and structure-activity relationship, molecular modeling, and NMR studies of a series of phenyl alkyl ketones as highly potent and selective phosphodiesterase-4 inhibitors. J Med Chem 51:7673–7688
82. Atkins RJ, Breen GF, Crawford LP et al (1997) Synthetic routes to quinoline derivatives: novel syntheses of 3-butyryl-8-methoxy-4-[(2-methylphenyl)amino]quinoline and 3-butyryl-8-(2-hydroxyethoxy)-4-[(2-methylphenyl)amino]quinoline. Org Process Res Dev 1:185–197
83. Steib AK, Thaler T, Komeyama K et al (2011) Highly diastereoselective iron-mediated C (sp^2)-C(sp^3) cross-coupling reactions between aryl grignard reagents and cyclic iodohydrine derivatives. Angew Chem Int Ed 50:3303–3307
84. Snider BB, Wu X (2006) Synthesis of (+)-myrtopsine, (+)-7,8-dimethoxymyrtopsine, and related 2,3-dihydro-3-hydroxy-2-(1-hydroxy-1-methylethyl)benzofuran natural products. Heterocycles 70:279–294

85. Gilman H, Soddy TS (1958) Organolithium compounds of quinoline and 2-phenylquinoline. J Org Chem 23:1584–1585
86. Biot C, Delhaes L, Maciejewski LA et al (2000) Synthetic ferrocenic mefloquine and quinine analogues as potential antimalarial agents. Eur J Med Chem 35:707–714
87. Comins DL, Nolan JM, Bori ID (2005) Regioselective lithium-halogen exchange and palladium-catalyzed cross-coupling reactions of 2,4-dihaloquinolines. Tetrahedron Lett 46:6697–6699
88. Sahin A, Cakmak O, Demirtas I et al (2008) Efficient and selective synthesis of quinoline derivatives. Tetrahedron 64:10068–10074
89. Tumambac GE, Wolf C (2004) Synthesis and stereodynamics of highly constrained 1,8-bis (2,2'-dialkyl-4,4'-diquinolyl)naphthalenes. J Org Chem 69:2048–2055
90. Tumambac GE, Wolf C (2005) Synthesis and stereodynamics of highly constrained 1,8-bis (2,2'-dialkyl-4,4'-diquinolyl)naphthalenes (2). J Org Chem 70:2930–2938
91. Mongin F, Fourquez J-M, Rault S et al (1995) Bromine-lithium exchange on bromo-8-hydroxyquinolines via formation of their sodium salts. Tetrahedron Lett 36:8415–8418
92. Sapountzis I, Dube H, Lewis R et al (2005) Synthesis of functionalized nitroarylmagnesium halides via an iodine–magnesium exchange. J Org Chem 70:2445–2454
93. Trecourt F, Mallet M, Mongin F et al (1995) Synthesis of substituted 8-methoxyquinolines by regioselective bromine-lithium exchange of 5,7-dihalo-8-methoxyquinolines and 7-bromo-8-methoxyquinoline. Synthesis 1159–1162
94. Baron O, Knochel P (2005) Preparation and selective reactions of mixed bimetallic aromatic and heteroaromatic boron-magnesium reagents. Angew Chem Int Ed 44:3133–3135
95. Sapountzis I, Lin W, Fischer M et al (2004) Functionalized arynes: preparation of polyfunctional arynes via 2-magnesiated diaryl sulfonates. Angew Chem Int Ed 43:4364–4366
96. Wommack JB Jr, Barbee TG Jr, Thoennes DJ et al (1969) Synthesis of quinoline- and isoquinolinecarboxaldehydes. J Heterocycl Chem 6:243–245
97. Issleib K, Haftendorn M (1970) Alkali phosphorus compounds and their reactive behavior. LXIII. 8-Quinolylphosphines. Z Anorg Allg Chem 376:79–86
98. Suggs JW, Pearson GDN (1980) Facile synthesis of 8-substituted quinolines. J Org Chem 45:1514–1515
99. Stradiotto M, Fujdala KL, Tilley TD (2001) Iridium(III) complexes of the new tridentate bis (8-quinolyl)silyl ('NSiN') ligand. Chem Commun 1200–1201
100. DiMauro EF, Mamai A, Kozlowski MC (2003) Synthesis, characterization, and metal complexes of a salen ligand containing a quinoline base. Organometallics 22:850–855
101. Lord A-M, Mahon MF, Lloyd MD et al (2009) Design, synthesis, and evaluation in vitro of quinoline-8-carboxamides, a new class of poly(adenosine-diphosphate-ribose)polymerase-1 (PARP-1) inhibitor. J Med Chem 52:868–877
102. Stang PJ, Olenyuk B, Chn K (1995) Preparation of nitrogen-containing bis(heteroaryl) iodonium salts. Synthesis 937–938
103. Staubitz A, Dohle W, Knochel P (2003) Expeditious functionalization of quinolines in positions 2 and 8 via polyfunctional aryl- and heteroarylmagnesium intermediates. Synthesis 233–242
104. Muchowski JM (2005) 1-Methylthio-3-lithioisoquinoline. The first simple isoquinoline derivative lithiated at the 3-position. ARKIVOC 470–479
105. Sainsbury M, Brown DW, Dyke SF et al (1970) 1,2-Dihydroisoquinolines. XIV. Alkylation. Tetrahedron 26:2239–2247
106. Denni-Dischert D, Marterer W, Bänziger M et al (2006) The synthesis of a novel inhibitor of B-raf kinase. Org Process Res Dev 10:70–77
107. Watson CY, Whish WJD, Threadgill MD (1998) Synthesis of 3-substituted benzamides and 5-substituted isoquinolin-1(2H)-ones and preliminary evaluation as inhibitors of poly(ADP-ribose)polymerase (PARP). Bioorg Med Chem 6:721–734
108. Tsou H-R, Otteng M, Tran T et al (2008) 4-(Phenylaminomethylene)isoquinoline-1,3 (2H,4H)-diones as potent and selective inhibitors of the cyclin-dependent kinase 4 (CDK4). J Med Chem 51:3507–3525

Top Heterocycl Chem (2013) 31: 131–170
DOI: 10.1007/7081_2012_95

Published online: 8 November 2012

Metalation of Pyrazine and Quinoxaline

Nelly Plé and Corinne Fruit

Abstract We report herein the main results, described during the last decade, dealing with the metalation methods of pyrazine and quinoxaline derivatives. These intermediate compounds play a key role in the synthesis of a wide range of building blocks used to elaborate more complex structures. Besides the deprotonative metalation carried out either with monometallic lithium bases or with alkali-metal complexes, other methods such as dehalometalation, oxidative addition, transmetalation, and cross-couplings are reported. The large amounts of results reveal the great interest to access such metalated heterocycles and to use them as synthetic tools.

Keywords Cross-coupling · Dimetallic complexes · Metalation · Monometallic lithium bases · Pyrazines · Quinoxalines · Transmetalation

Contents

N. Plé (✉)
Equipe de Chimie Hétérocyclique, COBRA-CNRS UMR 6014 & FR 3038, rue Lucien Tesnière, 76131 Mont-Saint-Aignan, France
e-mail: nelly.ple@insa-rouen.fr

C. Fruit
Equipe de Chimie Hétérocyclique, COBRA-CNRS UMR 6014 & FR 3038, rue Lucien Tesnière, 76131 Mont-Saint-Aignan, France

Université de Rouen, Place Emile Blondel, 76821 Mont-Saint-Aignan, France
e-mail: corinne.fruit@univ-rouen.fr

Abbreviations

AMMM	Alkali-metal-mediated metalation
Boc	*tert*-Butoxycarbonyl
CIPE	Complex-induced proximity effect
Dba	Dibenzylidene acetone
DMA	Dimethylacetamide
DMG	Directing metalation group
DNA	Desoxyribonucleic acid
HOMO	High occupied molecular orbital
LDA	Lithium diisopropyl amide
LTMP	Lithium tetramethylpiperidide
LUMO	Low unoccupied molecular orbital
THF	Tetrahydrofurane
TMEDA	N,N,N′,N′-Tetramethyl-ethylene-diamine
TMP	Tetramethylpiperidyl
TMSCl	Trimethylsilyl chloride
VEGFR	Vascular endothelial growth factor receptor

1 Introduction

Substituted pyrazines represent the core structure of a number of important natural as well as man-made heterocyclic compounds. Variously substituted pyrazines are flavor components in a wide variety of foodstuffs formed either thermally like alkylpyrazines present in coffee, meat, and potatoes [1] or biosynthetically. Significant examples are the biosynthetical 2-alkyl-3-methoxypyrazines associated with herbaceous flavour, which are found in wines such as Cabernet-Sauvignon and the 2 isopropyl-3-methoxypyrazine secreted by ladybirds which may be a defensive pheromone [2]. A more complex imidazopyrazine moiety is found in the scaffold of the Coelenterazine, a bioluminescent compound isolated from the jellyfish Aequorea vistoria [3–6] (Fig. 1). In addition, many substituted pyrazines possess important pharmacological activities such as cytotoxicity [7, 8] or as epithelial channel blockers [9].

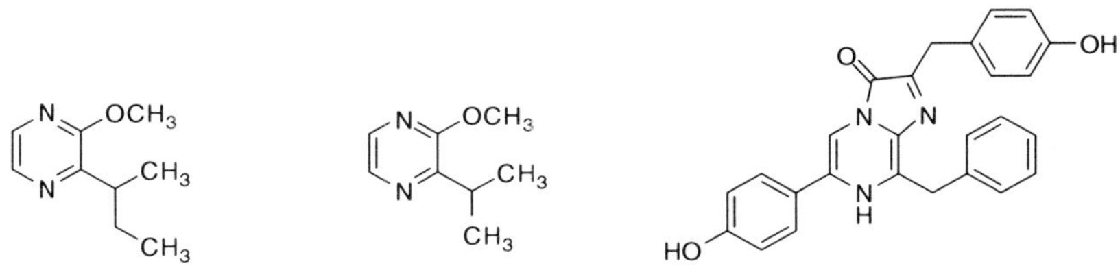

2-*sec*-butyl-3-methoxypyrazine (Cabernet-Sauvignon) 2-isopropyl-3-methoxypyrazine (Ladybirds) Coelenterazine (Jellyfish)

Fig. 1 Structure of some natural pyrazines [2–6]

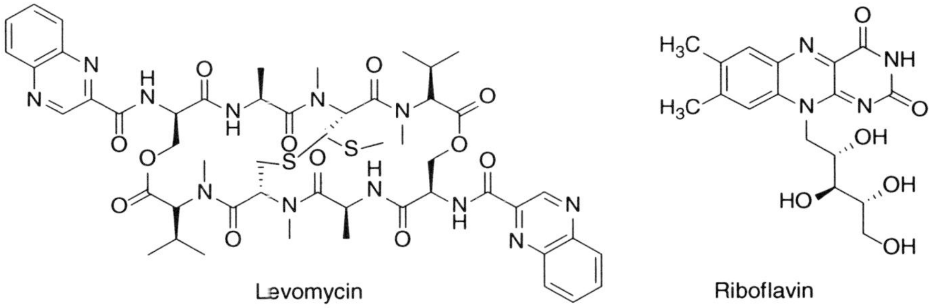

Fig. 2 Structure of some natural quinoxalines [10–15]

Functionalized quinoxalines, which belong to a class of azaheterocycles and are also named benzopyrazines, exhibit a wide range of biological activities. They are recognized in a number of naturally occurring compounds such as riboflavin (vitamin B2), flavoenzymes molybdopterines, and antibiotics of Streptomyces. Quinoxaline derivatives constitute also the basis of many insecticides, fungicides, and herbicides. Diversely substituted quinoxalines bearing various functional groups are important biological agents and are used for their antiviral, antibacterial (levomycin), anti-inflammatory, kinase inhibitory, and anticancer activities [10–15] (Fig. 2). Besides their medicinal uses, quinoxaline derivatives have found technical applications as dyes, organic semiconductors [16, 17], and as suitable ligands in coordination chemistry allowing their use as DNA cleaving agents [18].

Despite the importance of the polyfunctionalized pyrazines, only a limited number of methodologies have been reported to their synthesis. Pyrazines are generally synthesized by condensation of α-amino carbonyl compounds and the combination of α-diketones with vicinal diamines followed by dehydrogenation [19, 20], but generally these methods disappoint in the preparation of unsymmetrical substituted pyrazines. An alternative synthetic route is the use of cyclizing aza-Wittig reactions of two molecules of α-phosphazinyl-ketones or oxidation of dioxopiperazines [21]. In a general way, few regioselective syntheses have been reported.

Owing to their potential biological and technical interest a number of synthetic strategies have been developed to access quinoxalines. Quinoxaline derivatives are generally obtained by either condensation of an aryl-1,2-diamine with a 1,2-dicarbonyl compound by conventional one-pot or microwave synthesis methods or using an acyloin compound by tandem oxidation-condensations with catalysts [22–25]. As in the case of the pyrazines the lack of regioselectivity of these syntheses is a major impediment.

The pyrazine ring could be considered as an isoster of benzene or pyridine whereas quinoxaline could be compared to naphthalene or quinoline. The characteristics and the reactions of pyrazines and quinoxalines are determined by the presence of two nitrogen atoms at 1 and 4 positions in the ring. The replacement of the two opposite C–H groups in the benzene ring by nitrogen atoms has particular consequences on the stability and reactivity. The aromaticity and relative stabilities of azines have been compared to those of benzene and led to inconsistent conclusions. In some cases azines are considered more aromatic than benzene [26–28] or on contrary that "insertion of N atoms decreases the aromaticity" [28]. A more recent paper of Schleyer established that aromaticity of all azines is like that benzene, the same being true for azanaphtalene [29].

Examination and comparison of frontier orbitals reveal that contrary to benzene, in case of pyridine and pyrazine, the degeneracy of the π-frontier orbitals is removed. Calculations with DFT/B3LYP method have been performed and have highlighted that for pyridine as for pyrazine the HOMO orbital (~-0.263 a.u. or -7.15 eV) is not a π orbital but a σ one with the highest atomic coefficient on the nitrogen [30]. These results are in agreement with the preferential reactivity of the elctrophiles with the nitrogen atoms and with the very low reactivity of these heterocycles towards electrophilic substitution on the carbon atoms. Moreover, the CH$\rightarrow$N exchange induces a lowering of the energy of the LUMO making pyrazine and quinoxaline more reactive than pyridine or quinoline towards nucleophiles (addition and substitution reactions) (Fig. 3) [31, 32].

Nevertheless, quinoxalines are reactive to electrophiles as well as nucleophiles. Indeed the benzene ring allows SEAr reactions whereas nucleophilic substitutions occur in the diazine one, particularly if it is substituted by halogens or good leaving groups.

Due to these particularities, direct functionalization of pyrazine or quinoxaline rings cannot be carried out using usual synthetic methodologies with electrophiles. In another way a regioselective functionalization could be achieved if the electrophile could react with a specific metalated pyrazine or quinoxaline species. According to this strategy, a variety of synthetic methods have been intensively developed during the past decades to access metalated pyrazines or quinoxalines intermediates. This methodology includes dehydrometalation, dehalometalation, transmetalation, or oxidative addition (Scheme 1).

The dehydrometalation or direct metalation is one of the most useful and widely used methods to access a labile carbon–metal bond. Until recently, most organometallic reagents that facilitate this direct method have been mainly used with a strongly positive metal such as lithium to form highly polar and reactive

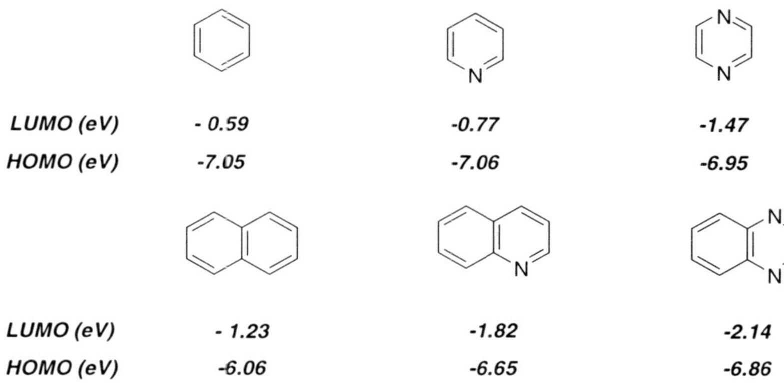

Fig. 3 HOMO and LUMO levels of azines compared to benzene and naphthalene and calculated by using DFT/B3LYP/6-311+G** (Dupas G, Plé N, Fruit C, unpublished results)

carbon–lithium bonds; this could be termed "conventional metalation." Nevertheless, the high reactivity of the lithio derivatives used or formed is incompatible with highly reactive functional groups and needs generally low reaction temperatures. More recently, several research groups have developed alternative metalation reagents which can be seen as composite molecules or mixture of several compound types (complex metalators) where the alkali metal is an essential component. This way could be considered as alkali-metal-mediated metalations (AMMMs).[1]

Among the other methods, the dehalometalation or non-deprotonative halogen–metal exchange has been widely developed in the pyridine series with bromo and iodo derivatives; however, in the case of pyrazines and quinoxalines the problem is deferred since the preparation of their bromo derivatives that could be used as substrates is not trivial. For these reasons, the dehydrometalation reaction has attracted the most attention due to the good accessibility of the starting materials which avoids the use of heavy halogenated heterocycles.

The transmetalation reaction could be carried out if M^1 is more electropositive than M^2. Usually, bases containing lithium (M^1 = Li), a highly electropositive metal, lead easily to lithioderivatives via the deprotonative metalation of substrates. These lithiated compounds are often chosen to access other orgnometallics by use of further transmetalation reactions.

The oxidative addition, which results from reaction of a metal such as magnesium with a halo-derivative, constitutes an old and well-known method to access metalated species. In heteroaromatic series, the presence of basic nitrogen ring atoms requires some "active" metals (M*) and does not allow to use this methodology easily.

[1] Mulvey introduced the term alkali-metal-mediated metalation to depict the reactions of ate bases because the reactivity ("synergy") they exhibit cannot be attained by homometallic compounds on their own.

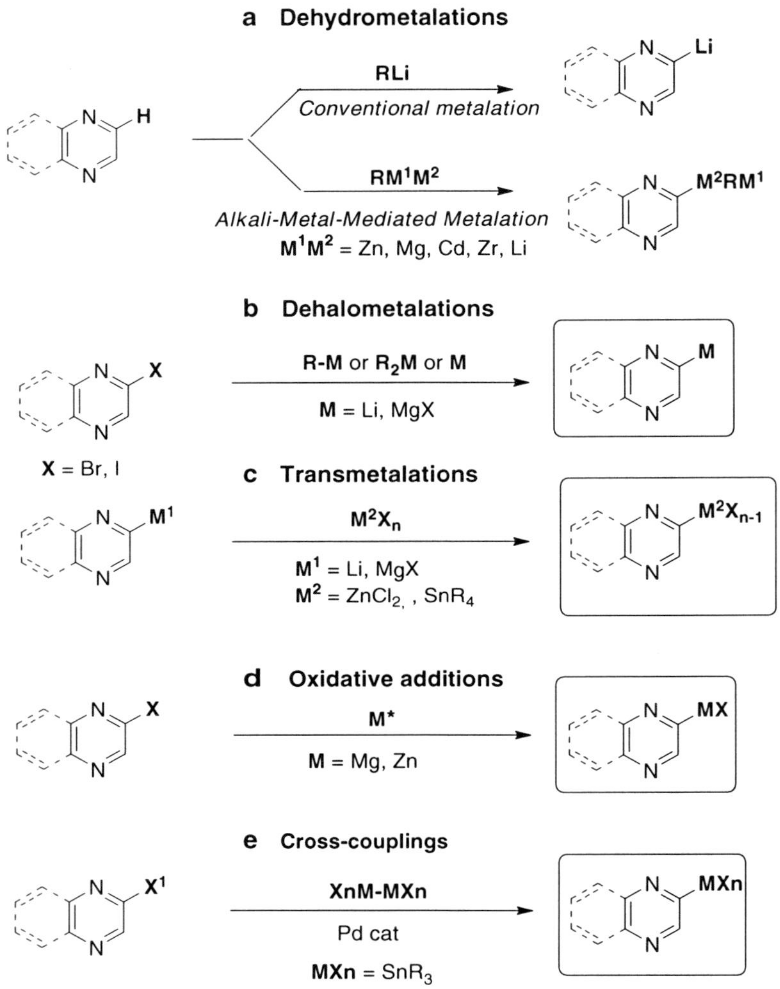

Scheme 1 Methods for the synthesis of organometallic intermediates in pyrazine and quinoxaline series

The palladium-catalyzed coupling reaction is a convenient method to induce formation of C–C bonds and involves reaction between a transition metallic compounds and a haloderivative. Stannylpyrazines have been often used to achieve such a reaction.

In this chapter we report the metalation of pyrazine and quinoxaline derivatives, according to the different methods considered to be suitable previously, to access structures with applicability in synthesis. The literature covered by this chapter begins mainly in 2002 because previous years have been comprehensively compiled [33, 34], although older works could be commented if necessary.

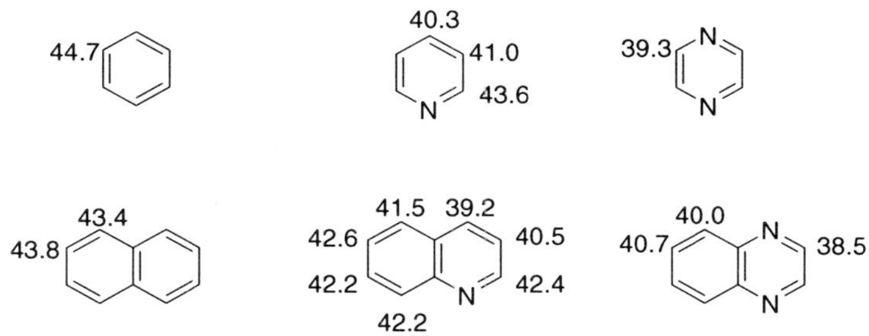

Fig. 4 Estimated p*K*a values for C–H bonds [33]

2 Dehydrometalation

2.1 *Conventional Metalation*

Considering the first methodology which is the dehydrometalation reaction, the general equation for the hydrogen-metal permutation is given in Eq. (1).

$$C_{Het} - H + Base - M \rightarrow C_{Het} - M + Base - H \qquad (1)$$

To achieve this reaction it is necessary that the thermodynamic acidity of the Base-H generated from metalated base Base-M should be lower than the acidity of the substrate C_{Het}-H. Using the standard ab initio B3LYP method, the p*K*a values of C–H bonds of numerous aromatic compounds have been recently reported [35]. Whereas the acidity of hydrogens in benzene, pyridine, naphthalene, and quinoline is very weak (p*K*a > 39), introduction of a second nitrogen atom into the ring enhances the acidities of protons. In pyrazine and quinoxaline the adjacent proton to nitrogen has the strongest acidity (Fig. 4), which is related to the less highly conjugated π orbitals compared to the pyridine and benzene series. It could be noted that contrary to pyridine or quinoline for which several values of p*K*a are observed for the different protons, involving a regioselectivity for the deprotonation, the symmetric structure of pyrazine and quinoxaline induces a unique site for metalation, in α position of the nitrogen atom.

In the azaheterocycles, the presence of electron lone pairs of the nitrogen atoms in the ring could cause coordination with the metal of the base and disaggregation of the latter, making it more reactive. Compared to pyridine (p*K*a = 5.2), pyrazine and quinoxaline (p*K*a ~ 0.4), which are the least basic in the diazines series, possess nitrogens which are less chelating. Nevertheless, since the ring hydrogens of pyrazine and quinoxaline are more acidic than those of pyridine or quinoline, deprotonation is possible.

The regiospecific dehydrometalation of aryl hydrogen by metallic bases generally requires the presence of substituents with heteroatoms (halogen, O, N, S). Due to their inductive electron-withdrawing effect and their ability to coordinate

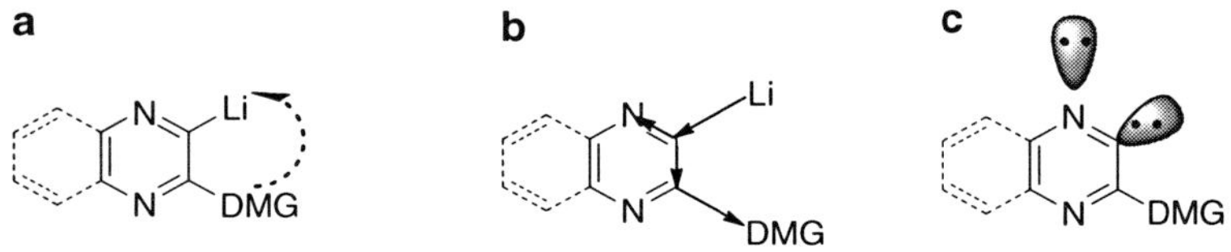

Fig. 5 Various factors influencing the stabilization of the lithiated derivatives

with the metals, such substituents designed as DMG (Directing Metalation Group) favor the deprotonation of the neighboring C–H bond.

The deprotonation by organolithium bases can be described as a two-step process in which the formation of a prelithiation complex brings the reactive group into proximity for directed metalation. This phenomenon was termed the complex-induced proximity effect (CIPE) [36–38].

For pyrazines and quinoxalines where the metalation occurs always at the adjacent position to the ring nitrogen, the metalated species is, on the one hand, stabilized by the electron-withdrawing effect of the ring nitrogen atom, but, on the other hand, destabilized by electronic repulsion between the carbanion and the lone pair of the adjacent nitrogen.

The electron-withdrawing effect of the nitrogen ring atoms decreases the energy level of the LUMO of the pyrazines and quinoxalines, making them more sensitive to nucleophilic addition reactions (See Fig. 3) [39–41]. As a consequence, "soft" alkyllithiums, which are strong bases (p*K*a ~ 40–50), have to be avoided since they can easily add to the π-deficient rings, even at low temperatures. The "harder" but less basic lithium diisopropylamide (LDA, p*K*a = 35.7) and lithium 2,2,6,6-tetramethylpiperidide (LTMP, p*K*a = 37.3) are relied on to effect deprotonation.

When lithium amides are used as the bases, the reaction is usually under thermodynamic control and the regioselectivity observed is the result of different effects such as stabilization by chelation of the metal with the DMG (a), stabilization by the electron-withdrawal of the adjacent ring nitrogen atom and the DMG (b), and destabilization by electronic repulsion between the carbanion and the lone pair of the adjacent ring nitrogen (c) (Fig. 5).

When no strong DMG is present, the solvent could have an efficient effect and in absence of a chelating solvent such as THF, the influence of the ring nitrogen is reinforced and could promote the deprotonation at the adjacent C–H group. Furthermore this reaction becomes more difficult with bare heterocycles, for which it can be suspected that lack of a DMG does not allow a possible coordination of the lithio derivative with the DMG making the lithio species less stable and more subject to side reactions [42].

According to the particularities of the pyrazines and quinoxalines, due to their facile competitive nucleophilic addition reactions [43–45], metalation of these scaffolds have been achieved first by the use of hindered lithium dialkylamides such as LDA or LTMP. Even if these less nucleophilic lithium amide reagents are more suitable for the metalation of these sensitive azaheterocycles, they suffer from

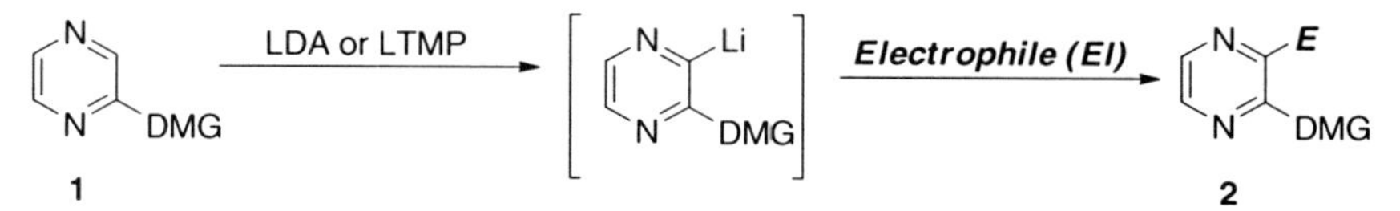

DMG : F, Cl,Br, I, OMe, SMe, SOMe, SO_2Ph,SOx^{t}Bu, NHCO^{t}Bu, CONH^{t}Bu

E : D, Cl,Br, I, CH(OH)R, CH(OH)Ar CHO, COOH, SPh, TMS....

Scheme 2 Functionalization of pyrazine via lithiation at adjacent position of DMG [33, 43, 47]

several limitations such as attack upon sensitive functional groups, low stabilities of the lithio derivatives which impose low reaction temperatures, short reaction times, or in situ trapping.

2.1.1 Pyrazines

The functionalization via the ortho-lithiation of chloropyrazine has been first reported in 1988 by Quéguiner and coworkers [46]. Subsequent work dealing with metalation of diazines and more particularly with pyrazine has been achieved by this team and reported in the literature [33, 43, 47]. A wide variety of DMG allowed the functionalization at the adjacent position leading to a wide range of 2,3-disubstituted pyrazines **2** generally in good yields (Scheme 2).

Halopyrazines

At first, it could be noted that due to the difficulties to obtain the bromopyrazine in appreciable amounts and to its instability, lithiation and functionalization of this compound have been achieved later using LDA as metalating agent with only few electrophiles (benzaldehyde, diphenylsulfide, and iodine). The 2,3-disubstituted bromo derivatives are obtained in moderate yields [48].

Metalation of fluoropyrazine **3** has been achieved with LTMP as metalating agent at low temperature (−75°C) with a short reaction time (5 min) leading to a wide range of 2,3-disubstituted fluoropyrazines. Because pyrazine derivatives are well known to be good electrophiles, above all when they are substituted by a good leaving group such as fluorine, a further nucleophilic substitution is observed by the released species coming from the electrophile used during the trapping step. So that, during functionalization of fluoropyrazine, formation of 2,3-diphenylthiopyrazine **5** and the 2-acetyl and 2-benzoyl-3-dimethylamino pyrazines **6** have been observed besides the expected disubstituted fluoropyrazine **4** when diphenylsulfide or *N,N*-dimethylacetamide or *N,N*-dimethylbenzamide have been used as electrophile (Scheme 3) [49].

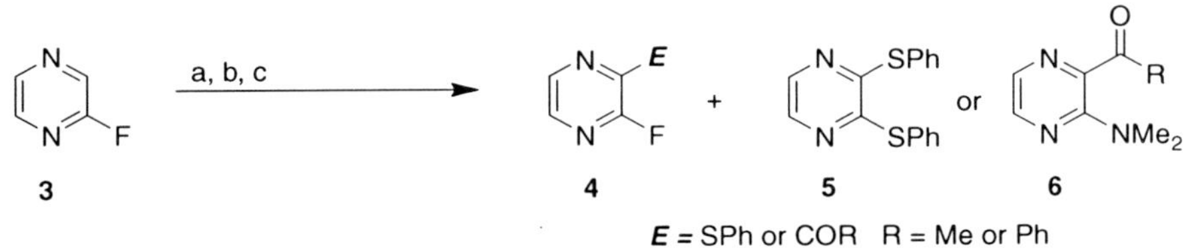

Scheme 3 (**a**) 1.2 equiv LTMP, THF, −75°C, 5 min (**b**) $(PhS)_2$ or $PhCONMe_2$ or DMA (**c**) H_3O^+ [49]

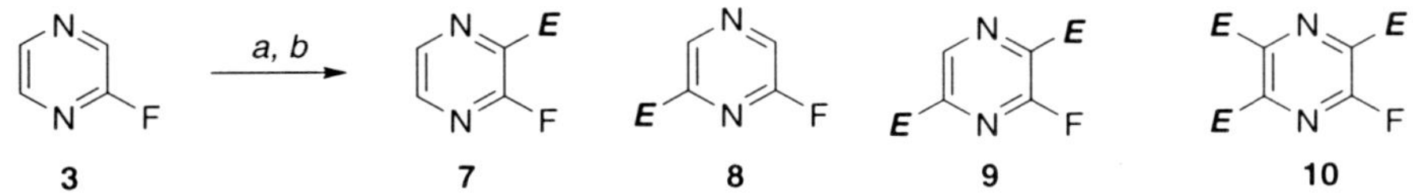

Scheme 4 (**a**) *n* equiv LTMP, n' equiv I_2 or Bu_3SnCl, THF, θ°C (**b**) EtOH/THF [51, 52]

Table 1 Lithiation of fluroropyrazine **3** with different equivalents of LTMP and electrophiles

E	*n*	*n'*	*θ* (°C)	**7**	**8**	**9**	**10**
I	1.1	1.1	−78	50	–	6	4
I	2.0	2.0	−78	11	–	35	15
I	4.0	4.0	−78	–	–	–	65
$SnBu_3$	1.1	1.5	−78	11	26	–	–
$SnBu_3$	1.1	1.5	−100	54	–	–	–
$SnBu_3$	2.0	1.5	−100	15	10	52	–
$SnBu_3$	4.0	1.5	−100	–	20	–	63

Recently, this strategy has been developed with the chloropyrazine and dimethylsulfide as electrophile to access in one-pot 2,3-bis(methylthio)pyrazine in good yield (91%) [50].

Starting from fluoropyrazine, a study of regioselective synthesis of iodo- and tributylstannylfluoropyrazines has been investigated. Lithiation of fluoropyrazine with stoichiometric amounts of LTMP and iodine afforded the 2-fluoro-3-iodopyrazine as sole product whereas use of various equivalents of metalating agent and iodine afforded mono-, di-, and triiodo derivatives (Scheme 4, Table 1) [51]. In the same manner, use of tributyltin chloride as electrophile led to mono- and di-stannylpyrazines [52].

Formation of compounds **8**, **9**, and **10** is a result of metalation at the position adjacent to the nitrogen atom without assistance of the fluorine atom as DMG and could be imputed to the influence of the nitrogen ring.

Such a metalation without a DMG has been previously reported during direct metalation of bare pyrazine **11** by use of an excess of LTMP (4 equiv) with very short reaction time (5 min) at low temperature (−78°C), [42] with acetaldehyde, iodine, and benzaldehyde as the electrophile leading to compounds **12** in moderate yields, when benzaldehyde is used in excess the 2,5-disubstituted pyrazine **13** was obtained (Scheme 5).

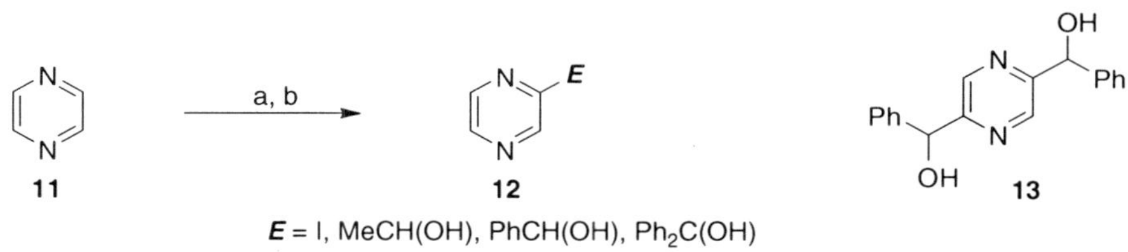

Scheme 5 (**a**) 4 equiv LTMP, THF, −78°C, 5 min (**b**) Electrophile then H_3O^+ [42]

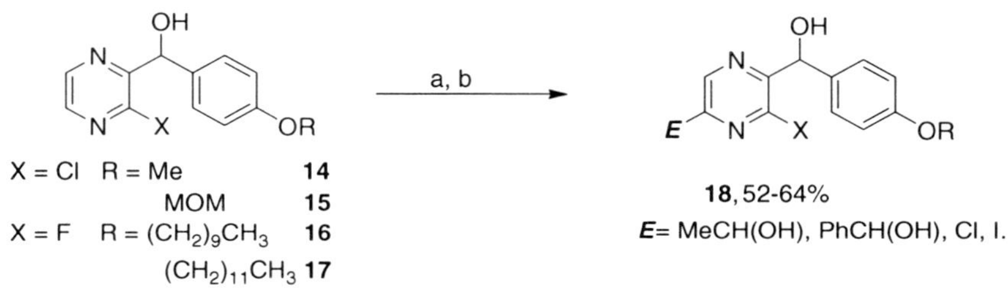

Scheme 6 (**a**) 3 equiv LTMP, THF, −78°C (**b**) Electrophile then H_3O^+ [51, 53]

It could be mentioned that deprotonative metalation of 2-chloro- and 2-fluoro-(1-hydroxy-4-alkoxyphenylmethyl)pyrazines with LTMP in excess (3 equiv) in THF at −75°C occurred at the C6 position (Scheme 6). With iodine as the electrophile, the 6-iodo derivative obtained has been used to induce Negishi reaction and to access a natural product, Septorin, the main agent of a wheat disease impeding growth [51, 53].

When alkylamides such as LTMP are used as metalating agent, the deprotonation is considered as thermodynamically controlled, and then it could be assumed that the heats of formation of the lithio compound could be examined as a simple approach for the regioselectivity. An explanation to this regioselectivity has been found by using semiempiric calculations Li/PM3 to determine the heat of formation of the intermediate lithio derivatives **19a** and **19b**. Considering that the hydrogen of the hydroxyl group is first abstracted by LTMP, a lithium atom could form a chelate between the oxygen of the alcoholate and the neighboring nitrogen. The lithium at the C6 position may coordinate with the adjacent free nitrogen N1 whereas such coordination cannot be observed when the lithium is at the C5 position since the nitrogen N4 is already chelated (Fig. 6).

This assumption is in agreement with the calculation of the heat of formation of the two lithio derivatives by Li/PM3, the 6-lithioderivative **19a** being more stable than the 5-lithio one **19b** with a difference of $\Delta(\Delta H_f) = 6.8$ kcal/mol. A regioselective functionalization at the C6 position has also been achieved when the 2-fluoro-3-phenylpyrazine **20** reacted with an excess of LTMP (3 equiv) followed by reaction with various electrophiles [51] leading to compounds **21** (Scheme 7).

19a ΔH_f = - 5.37 kcal/mole

19b ΔH_f = + 1.43 kcal/mole

Fig. 6 Theoretical heat of formation of the lithiated intermediates **19a** and **19b**, determined by semiempirical Li/PM3 calculations

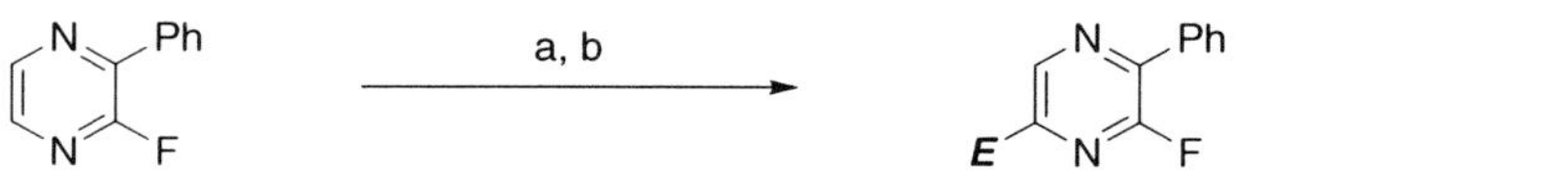

20

21, 59-77%
E = MeCH(OH), PhCH(OH), $SiMe_3$, $SnBu_3$

Scheme 7 (**a**) 3 equiv LTMP, THF, −78°C, 5 min (**b**) Electrophile then H_3O^+ [51]

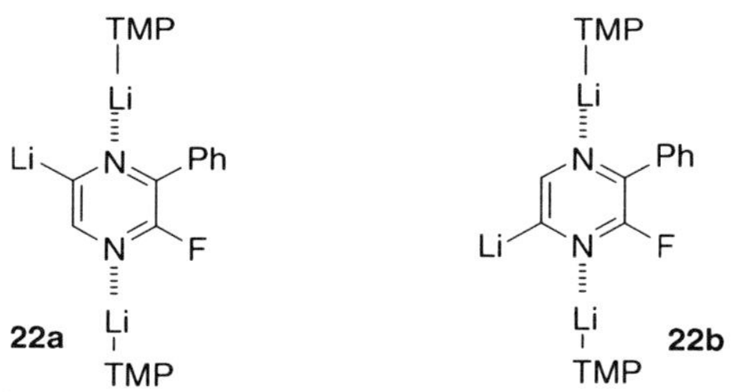

ΔH_f = + 0.70 kcal/mol

ΔH_f = - 4.26 kcal/mol

Fig. 7 Calculated heat of formation of the lithiated intermediates **22a** and **22b** [51]

To explain this regioselectivity which occurs exclusively at the C6 position, theoretical calculations using Li/PM3 semiempirical method have been performed [51]. A complexation between the lithium of LTMP and the two nitrogen atoms of the pyrazine moiety, which behaves as a complexing agent, has been taken into account (Fig. 7). The values calculated indicated that the 6-lithio derivative **22b** is clearly more stable than the 5 one **22a**. This result could explain the complete regioselectivity at C6 position.

Pyrazinethiocarboxamides

While some DMGs such as carboxamides have been studied in induced metalation of the pyrazine moiety, the regioselectivity of this reaction has highlighted to be dependent on the experimental conditions. With electrophiles at low temperature

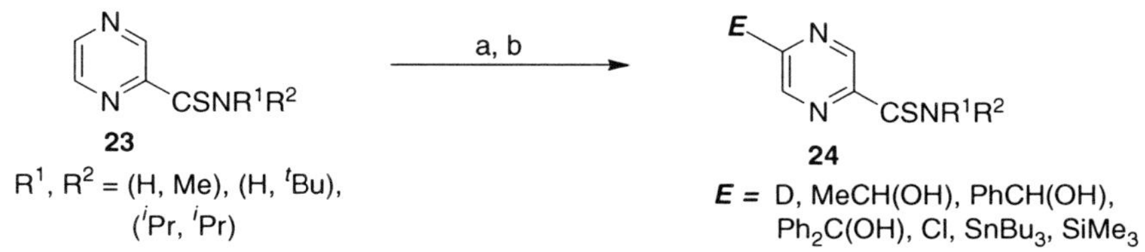

Scheme 8 (a) LTMP, THF, −78°C (b) Electrophile then H_3O^+ [54]

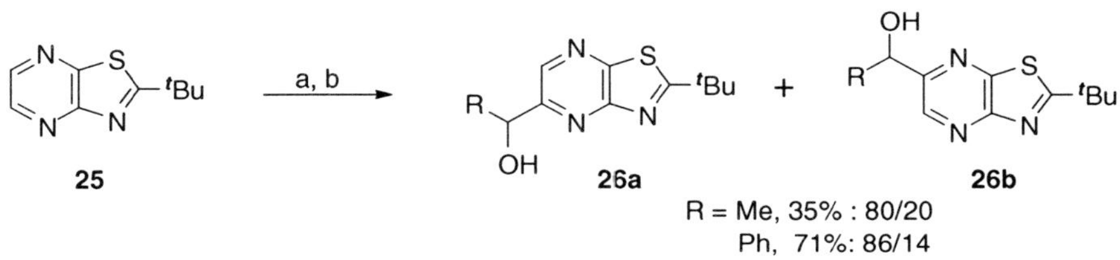

Scheme 9 (a) 4.1 equiv LTMP, THF, −78°C, 5 min (b) RCHO, −78°C, 45 min then H_3O^+ [55]

(<−80°C), the C5 compounds are in majority (kinetic isomer) whereas at higher temperature, the metalation occurred at the neighboring position of the DMG. When thiocarboxamides are used as DMG, metalation of compounds **23** with various electrophiles gave only 2,5-disubstituted pyrazines **24**, this regioselectivity could be due to the sulfur atom (Schemes 8) [54].

Metalation of 6-tert-butylthiazolo[4,5-b]pyrazine **25** has been achieved with LTMP at −75°C and a short time (5 min) [55]. The ratio between the two isomers **26a** and **26b** is depending on the amounts of metalating agent. With one equivalent of LTMP and acetaldehyde as the electrophile the **26b** compound is obtained as sole product in moderate yield (48%) whereas an excess of LTMP led to a mixture of **26a** and **26b** in better yields, with **26b** as major product (Scheme 9).

Pyridinylpyrazine

During the metalation reaction, the assistance of the DMG is inducing various effects due to its electron-withdrawing effect and to its coordination to the lithium atom. Among them is the CIPE of the base which favors the deprotonation and the stability of the lithio derivative. According to their π-deficient character, azines such as pyridine and pyrazine could play the role of DMG with their ring nitrogen atom. The site influenced by the ring nitrogens and the heats of formation of monolithio pyridine-2-ylpyrazines (determined by semiemperical Li/PM3 method) are given in Fig. 8.

Metalation of **27** has been achieved with LTMP (4 equiv.) at −100°C followed by reaction with various electrophiles. A high regioselectivity has been observed on the pyrazine ring at the ortho position of the linkage between the two cycles (Scheme 10) [56].

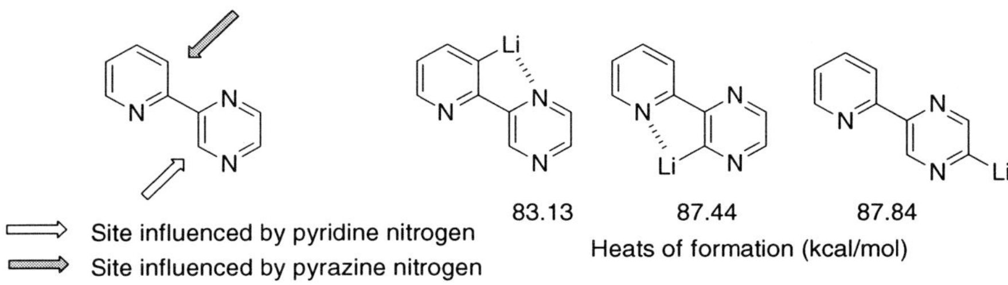

Fig. 8 Calculated heat of formation of the lithiated pyridin-2-ylpyrazines [56]

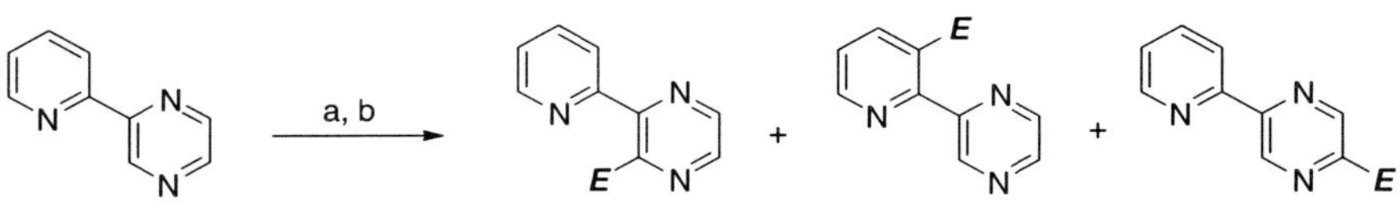

27 **28** **29** **30**

E	28 (%)	29 (%)	30 (%)
D	81	21	16
SPh	59	-	-
SnBu3	72	-	-
*p*OMePhCH(OH)	51	8	-

Scheme 10 (**a**) 4 equiv LTMP, THF, −100°C, 15 min (**b**) Electrophile, −100°C, 30 min then H_3O^+ [56]

2,5- and 2,6-Disubstitued Pyrazines

Other ortho-directing groups have been used to allow the functionalization via metalation, among them we can notice protecting groups such as ketals or *tert*-butoxycarbonyl (Boc). Lithiation of different 2,5-disubstituted pyrazines has been disclosed using Boc or neopentylglycol ketal as DMGs [57]. The 3,6-diiododerivatives obtained with the diketal compounds and the 3,6-distannylpyrazines coming from 2,5-di-*tert*-butoxycarbonylpyrazine are used to synthesize pyrazine ladder polymer through Pd/Cu-catalyzed couplings [57].

Deprotonative functionalization of 2,6-disubtituted pyrazines **31** has been performed. When the two DMGs are identical no problem of regioselectivity could appear, at most a competition between mono- and difunctionalization could be observed (Scheme 11) [43, 45, 47].

The twofold deprotonation of the 2,6-dichloropyrazine **33** has been performed with 2.5 equiv of LTMP at −100°C followed by a subsequent addition of two different electrophiles leading to tetrasubstituted compounds **34** (Scheme 12) [58].

When the synthesis of pyrazine C-ribosides via direct metalation of 2,6-disubstituted pyrazines bearing two different DMG (Cl, OMe) or (Cl, OBn) has been carried out, a complete regioselectivity of the functionalization has been observed at the ortho position of the alkoxy group [58].

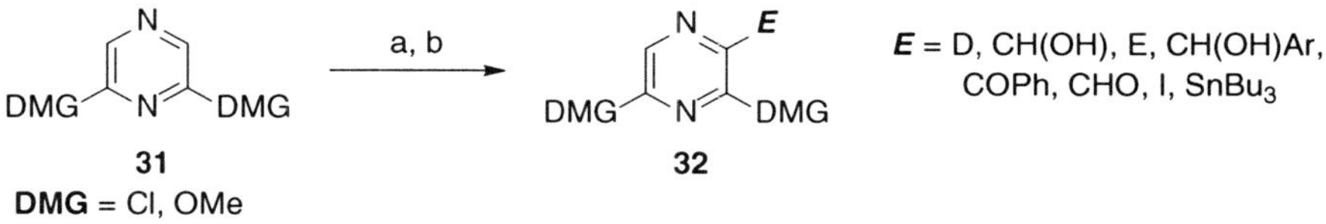

Scheme 11 Functionalization of 2,6-disubstituted pyrazines (**a**) LTMP (**b**) Electrophile [43, 45, 47]

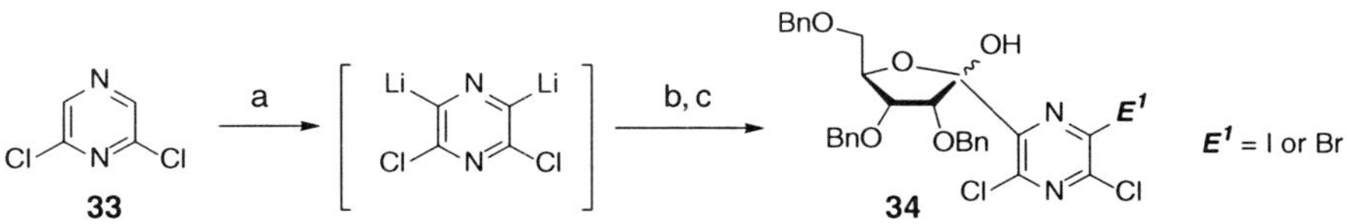

Scheme 12 (**a**) 3 equiv LTMP, THF, −100°C (**b**) Electrophile 1 (**c**) Electrophile 2 then hydrolysis [58]

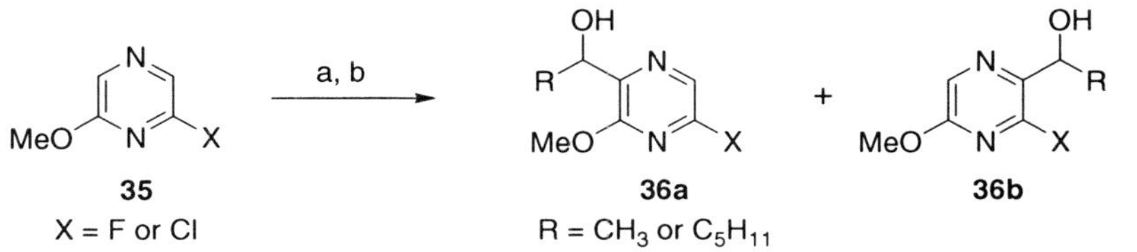

Scheme 13 (**a**) 2.2 equiv alkylamide, THF, −78°C, 5 min (**b**) RCHO then hydrolysis [59]

To establish a comparison between the ortho directing power of fluoro, chloro, and methoxy groups, the lithiation of 2-halo-6-methoxypyrazines **35** has been reinvestigated using various alkylamides as metalating agent at −78°C with a short reaction time (5 min) (Scheme 13, Table 2) [59].

The main isomer has the substituent at the ortho position relative to the fluorine atom as in **36b**, contrary to what is observed with the chlorine atom as in **36a**. When LDA is used as metalating agent, the metalation is more regioselective than with more bulky bases such as LTMP or LB. These results allowed to estimate the relative ortho directing power of fluorine, chlorine, and methoxy which has been established as F > OMe > Cl (Fig. 9).

2.1.2 Quinoxalines

Due to its low-LUMO energy level (−2.14 eV for quinoxaline), the quinoxalines are very sensitive towards nucleophiles making their functionalization via metalation more difficult, due to the competitive reaction of nucleophilic addition.

Two quinoxaline ketones have been prepared by metalation of the 2-methoxy and the 2-thiomethylquinoxalines followed by reaction with *N*-methoxy-*N*-methylbenzamide [60]. Later, the functionalization via metalation of 2-chloro,

Table 2 Lithiation of 2-halo-6-methoxypyrazines **35** with different alkylamides

X	R	Alkylamide	Yield (%)	**36a:36b**
F	Me	LDA	72	4:96
F	Me	LTMP	78	12:88
F	Me	LB	71	12:88
F	C_5H_{11}	LDA	74	4:96
F	C_5H_{11}	LTMP	65	14:86
F	C_5H_{11}	LB	80	14:86
Cl	Me	LDA	91	88:12
Cl	Me	LTMP	80	68:32
Cl	Me	LB	71	62:38
Cl	C_5H_{11}	LDA	74	88:12
Cl	C_5H_{11}	LTMP	82	67:33
Cl	C_5H_{11}	LB	79	61:39

LDA lithium di*iso*propylamide, *LTMP* lithium 2,2,6,6,-tetramethylpiperidine, *LB* lithium *tert*-butyl-(1-*iso*propylpentyl)amide

Fig. 9 Regioselectivity of the functionalization of 2-halo-6-methoxypyrazines

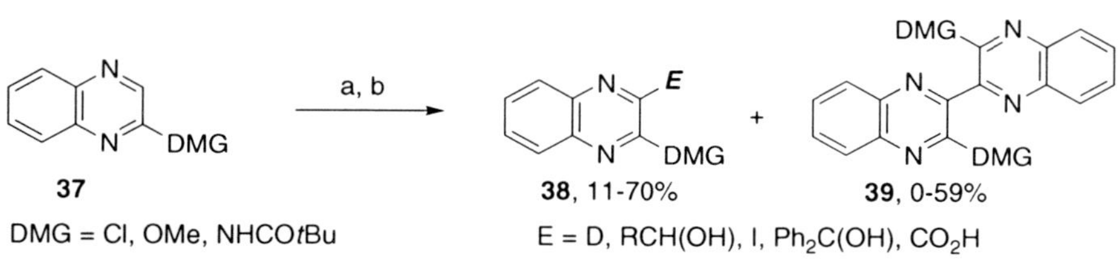

Scheme 14 (**a**) LTMP, THF, −78°C (**b**) Electrophile then hydrolysis [32]

2-methoxy, and 2-pivalolaminoquinoxalines **37** has been successfully achieved with various electrophiles leading to 2,3-disubstituted quinoxalines **38** [32]. In the case of 2-chloro and 2-methoxyquinoxaline, formation of dimers **39** has been observed besides the expected compounds **38** (Scheme 14). The dimers result from a nucleophilic addition of the lithio derivative on the quinoxaline **37**, followed by a further aromatization.

It could be noticed that contrary to the sulfonylpyrazine series, metalation of 2-*tert*-butylsulfonyl quinoxaline has been unsuccessfully tested, leading to either starting material or untreatable tarry products [61].

When the pyrazine ring is disubstituted, the metalation occurs on the benzene ring; the regioselectivity is influenced by the presence of a DMG group. Treatment of 6-chloro-2,3-dimethoxyquinoxaline with LTMP (4 equiv) in THF at −75°C,

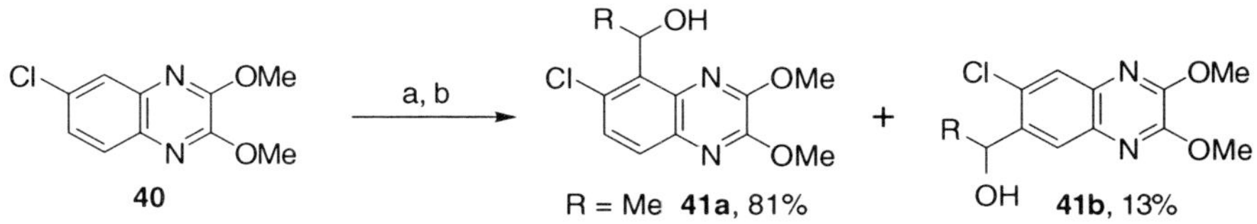

Scheme 15 (**a**) 4 equiv LTMP, THF, −78°C (**b**) RCHO then hydrolysis [62]

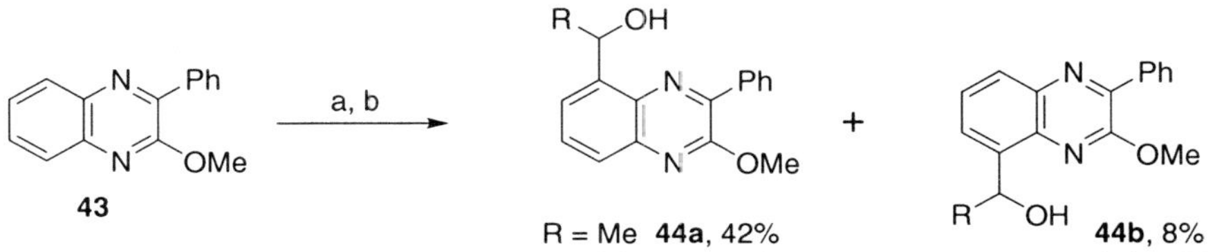

Scheme 16 (**a**) 4 equiv LTMP, THF, 0°C, 30 min (**b**) RCHO then hydrolysis [62]

followed by reaction with various electrophiles afforded the 5-substituted quinoxalines as major products (Scheme 15) [62]. This high regioselectivity at the C5 position is due to the additive effect of the ortho-directing chlorine atom and of the *peri* ring nitrogen atom.

Metalation of 2-methoxy-3-phenylquinaxoline **43** has been performed with an excess of LTMP in THF at 0°C. In this case a competitive lithiation could undergo on the benzene ring. In this case a mixture of 5-substituted **44** (as major compound) and 8-substituted quinoxalines **45** has been obtained besides starting material (Scheme 16) [62].

When TMSCl was used as the electrophile under in situ trapping conditions, the 5-substituted compound was obtained in 65% yield besides a small amount (10%) of 5,8-disubstituted quinoxaline. As previously, the relative amounts of the two isomers could be explained by the heats of formation of the lithiated species calculated with the semiempirical Li/PM3 method, which indicates that the C5 lithiated compound is more stable than the C8 one with a difference of 2.7 kcal/mol [62].

2.2 *Alkali-Metal-Mediated Metalation*

Use of hindered lithium amides such as LDA or LTMP allows deprotonation of a wide range of pyrazine and quinoxaline derivatives. However, to avoid side reactions, due to the high reactivity of the generated lithio intermediates, these reactions generally require strict experimental conditions (low temperatures, in situ electronic interceptions, short reaction times, limited functional group tolerance, etc.) which could impose limitations in their applications.

To overcome these drawbacks, over the past few years, a new generation of multimetallic reagents has been developed, allowing deprotonation of functionalized aromatic and heteroaromatic substrates with a wider functional group tolerance and often with milder reaction conditions. Owing to a variable central metal (magnesium, zinc, manganese, aluminium, or cadmium), variable ligands (both in their nature and in their number), and a variable second metallic center (an alkali metal such as lithium or sodium), deprotonative reactions using these organometallic "ate" complexes has opened up a new pathway to the functionalization of aromatic substrates [63, 64] and in particular with sensitive π-deficient heterocycles such as pyrazine and quinoxaline.

Among these multimetallic type compounds we can give as example $R_2Zn(TMP)$ Li(TMEDA) (R = *t*Bu, Bu or Me) described by the teams of Kondo, Uchiyama, Mulvey and Hevia, $(TMP)_2Zn{\cdot}2MgCl_2{\cdot}LiCl$, TMPZnCl·LiCl and $Al(TMP)_3{\cdot}3LiCl$ are reported by Knochel, $(Me_3SiCH_2)_2Mn(TMP)Li{\cdot}TMEDA$ used by Mulvey, MeCu $(TMP)(CN)Li_2$ described by Uchiyama and Wheatley and $(TMP)_3Cd$ Li by Mongin. In all these multimetallic compounds, the anion of the 2,2,6,6-tetramethylpiperidine (TMP) is generally found and could play the main role due to its steric and electronic features, such as a low nucleophilicity and a high Brøensted basicity. These multicomponent systems including TMP present different reactivity profiles since each component in the system such as alkali-metal, softer metal, supporting anions, solvent, and ligands can influence reactivity (optimally at ambient temperature) or selectivity (an ability to deprotonate at positions inaccessible via other reagents). Other experimental factors such as stoichiometry, solubility, nature of the solvent, concentration, and temperature are also of main importance.

Among the wide range of examples of deprotonative reactions using mixed-metal (designed as ate-complex) given in the literature, Kondo and Uchiyama, in 1999, reported a notable systematic study of the monoamido-dialkylzincate reagent "LiZn*t*Bu_2(TMP)" as a complex zincator carrying out deprotonation reactions of a series of functionalized aromatics with high levels of chemo- and regioselectivity. Since these pioneering works, the use of mixed-metal complexes has experienced an important development making them interesting bases of high importance in the synthetic routes via metalation reactions. Nevertheless it could be noted that examples with pyrazine and quinoxaline substrates remain limited.

By a classification dealing with the nature of the soft metal used in such multicoponent bases, we report herein the examples of deprotonation and cross-coupling reactions exclusively with pyrazine and quinoxaline substrates.

2.2.1 Alkali-Magnesium Reagents

Pyrazine

With the aim to develop the metalation and functionalization of heteroaryl derivatives with other metalating agents than lithium alkylamides which require low temperatures, use of magnesium amides R2NMgCl or R2NMgR′ is an

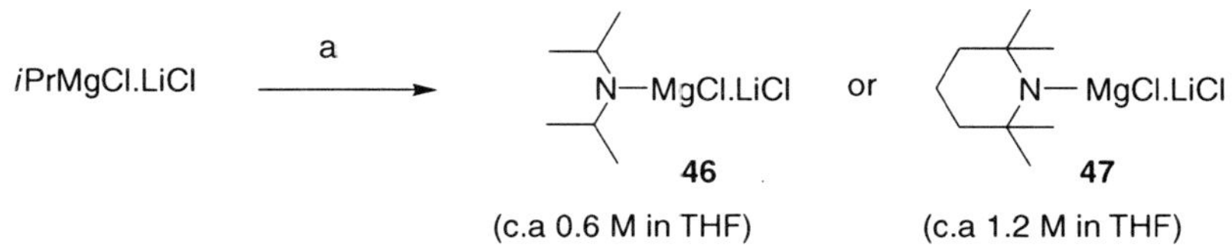

Scheme 17 (a) DIPA or TMPH, 25°C, 1–24 h [65]

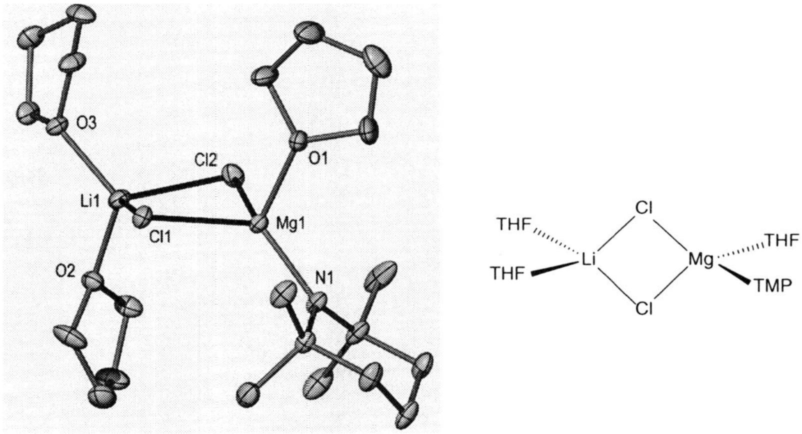

Fig. 10 Structural data of (TMP)MgCl·LiCl **47** [66]

interesting alternative; nevertheless, theses bases which suffer from a low solubility and usually for which a large excess of base is necessary to achieve high conversions have found fewer applications. A new class of magnesium bases of type $R_2NMgCl{\cdot}LiCl$ by reacting $i\mathrm{Pr}_2MgCl{\cdot}LiCl$ with di*iso*propylamine (DIPA) or 2,2,6,6-tetramethylpiperidine (TMPH) in THF has been described by Knochel [65], leading to $i\mathrm{Pr}_2NMgCl{\cdot}LiCl$ **46** or (TMP)MgCl·LiCl **47** (Scheme 17).

These bases display an excellent solubility in THF (0.6–1.2 M) as well as an enhancement of the kinetic basicity that has allowed a selective magnesiation of various aryl or heteroaryl derivatives substituted by sensitive functional groups such as ester, nitrile, or even ketone. The molecular structure of the base (TMP) MgCl·LiCl **47** has been established by NMR spectra and crystal data, highlighting that the active basic ligand TMP is bonded to the magnesium, that NMg–N(TMP) interaction is terminal and thus only one bond needs to be broken to release the active base (Fig. 10) [66]. These structural data could justify the magnesiating power of these bases.

Thus, starting from dichloropyrazines, the base (TMP)MgCl·LiCl **47** allowed the magnesiation and the selective functionalization of all positions of the pyrazine ring after trapping with various electrophiles [67]. Treatment of 2,3-dichloropyrazine **48** with (TMP)MgCl·LiCl (**47**, 1.1 equiv, 25°C, 15 min) led to 5-magnesiated

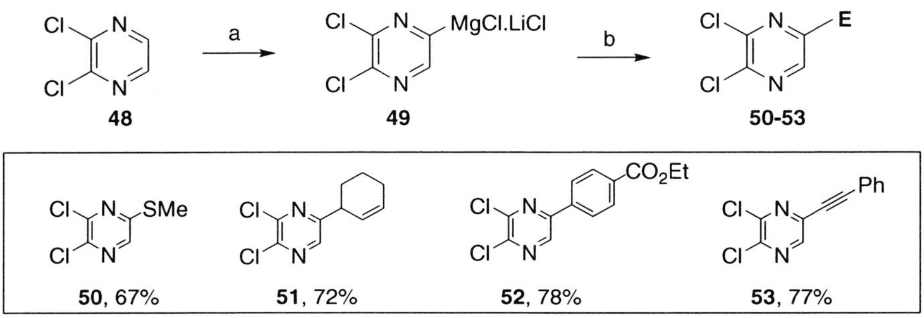

Scheme 18 (a) 1.1 equiv TMPMgCl.LiCl **47**, THF, 25°C, 15 min (b) Electrophile [67]

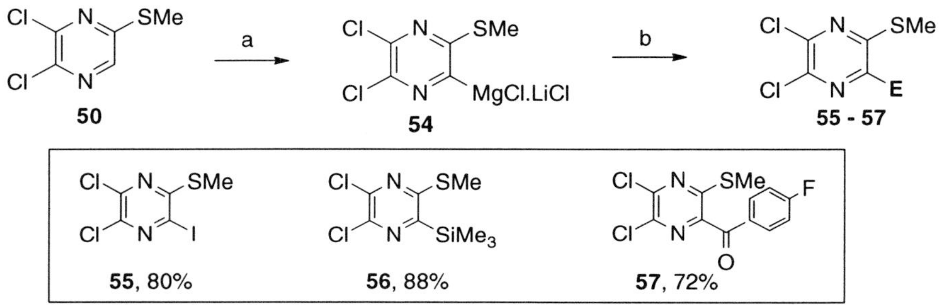

Scheme 19 (a) 1.1 equiv TMPMgCl.LiCl **47**, THF, −40°C, 60 min (b) Electrophile [67]

pyrazine **49**, which was trapped by electrophiles such as $MeSO_2SMe$ and 3-bromocyclohexene (after transmetalation with $ZnCl_2$ and addition of catalytic amount of CuCN.2LiCl), leading to the expected pyrazines **50** and **51** in good yields (Scheme 18).

The further formation of a new carbon–carbon bond was performed by a Negishi cross-coupling or a Sonogashira reaction of the in situ generated 2,3-dichloro-5-iodopyrazine giving compounds **52** and **53** in 78% and 77% yield.

Starting from compound **50**, a further magnesiation was carried out at the last position by the addition of (TMP)MgCl·LiCl (**47**, 1.1 equiv) leading to the magnesiated species **54** which reacted with iodine and with TMSCl to give the compounds **55** and **56**, respectively. After transmetalation of **54** with CuCN.2LiCl, reaction with 4-fluorobenzoyl chloride has furnished the ketone **57** in 72% yield (Scheme 19).

An alternative metalation method of the methylsulfanylpyrazine **58** involving a frustrated Lewis pair TMPMgCl BF_3·LiCl **60** resulting from the Lewis base (TMP) MgCl·LiCl **47** with the strong Lewis acid $BF_3·OEt_2$ has been reported by Knochel [68], leading to the iododerivative **59** in 81% yield after iodolysis (Scheme 20).

The highly reactive but less stable base $(TMP)_2Mg·2LiCl$ **61** resulting from addition of 1 equiv of LTMP to **47** (Fig. 11) has been used for the metalation of bare pyrazine [69].

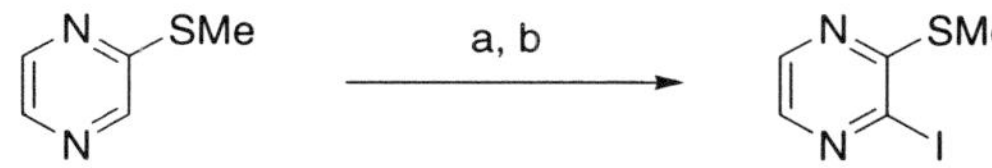

Scheme 20 (**a**) 1.1 equiv TMPMgCl.BF_3.LiCl **65**, THF, −40°C, 10 min (**b**) 1.5 equiv I_2, −40 to 25°C, 10 min [68]

(TMP)MgCl.LiCl + LiTMP ⟶ $(TMP)_2$Mg.2LiCl

47 **61**

Fig. 11 Synthesis of $(TMP)_2$Mg.2LiCl **61** [69]

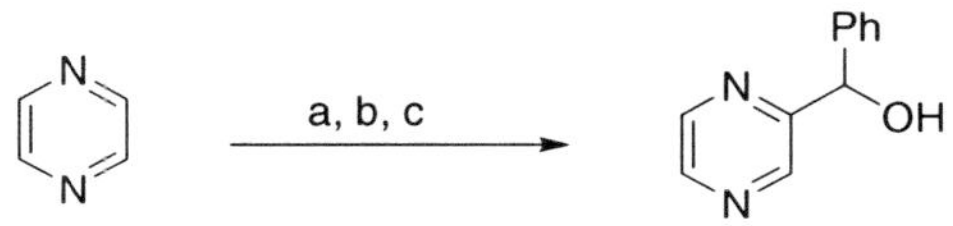

Scheme 21 (**a**) 0.5 equiv $ZnCl_2$ (**b**) 1.1 equiv **61**, 25°C, 15 min (**c**) PhCHO, 30 min [69]

It was shown that a former treatment of the pyrazine with $ZnCl_2$ before the addition of **61** gave the best result. By using the $ZnCl_2$·$(TMP)_2$Mg.LiCl protocol, the metalation of pyrazine **11** was achieved within 30 min. Further reaction with benzaldehyde gave the corresponding alcohol **62** in 60% yield (Scheme 21) [69].

Quinoxaline

Investigation to metalate quinoxaline **63** with $(TMP)_2$Mg·2LiCl **61** at 25°C in THF gave only traces of desired quinoxalyl iodide **64** after iodolysis; while the major product the dimeric heterocycle **65** was isolated in 34% yield. In contrast, treatment of **63** with $ZnCl_2$ (0.5 equiv) in THF, followed by addition of $(TMP)_2$Mg.2LiCl gave the 2-iodoquinoxaline **64** in 94% yield besides only traces of dimer (Scheme 22) [69, 70].

Under similar reaction conditions, the 6-bromoquinoxaline **66** was functionalized at the 5 position after reaction with various electrophiles leading to compounds **67–69** (Scheme 23). Compound **70** was iodated on the pyrazine ring by iodolysis to give **71** (Scheme 24). It could be noticed that compounds **68** and **69** were obtained with a further transmetalation using CuCN·2LiCl (1.1 equiv).

The authors have also investigated the preparation of other organometallics by performing dehydrometalation with **61** in the presence of stannyl compounds. When $(TMP)_2$Mg·2LiCl **61** was added to a mixture of quinoxaline **63** and n-Bu_3SnCl at −78°C, stannane **72** was isolated in good yield (Scheme 25) [69].

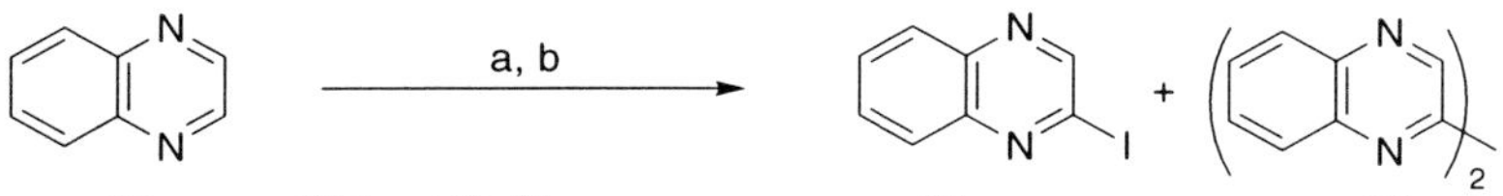

		64	65
63	Without $ZnCl_2$	traces	34%
	With $ZnCl_2$ (0.5 equiv)	94%	traces

Scheme 22 (**a**) 0.55 equiv $(TMP)_2Mg{\cdot}2LiCl$ **61**, 25°C, 2 h (**b**) I_2, 0–25°C, 1 h [69]

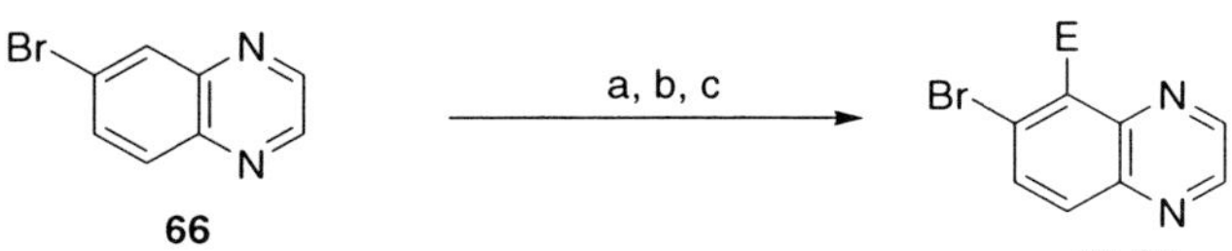

66

67-69

E = Br **67**, 70%
= CO^tBu **68**, 81%
= COPh **69**, 86%

Scheme 23 (**a**) 0.5 equiv $ZnCl_2$ (**b**) 0.55 equiv $(TMP)_2Mg{\cdot}2LiCl$ **61**, 25°C, 2 h (**c**) $(BrCl_2C)_2$ or CuCN.LiCl then RCOCl [69]

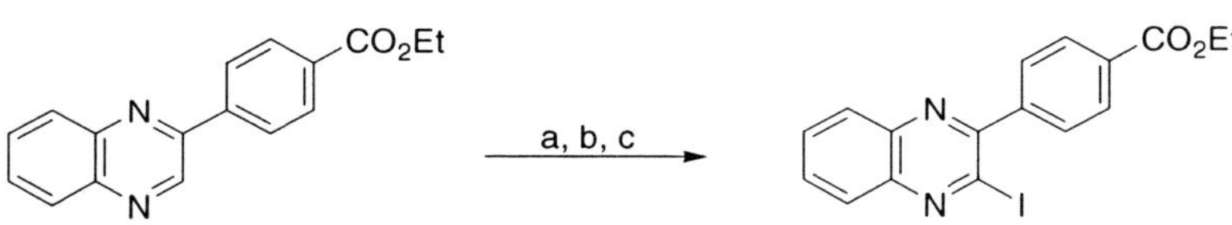

70 **71**, 73%

Scheme 24 (**a**) 0.5 equiv $ZnCl_2$ (**b**) 0.55 equiv $(TMP)_2Mg{\cdot}2LiCl$ **61**, 25°C, 2 h (**c**) I_2, 0–25°C, 20 min [69]

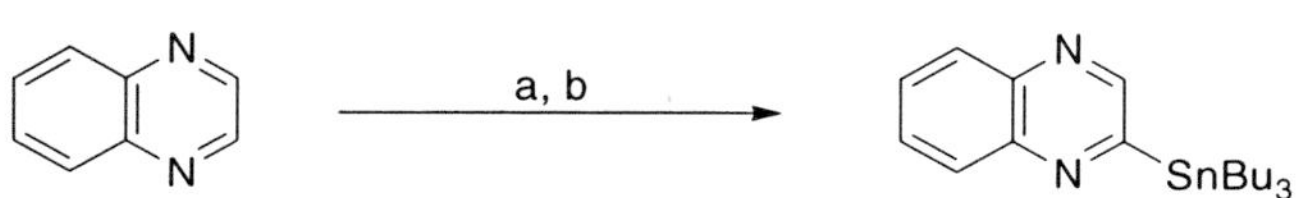

63 **72**, 73%

Scheme 25 (**a**) 1.1 equiv $n\text{-}Bu_3SnCl$ (**b**) 1.1 equiv $(TMP)_2Mg{\cdot}2LiCl$ **61**, −78°C, 30 min [69]

The quinoxaline **73** bearing a phosphorodiamide group as DMG was magnesiated with $(TMP)_2Mg{\cdot}2LiCl$ **61** at −50°C in 1.5 h without any dimerization side reaction. After a transmetalation with $ZnCl_2$, it underwent a Negishi cross-coupling in the presence of $Pd(dba)_2$ (5% mol) and $P(2\text{-furyl})_3$ (10% mol) with either 4-chloroiodobenzene or ethyl-4-iodobenzoate leading to the 2-aryled quinoxalines **75** and **76** in up to 79% yield. Treatment of the intermediate quinoxalylzinc **74** with methallyl bromide in presence of $CuCN{\cdot}2LiCl$ (10% mol) led to the allylated quinoxaline **77** in 71% yield (Scheme 26) [71].

The metalation and cross-coupling reaction obtained in the presence of $ZnCl_2$ involved a transmetallation; nevertheless, a direct zincation has been achieved by the use of a zincate base, analogous to the magnesiate one, with a zinc atom instead of magnesium.

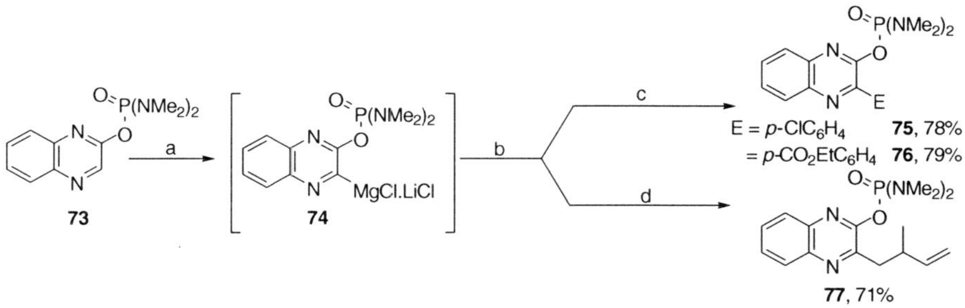

Scheme 26 (**a**) 1.5 equiv $(TMP)_2Mg{\cdot}2LiCl$ **61**, −30°C, 1.5 h (**b**) 1.6 equiv $ZnCl_2$ (**c**) Pd-catalyzed cross-coupling reaction with 5 mol% $Pd(dba)_2$ and 10 mol% $P(2\text{-furyl})_3$ (**d**) CuCN.2LiCl 10 mol% [71]

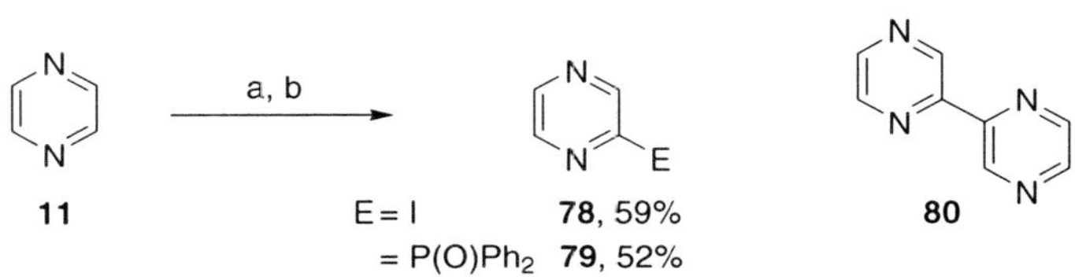

Scheme 27 (**a**) 0.5 equiv $ZnCl_2{\cdot}TMEDA$, 1.5 equiv LTMP, THF, rt, 2 h (**b**) I_2 or Ph_2PCl [44]

2.2.2 Alkali-Zinc Reagents

Deprotonation of pyrazine and quinoxaline using an in situ mixture of $ZnCl_2{\cdot}TMEDA$ (0.5 equiv) and LTMP (1.5 equiv) at room temperature was first reported by Mongin and coworkers [44, 72]. With THF as solvent, trapping with iodine after 2 h afforded the iodide **78** in 59% yield. When hexane was used as solvent, formation of appreciate amounts of dimer **80** was observed. With the use of chlorodiphenylphosphine as electrophile and THF without cosolvent, the phosphine oxide **79** was obtained in 52% yield (Scheme 27).

Metalation of quinoxaline **63** followed by quenching with iodine as the electrophile leads to the formation of the monoiodide **64** and the diiodide **81**, obtained in low yields (25% and 17%) besides the dimer **65** (Scheme 28).

If the use of mixed Mg/Li bases has highlighted a great interest due to their ability to allow metalation of sensitive substrates and a compatibility with functional groups, use of bases which could tolerate more sensitive functionality such as an aldehyde or a nitro group would expand the scope of direct metalations. Therefore, the use of the chemioselective zinc base $TMP_3Zn{\cdot}2MgCl_2{\cdot}2LiCl$ has been reported; however, this reagent was unsatisfactory for metalation of electron-poor functionalized arenes and heteroarenes [73]. Thus, a mild and selective complex base TMPZnCl·LiCl **82** resulting from treatment of TMPH with *n*-BuLi (1 equiv) followed by the addition of $ZnCl_2$ (1.1 equiv) at −30°C for 30 min provides a ca. 1.3 M solution of **84** stable at room temperature (Scheme 29), allowing zincation of diazines especially pyrazines [74].

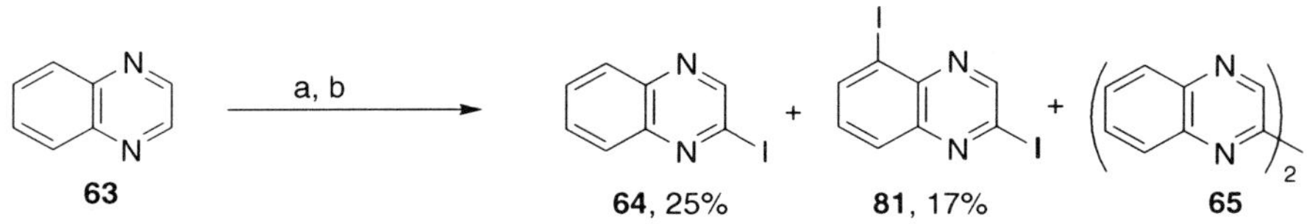

Scheme 28 (**a**) 0.5 equiv $ZnCl_2$·TMEDA, 1.5 equiv LTMP, THF, rt, 2 h (**b**) I_2 [72]

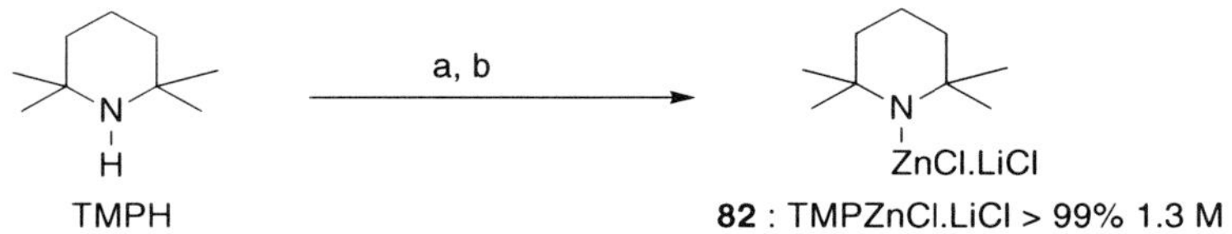

Scheme 29 (**a**) 1.1 equiv *n*-BuLi, THF, −40°C to −10°C, 1 h, (**b**) 1.1 equiv $ZnCl_2$, THF, −10°C, 30 min then 25°C, 30 min [74]

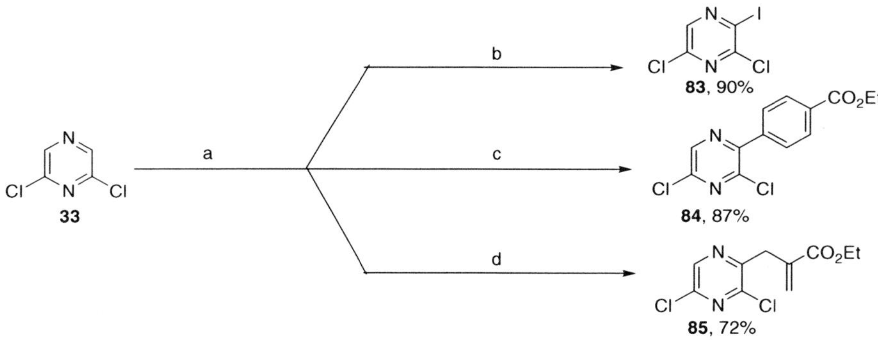

Scheme 30 (**a**) 1.1 equiv (TMP)ZnCl·LiCl **82**, THF, 25°C, 0.5 h (**b**) I_2 (**c**) Ethyl-4-iodobenzoate, Pd-catalyzed cross-coupling reaction with 5 mol% $Pd(dba)_2$ and 10 mol% $P(2\text{-furyl})_3$ (**d**) Ethyl 2-bromomethylacrylate, CuCN.2LiCl 10 mol% [74]

The 2,6-dichloropyrazine **33** was zincated quantitatively with TMPZnCl·LiCl **82** (1.1 equiv, 25°C, 30 min) and reacted with iodine to give **83**. A further Negishi cross-coupling with ethyl 4-iodobenzoate or an allylation with ethyl 2-bromomethylacrylate (after addition of CuCN·2LiCl) afforded the expected products **84–85** in 72–90% yields (Scheme 30) [74].

Regio- and chemoselective multiple functionalization of dichloropyrazine derivatives have been carried out by successive metalations using TMPZn·LiCl and TMPMg·LiCl followed by trapping with various electrophiles (Scheme 31) [67].

This methodology has been applied to carry out the total synthesis of coelenterazine in nine steps and 9% overall yield [67].

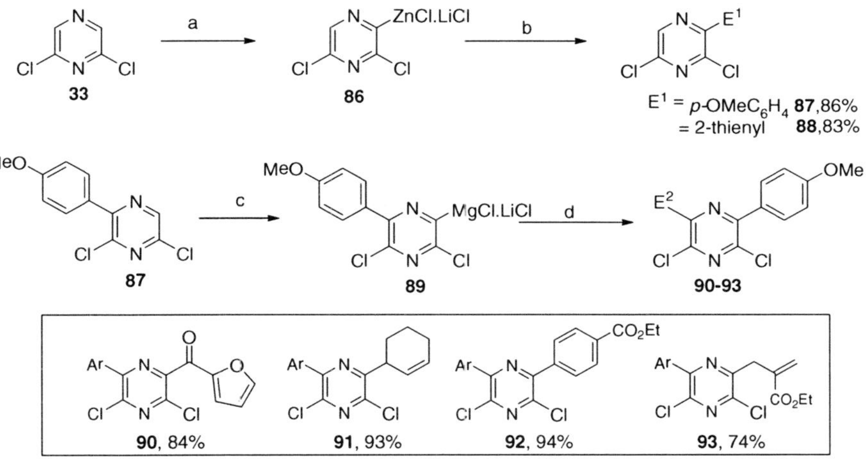

Scheme 31 (**a**) 1.1 equiv TMPZnCl·LiCl **82**, THF, 25°C, 0.5 h (**b**) Pd-catalyzed cross-coupling reaction with 4-iodoanisole or 2-thiophene and 3 mol% $Pd(dba)_2$ and 6 mol% $P(2\text{-furyl})_3$ (**c**) 1.1 equiv TMPMgCl·LiCl **47**, THF, −40°C, 1 h (**d**) Pd-catalyzed cross-coupling reaction with ethyl-4-iodobenzoate and 3 mol% $Pd(dba)_2$ and 6 mol% $P(2\text{-furyl})_3$; or furoyl chloride or 3-bromocyclohexene or ethyl 2-(bromomethyl)acrylate in the presence of CuCN.2LiCl [67]

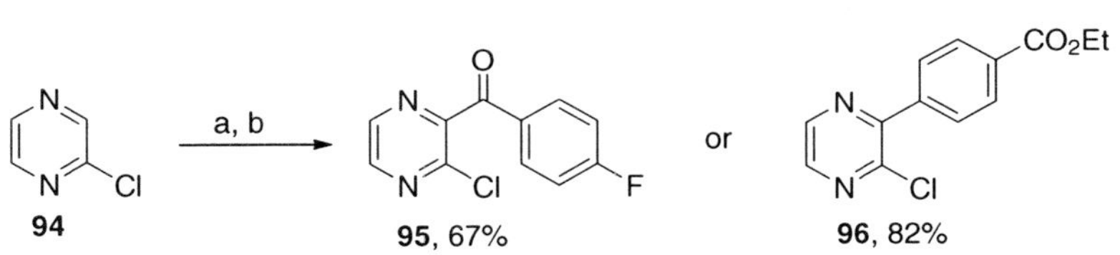

Scheme 32 (**a**) 1.1 equiv TMPZnCl·LiCl **82**, THF, 70°C, 45 min under microwave irradiation (**b**) Pd-catalyzed cross-coupling reaction with 4-fluoro-iodobenzene and $Pd(PPh_3)_4$ or ethyl-4-iodobenzoate and 3 mol% $Pd(dba)_2$ and 6 mol% $P(2\text{-furyl})_3$ [75]

Due to the thermal stability of the organo zinc reagents and their tolerance towards functional groups, use of TMPZnCl·LiCl and microwave irradiation allowed the high-temperature zincation of pyrazine derivatives. Thus, 2-chloropyrazine **94** reacted with **82** under microwave irradiation (70°C, 45 min) (Scheme 32). Acylation and Negishi cross-coupling gave pyrazines **95** and **96** in good yields [75].

Compound **87** was zincated at 100°C within 1 h. Negishi coupling and acylation with furoyl chloride (after transmetalation with CuCN·2LiCl) afforded the fully substituted pyrazines **97** and **90** (Scheme 33) [75].

Additionally, the metalation of more sensitive substrates has been accomplished by using the zinc base $TMP_2Zn{\cdot}MgCl_2{\cdot}2LiCl$ **98**. Thus, the larger-scale zincation of quinoxaline **63** (13.5 g) has been achieved within 3 h using **98** (0.44 M in THF, 114 mL); subsequently Pd-catalyzed cross-coupling reaction with iodoanisole (1.0 equiv) using $Pd(dba)_2$ (0.5 mol%) and $(o\text{-fur})_3P$ (1 mol%) as catalytic system has been performed leading to the arylated quinoxaline **99** in 82% yield (Scheme 34) [76].

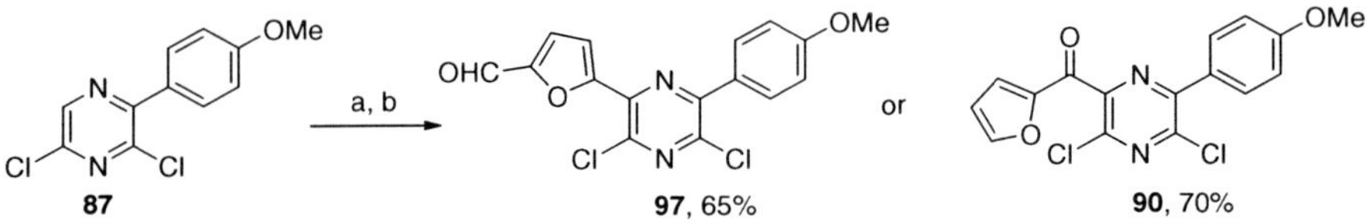

Scheme 33 (a) 1.1 equiv TMPZnCl·LiCl **82**, THF, 70°C, 45 min under microwave irradiation (b) Pd-catalyzed cross-coupling reaction with 5-iodofurfural and 3 mol% $Pd(dba)_2$/6 mol% $P(2\text{-furyl})_3$ or furoyl chloride with 1 equiv of CuCN.2LiCl [75]

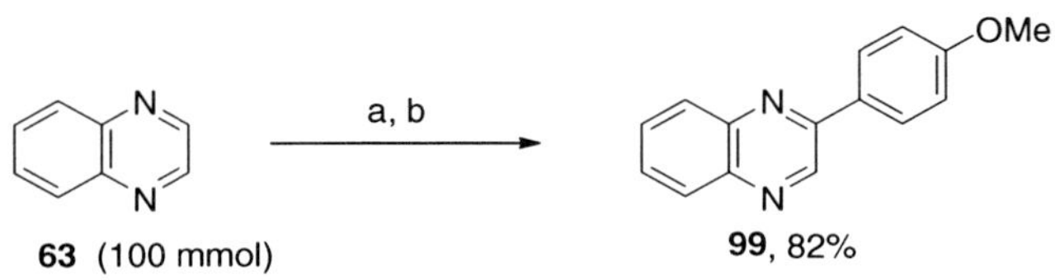

Scheme 34 (a) 0.5 equiv $TMP_2Zn{\cdot}MgCl_2{\cdot}2LiCl$ **98**, THF, 25°C, 2 h (b) Pd-catalyzed cross-coupling reaction with 4-iodoanisole and 0.5 mol% $Pd(dba)_2$/1 mol% $P(2\text{-furyl})_3$ [76]

Recently, the deprotonative zincation reaction of unsubstituted pyrazine has been reported with the lithium zincate base $[Li(TMP)Zn(^tBu)_2]$ **100** leading to a 2,5-dizincated pyrazine molecule **101** for which the structure has been confirmed by ^1H and 13C experiments and by crystal structure analysis. After quenching of **101** with iodine the 2,5-diiodopyrazine **102** was isolated in a 68% yield [77]. The homoleptic alkyl zincate $[(PMDETA)LiZn^tBu_3]$ **103** has promoted chemoselective addition of the ^{t}Bu group to one α-C of the heterocycle at room temperature causing its dearomatization to give **104** for which the crystal structure provided confirmation of addition on the C=N bond of pyrazine in positions 1 and 2. A further aerobic oxidation furnished the *tert*butylpyrazine **105** (Scheme 35).

Use of **100** as zincate in the presence of 2,6-dichloropyrazine **33** followed by reaction with iodine led to the trihalogenoderivative **83** in 46% yield. When **100** reacted with quinoxaline **63** at 0°C for 1 h, the iodoquinoxaline **64** was obtained in moderate yield (50%) [78] (Scheme 36).

2.2.3 Alkali-Cadmium Reagents

The deprotonative metalation of pyrazines has been described by Mongin using a new mixed lithium-cadmium ate base: TMP-cadmiate $(TMP)_3CdLi$ **106** resulting from treatment of LTMP (3 equiv) with $CdCl_2$.TMEDA at 0°C (Scheme 37) [79]. The latter proved to be compatible with reactive functional groups such as nitrile and sensitive π-deficient heteroaromatics even at room temperature (Scheme 38) [80, 81].

It could be noticed that the deprotonation and the subsequent trapping with iodine of the cyanopyrazine **108** gave a mixture of monoiodide and unidentified

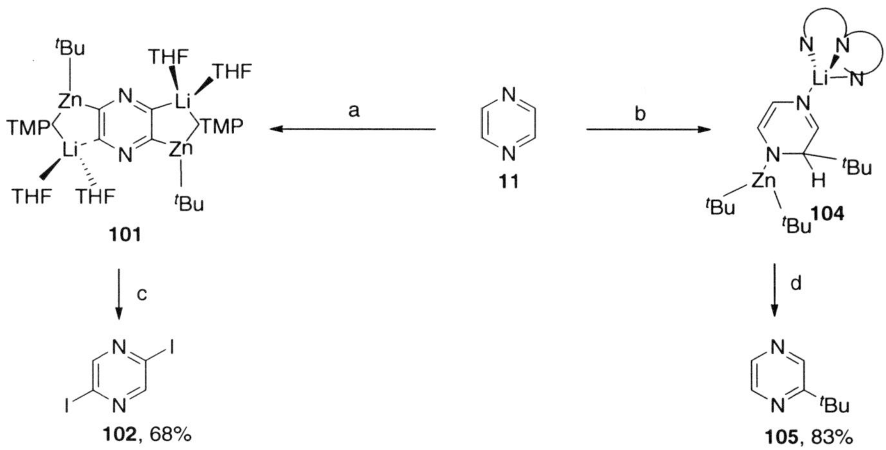

Scheme 35 (**a**) 2[(THF)LiZn(TMP)(tBu)$_2$] **100**, THF, 0°C (**b**) [(PMDETA)LiZntBu$_3$] **103**, hexane 25°C (**c**) I_2 (**d**) aerobic oxidation [77]

Scheme 36 (**a**) 1 equiv Li(TMP)Zn(tBu)$_2$] **100**, THF, 0°C, 1 h (**b**) I_2 [78]

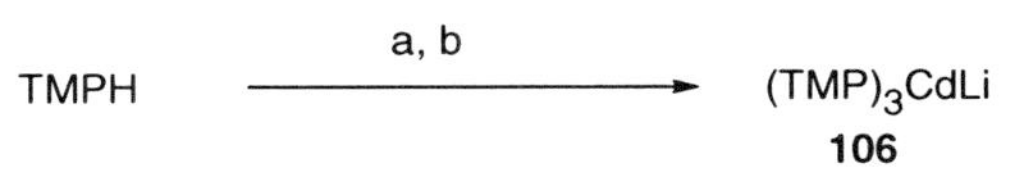

Scheme 37 (**a**) 1 equiv n-BuLi, THF, (**b**) 0.33 equiv $CdCl_2$.TMEDA, THF, 0°C [79]

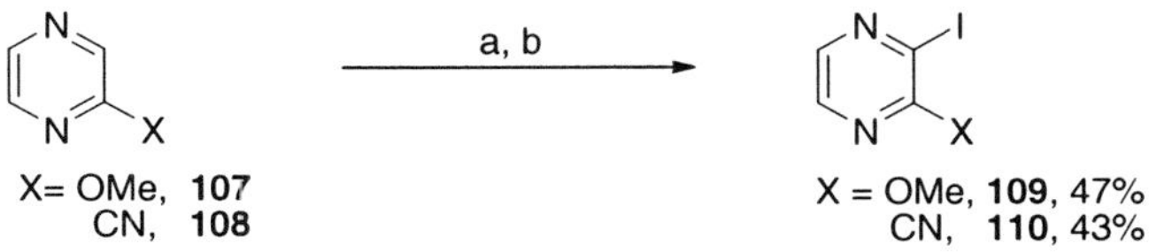

Scheme 38 (**a**) 0.5 equiv (TMP)3CdLi **106**, THF, rt, 2 h (**b**) I_2 [79–81]

diiodide in a 75:25 ratio. The metalation of unsubstituted pyrazine was also performed at room temperature in THF using $(TMP)_3CdLi$ **106** and iodine as electrophile [72, 82]. In this case, bare pyrazine which presents two activated metalation positions could lead to the 2,5-dihalogeno compound. It was shown that di-iodopyrazine could be obtained depending on the equivalents of **106**. For example, iodopyrazine **78** was isolated in 63% yield using 0.33 equiv of **106** while 0.5 equiv led to iodopyrazine **78** in 59% yield and concomitant formation

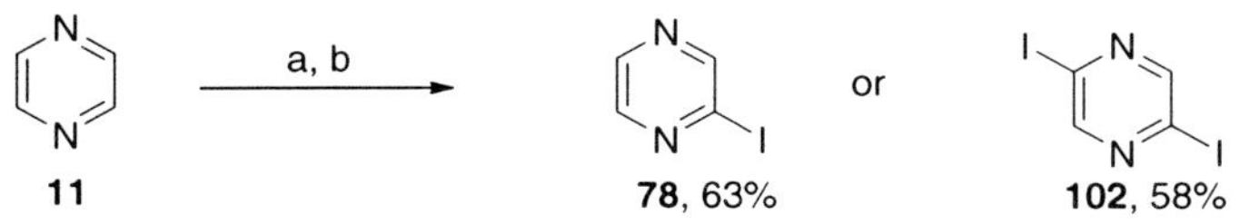

Scheme 39 (**a**) 0.33–1 equiv $(TMP)_3CdLi$ **106**, THF, rt, 2 h (**b**) I_2 [72, 82]

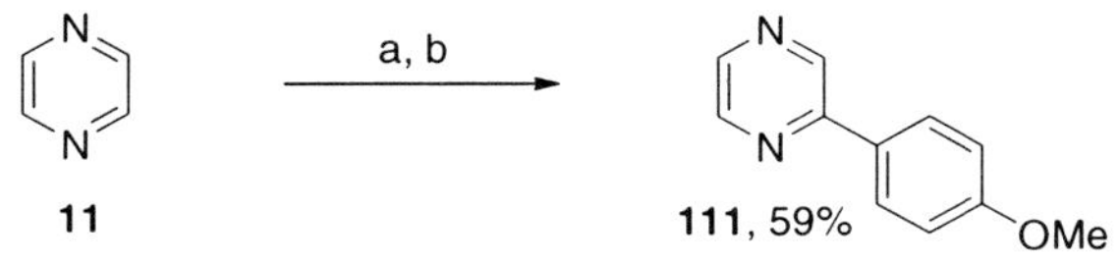

Scheme 40 (**a**) 0.5 equiv $(TMP)_3CdLi$ **106**, THF, rt, 2 h (**b**) Pd-catalyzed cross-coupling reaction with 4-bromoanisole and 6 mol% $Pd(OAc)_2$/2 mol% dppf, reflux, 18 h [80]

of 2,5-diiodopyrazine **102** in 20% yield (Scheme 39) [82]. The yield of **102** could be improved by using 1 equiv of **106**. In this case, only the diiodide **102** was isolated in 58% yield (in 40% yield when a 25 mmol scale was done).

Alternative trappings of cadmiates intermediates have been involved in palladium-catalyzed cross-coupling reaction with aromatic bromide. From bare pyrazine, the expected biaryl compound **111** was isolated in mediocre yield due to the concomitant formation of homocoupling products (Scheme 40) [79, 80]. Compared to the previously described methods for the synthesis of this biaryl compound, this procedure has the advantage of being "one pot."

Due to the toxicity of cadmium compounds, catalytic quantities of cadmium salts are preferred rather than stoichiometric ones.

2.2.4 Alkali-Zirconium Reagents

In the course of development of amide-type bases of transition metal complexes derived zirconium is attractive because of its low toxicity (specially compared to cadmium-complexes) and its low cost. A THF-soluble kinetically highly active zirconium amide was described recently by Knochel. Indeed, $TMP_4Zr{\cdot}4MgCl_2{\cdot}6LiCl$ (abbreviated as tmp_4Zr **112**) allows a direct zirconation of functionalized heterocycles such as methylsulfanylpyrazine **58** [83]. The subsequent trapping of the zirconium intermediate with ethyl 2-(bromomethyl)acrylate in the presence of CuCN.2LiCl afforded compound **113** in 81% yield (Scheme 41). It could be noticed that the aryl zirconated species display a better reactivity toward electrophile than the other organometallic complexes.

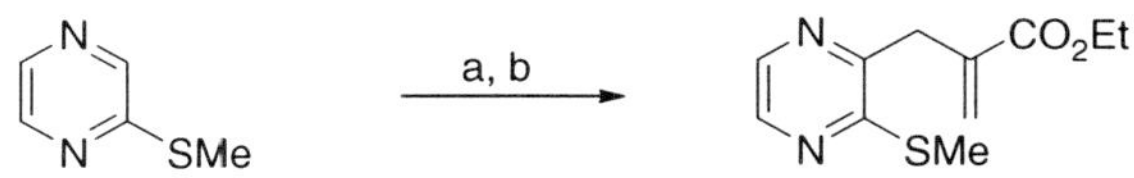

58 — 113, 81%

Scheme 41 (**a**) 0.25 equiv TMP_4Zr **112**, THF, −35°C, 20 min (**b**) CuCN.2LiCl, Ethyl 2-(bromomethyl)acrylate [83]

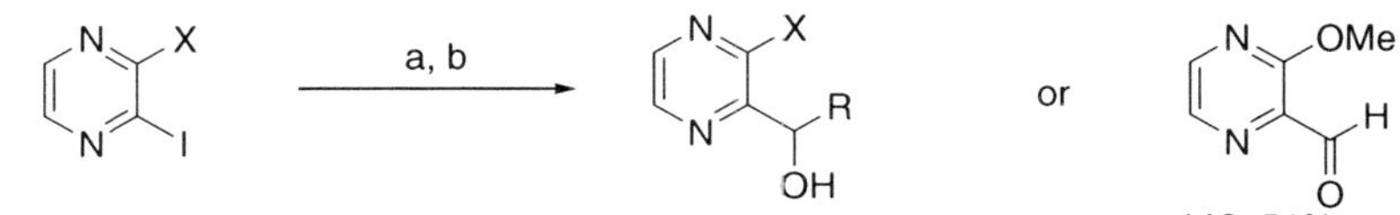

78, 109, 114, 115 X = H, OMe, Cl, I — **62, 116-118, 120-122**, 11-59% R = Ph, C_5H_{11} — **119**, 51%

Scheme 42 (**a**) 1 equiv *n*-BuMgCl or *i*-PrMgCl, 0.5 h, 0°C; (**b**) RCHO, 2 h, 0°C followed by brine [84]

3 Dehalometalation

3.1 Pyrazines

Since electron-deficient heteroaromatics are generally reactive towards nucleophiles, reaction of halogenated pyrazines with alkyllithium proceeds with side reactions related to the halide being a good leaving group. Nevertheless, the dehalometalation could occur with less nucleophilic organometallics. The preparation of Grignard derivatives of pyrazines is described using an iodine–magnesium exchange reaction (Scheme 42). This convenient method allows the functionalization of halogeno- or methoxypyrazine at 0°C using successfully *i*-PrMgCl or the least nucleophilic *n*-BuMgCl. The Grignard derivatives were trapped with aldehydes or *N,N*-dimethylformamide to give the corresponding alcohols or aldehyde in modest yields [84] (Table 3).

When 3-chloro-2-iodopyrazine, known to be really sensitive to nucleophilic attack, was used as starting material, *n*-Bu_2Mg/NEt_3 gave the best result (Scheme 43).

More recently, the metal–halogen exchange reaction was described with lithium tri-*n*-butylmagnesate. The reaction was performed exclusively with 2-iodopyrazine using 0.35 equivalent of lithium-magnesate at −10°C. When 2-chloro or 2-bromopyrazine is used as starting material, the reaction was unsuccessful. The in situ formed magnesate intermediate was trapped with various electrophiles at room temperature (Scheme 44). [85]

Another way to perform metal–halogen exchange is by use of lithium alkylamide such as LTMP. In this case, the dehalogenation reaction is described as a halogen–lithium exchange well known as a "halogen-dance" in pyridine series [86, 87]. In presence of an excess of LTMP, the 2-fluoro-6-iodo-3-phenylpyrazine **127** underwent a further isomerization involving an iodine atom migration leading to 5-iodo derivative **128** (Scheme 45) [51].

Table 3 Synthesis of pyrazine derivatives

-X	RMgCl	Electrophile	Product	Yield (%)
-H	*n*-BuMgCl	PhCHO	**62**	59
-H	*n*-BuMgCl	$C_5H_{11}CHO$	**116**	33
-OMe	*i*-PrMgCl	PhCHO	**117**	56
-OMe	*i*-PrMgCl	$C_5H_{11}CHO$	**118**	50
-OMe	*i*-PrMgCl	Me_2NCHO	**119**	51
-Cl	*n*-BuMgCl	PhCHO	**120**	24
-Cl	*n*-BuMgCl	$C_5H_{11}CHO$	**121**	20
-I	*n*-BuMgCl	PhCHO	**122**	11

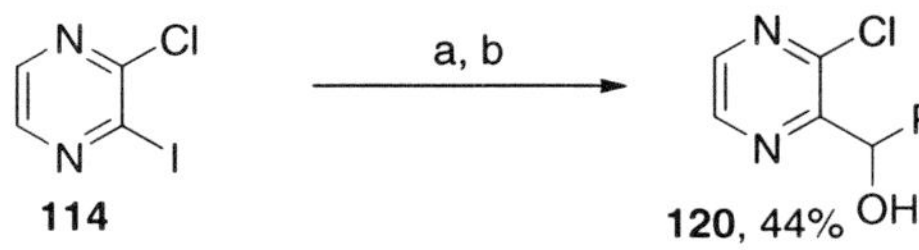

114 → **120**, 44%

Scheme 43 (**a**) 1 equiv *n*-Bu_2Mg/NEt_3, 0.5 h, 0°C; (**b**) PhCHO, 2 h, 0°C followed by brine [84]

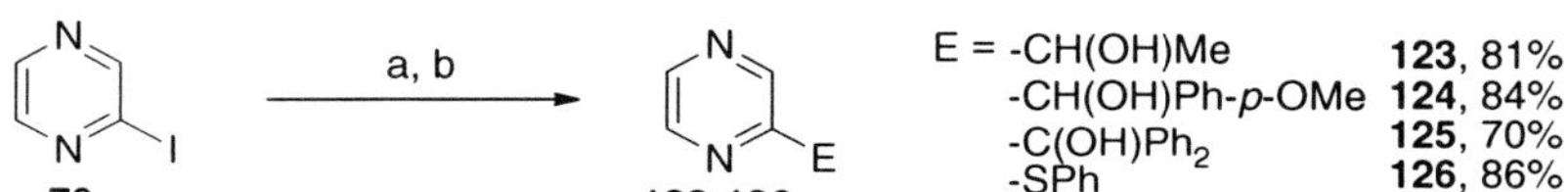

78 → **123-126**

Scheme 44 (**a**) 0.35 equiv *n*-Bu_3MgLi, THF, −10°C, 2.5 h; (**b**) Electrophile, −10°C to rt, 18 h followed by hydrolysis [85]

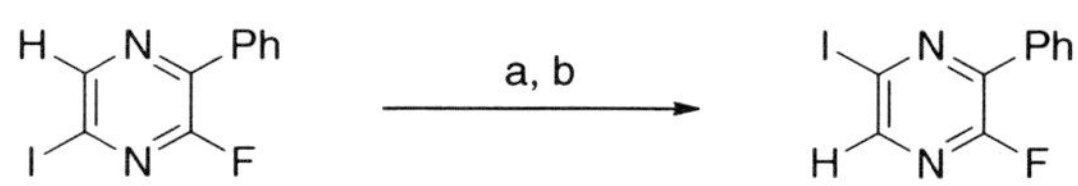

127 → **128**, 89%

Scheme 45 (**a**) 3.1 equiv LTMP, THF, −78°C, 0.5 h; (**b**) HCl, EtOH [51]

The reaction of 5,6-diiodo derivative **129** with LTMP under similar conditions, followed by reaction either with HCl or with aldehyde, highlighted that the iodine–lithium exchange occurred exclusively at the C_6 position (Scheme 46).

Another successful route to lithio-derivatives by a dehalogenation reaction involved the Barbier type reaction. In this case, the reaction could be performed at room temperature. The main drawback of the Barbier reaction is usually the slow attack of the metal by the halogen derivative which could be solved by the use of sonication. Indeed, sonication was known to afford an enhancement in the preparation of lithium derivatives from metal powder. Barbier type reaction with lithium metal was proved to be efficient under sonication in the pyrazine series, even with halogen substituents (Scheme 47, Table 4) [88]. This very convenient method allows a very fast and smooth functionalization of these compounds at room temperature with a short reaction time.

The obtained yields revealed a good compatibility of this reaction with halogen substituents.

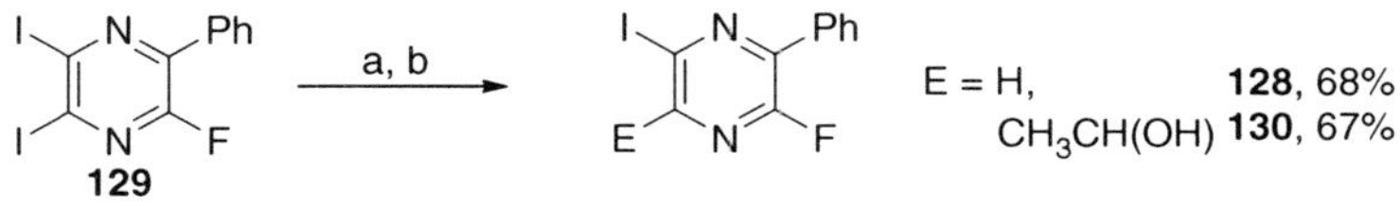

Scheme 46 (**a**) 3.1 equiv LTMP, THF, −78°C, 10 min; (**b**) HCl, EtOH or CH_3CHO [51]

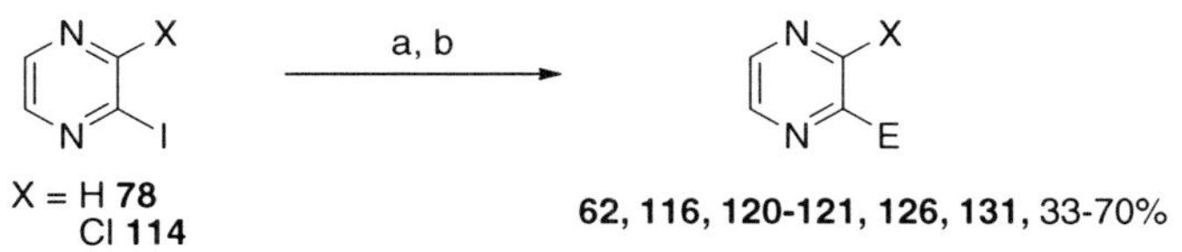

Scheme 47 (**a**) 2.2 equiv Li, 1.1 equiv Electrophile, THF, rt,)))), 0.5 h; (**b**) EtOH [88]

Table 4 Synthesis of pyrazine derivatives using Barbier type reaction under sonication

-X	Compound	Electrophile	Product	Yield (%)
-H	**78**	PhCHO	**62**	70
-H	**78**	$C_5H_{11}CHO$	**116**	64
-H	**78**	PhSSPh	**126**	44
-Cl	**114**	PhCHO	**120**	70
-Cl	**114**	$C_5H_{11}CHO$	**121**	60
-Cl	**114**	PhSSPh	**131**	33

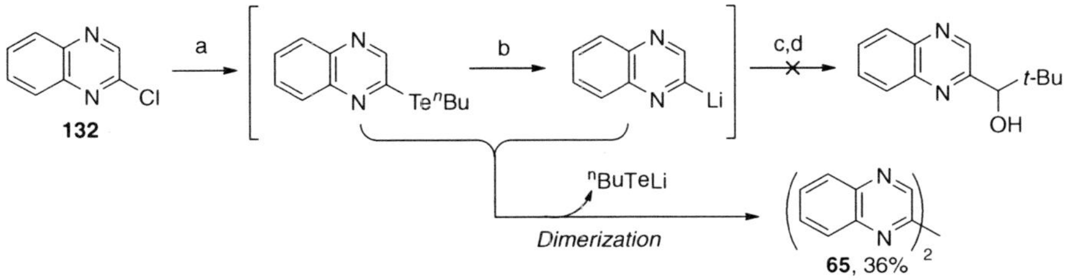

Scheme 48 (**a**) 1.1 equiv *n*-BuTeLi, THF, rt; (**b**) 1.1 equiv *n*-BuLi, THF, −78°C, 10 min; (**c**) 5 equiv pivalaldehyde, −78°C to rt, 10 min; (**d**) HCl 1N, rt [89]

3.2 Quinoxalines

The lithiation of 2-chloroquinoxaline was considered using a halogen–tellurium exchange followed by a transmetalation of the teluride intermediate with alkyllithium (Scheme 48). This method was presumed to allow the conversion of π-deficient chloroheteroaromatics, which are not active in the halogen–lithium exchange reaction, into the lithio adduct. Nevertheless, the halogen–tellurium exchange took place but dimerization occurred when 2-chloroquinoxaline **132** was used as substrate [89].

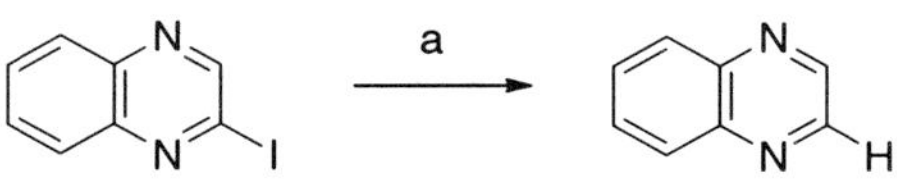

64 63, 44%

Scheme 49 (**a**) 1 equiv Indium powder (100 mesh), H_2O, reflux, 2 h [90]

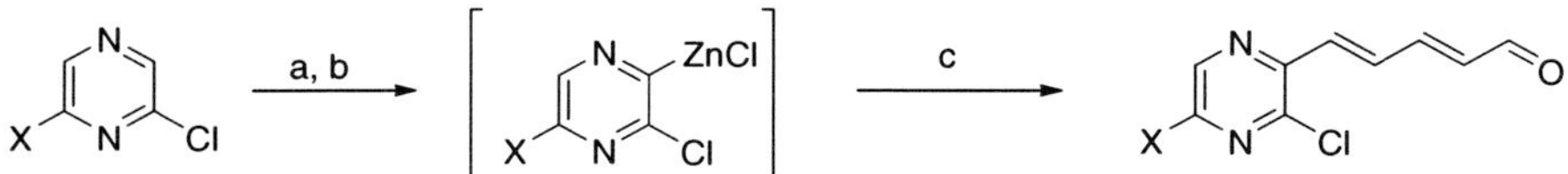

X = H **94**
X = Cl **33**
X = H **133**
X = Cl **134**
X = H **135**, 72%
X = Cl **136**, 46%

Scheme 50 (**a**) 1.3 equiv LTMP, THF, −78°C, 0.5 h; (**b**) 3 equiv $ZnCl_2$, −78°C to rt; (**c**) (2*E*,4*E*)-5-bromopentadienal, 5 mol% $Pd(PPh_3)_4$, THF, rt, 2 h [91]

More recently, the same author described a novel indium-mediated deiodination of 2-iodoquinoxaline **64** in water [90]. This method using indium powder is facile and safe and allowed the synthesis of quinoxaline **63** in 44% yield (Scheme 49).

4 Transmetalation

4.1 Pyrazines

Among transmetalation reaction for generating organometallic compounds, which are attractive tools in organic synthesis, lithium/metal exchange has proved to be an efficient methodology. This methodology allowed a further cross-coupling reactions such as Negishi, Kumada, or Stille reactions. Lithium/zinc exchange was described on easily available 2-chloro- or 2,6-dichloro-3-lithiopyrazines using zinc chloride. This transmetalation was described as the key step for the synthesis of various push–pull molecules with a central pyrazine unit [91]. Indeed, the transmetalation was followed by a Negishi cross-coupling reaction of the organozinc derivative **133** or **134** with (2*E*,4*E*)-5-bromopentadienal (Scheme 50).

The main advantage of this lithium/zinc exchange is to generate an organometalic intermediate allowing a further cross-coupling reaction without protection of the aldehyde group of the coupling partner. Nevertheless, this aldehyde could be protected in order to generate the C6-lithiated derivative submitted to a further lithium/zinc exchange using zinc chloride (Scheme 51). The trisubstituted pyrazines were obtained via a Negishi coupling reaction between the organo zinc **138** and 1-bromo-6-arylhexatrienes as a mixture of isomers (1*E*,3*E*,5*Z*) and (1*E*,3*E*,5*E*). It must be noticed that the yield of these compounds depends on the amount of LTMP, a wide excess of the lithium amide warranted an appreciable yield [92].

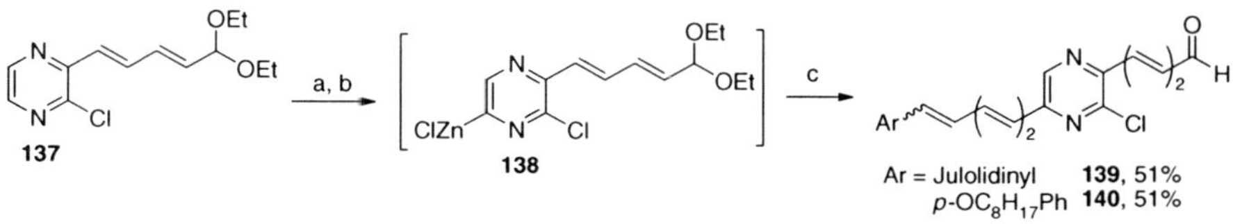

Scheme 51 (**a**) 4.1 equiv LTMP, THF, −78°C, 2 h; (**b**) 3 equiv $ZnCl_2$, −78°C to rt; (**c**) Ar $(CH{=}CH)_3$-Br, 5 mol% $Pd(PPh_3)_4$, THF, rt, 4–18 h then HCl/H_2O [92]

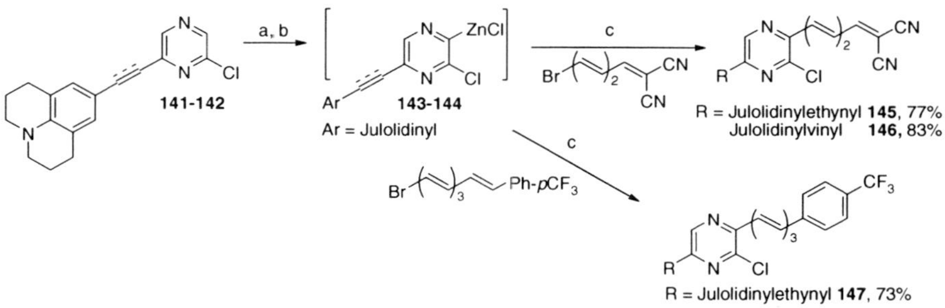

Scheme 52 (**a**) 1.2 equiv LTMP, THF, −78°C, 0.5 h; (**b**) 3 equiv $ZnCl_2$, −78°C to rt; (**c**) 5 mol% $Pd(PPh_3)_4$, THF, rt, 4 h then HCl/H_2O [92]

The authors have shown that zinc transmetalation of lithiated pyrazines is a key step in the synthesis of new rod-like conjugated molecules having a pyrazine core moiety as electron-withdrawing group. The light emitting properties of these molecules are investigated in terms of optical absorption and emission spectra. In order to appreciate the influence of the optical properties of the length and of the nature of the conjugated chain at C6 and C3 positions of the pyrazine ring, compounds **145–147** have been synthesized using a lithium/zinc exchange (Scheme 52).

In addition, transmetalation reaction using metal/lithium exchange could be also an effective route for preparing orgnanolithium reagents. Due to their highly reactivity toward nucleophiles, examples of metal/lithium exchange are sparse in pyrazine and quinoxaline series. Tin/lithium exchange was described on 2-halogeno-3-tributylstannylpyrazines allowing an original synthesis of its 2-halogeno-6-tributylstannylpyrazine isomers. When metalation of 2-fluro-3-tributylstannylpyrazine **7** was performed with 2.1 equiv of LTMP with a short reaction time and was followed by protonation of the lithio derivative, compound **8**, resulting from intramolecular tin/lithium exchange, was isolated in 63% yield (Scheme 53) [52].

The mechanism of the isomerization of 2-fluro-3-tributylstannylpyrazine **7** via migration of the tributylstannyl group was shown in Scheme 54.

The 2-chloro-6-tributylstannylpyrazine, resulting from the intramolecular transmetalation, was proved to be an effective intermediate in the synthesis of natural pyrazine alkaloids such as Botrillazines [93]. Starting from 2-chloropyrazine, the

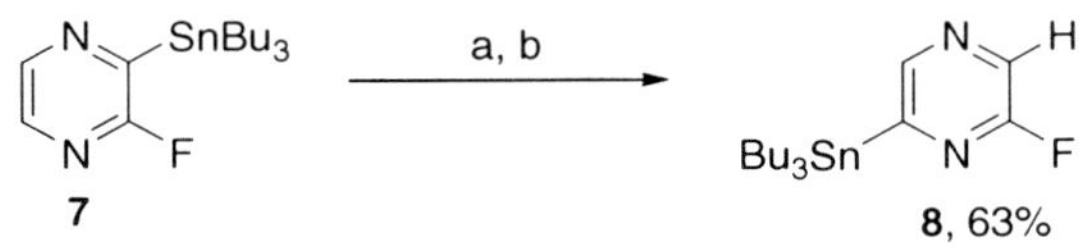

Scheme 53 (a) 2.1 equiv LTMP, THF, −78°C, 5 min; (b) EtOH, THF, −78°C [52]

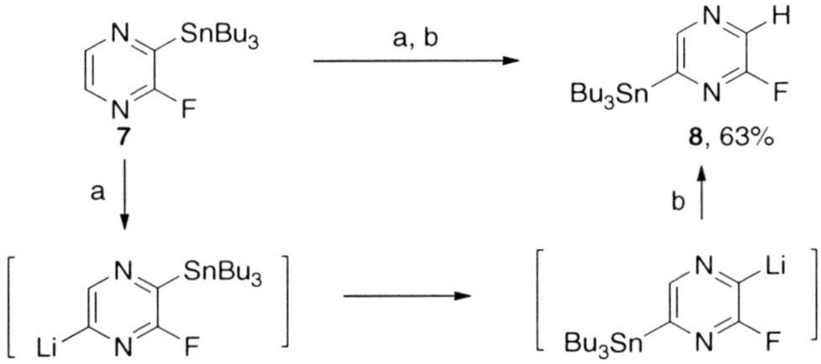

Scheme 54 (a) 2.1 equiv LTMP, THF, −78°C, 5 min; (b) EtOH, THF, −78°C [52]

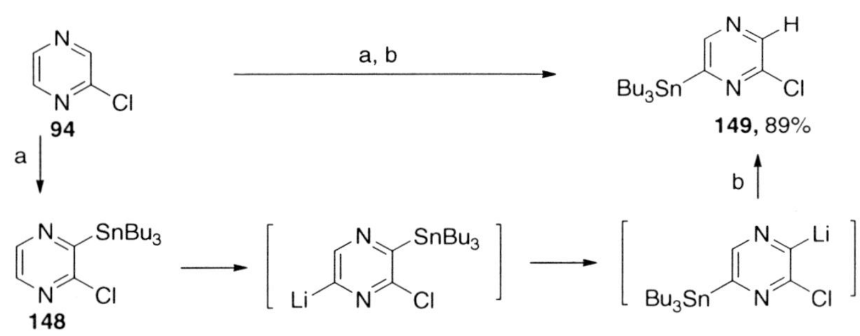

Scheme 55 (a) 3.1 equiv LTMP, 1 equiv $SnBu_3Cl$, THF, −100°C to −40°C, 2.5 h; (b) H_2O, HCl, THF, −40°C to rt [93]

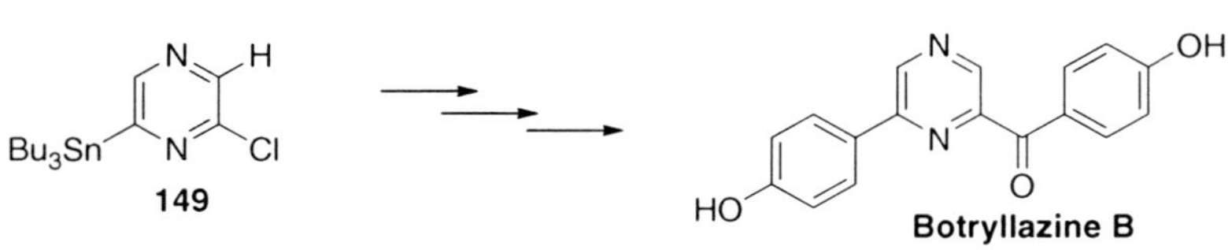

Scheme 56 Synthesis of Botryllazine B from 2-chloro-6-tributylstannylpyrazine intermediate

2-chloro-3-lithio-6-tributylstannylpyrazine intermediate was similarly obtained via migration of the tributylstannyl group in the presence of an excess of LTMP (Scheme 55).

Compound **149**, resulting from the hydrolysis of the 2-chloro-3-lithio-6-tributylstannylpyrazine intermediate, was the key intermediate in the synthesis of Botrillazine B and its analogues (Scheme 56) [93, 94].

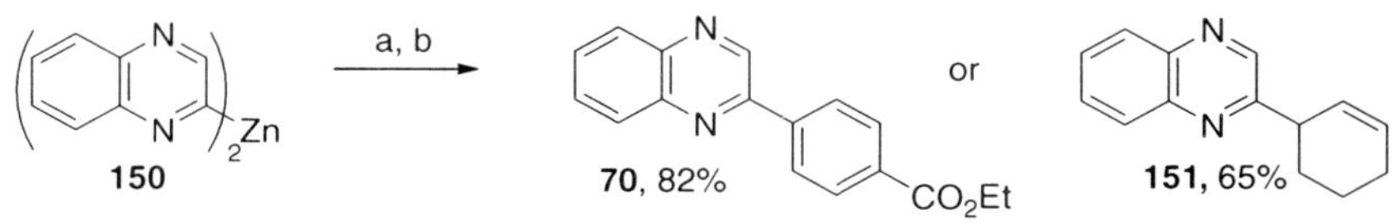

Scheme 57 (**a**) 1.1 equiv CuCN.2LiCl, THF, −40°C to rt (**b**) Pd-catalyzed cross-coupling reaction with ethyl-4-iodobenzoate and 3 mol% Pd(dba)$_2$ and 6 mol% P(2-furyl)$_3$ or 3-bromocyclohexene, −40°C to rt, overnight [69]

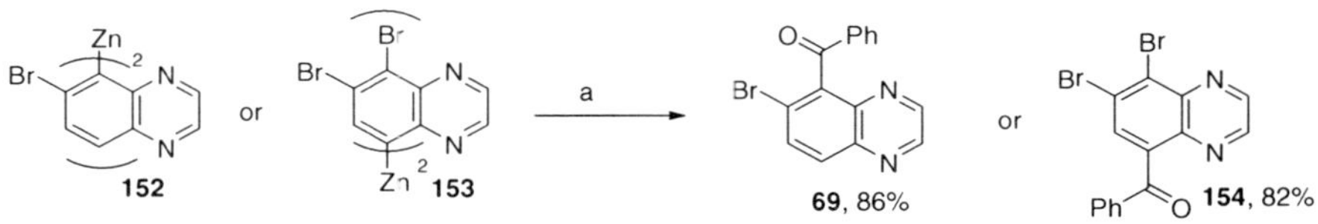

Scheme 58 (**a**) 1.1 equiv CuCN.2LiCl, THF, benzoyl chloride, −40°C to rt, overnight [69]

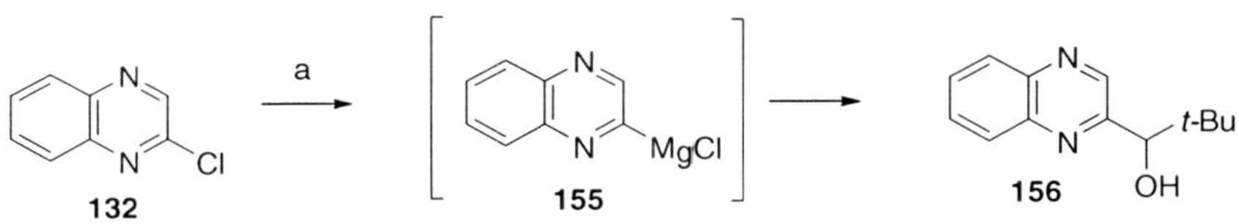

Scheme 59 (**a**) 4 equiv Mg, THF, pivalaldehyde, −40°C, 5 min or −20°C, 10 min, then rt [95]

4.2 Quinoxalines

The in situ transmetalation of zincate complex **150**, preformed with magnesium base **61** in the presence of ZnCl$_2$ and quinoxaline, was described by Knochel using CuCN.2LiCl [69] (Scheme 57). The trapping of cuprate intermediate with ethyl-4-iodobenzoate or 3-bromocyclohexene gave compounds **70** or **151**.

Zinc/Copper exchange on the 6-bromo or 5,6-dibromoquinoxaline followed by the reaction of benzoyl chloride gave access to the corresponding ketone **69** or **154** in 86% and 82% yield, respectively (Scheme 58) [69].

5 Oxidative Addition

The only example in oxidative addition concerned the quinoxaline series. Sugimoto described the useful oxidative magnesiation of chloroquinoxaline using active magnesium under Barbier conditions [95]. Indeed, the oxidative magnesiation of 2-chloroquinoxaline **132** in the presence of pivalaldehyde gave the corresponding alcohol **156** in modest yield without usual extra cooling such as −78°C (Scheme 59).

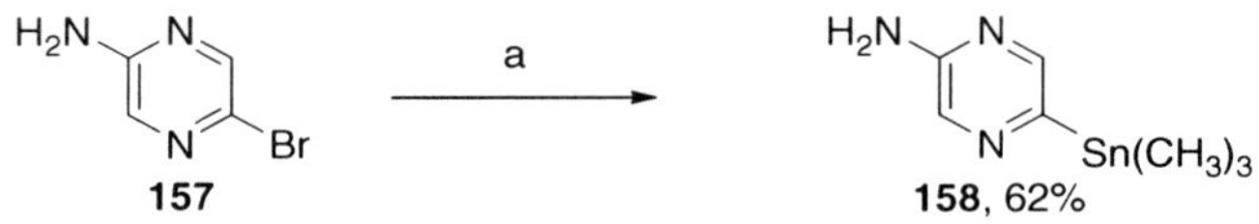

Scheme 60 (**a**) Me_6Sn_2, $Pd(PPh_3)_4$, toluene [100]

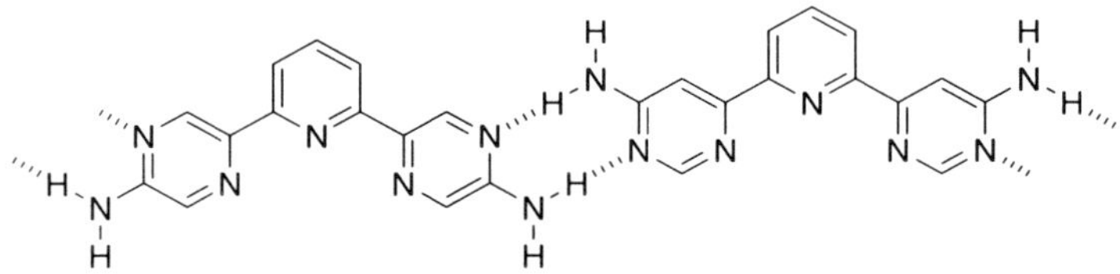

Fig. 12 Terpyridine-like ligands with aminopyrazine unit in the metal-complex conformation exhibiting complementary hydrogen bonds [100]

Since chloroderivative **132**, which was considered as an inert compound towards metalating reagent, was magnesiated at −20°C, this simple method could be really useful to introduce electrophilic substituents to the quinoxaline moiety.

6 Cross-Coupling Reaction

Although the organometallic species are numerous in the pyrazine series, the organostannane derivatives are still attractive as key intermediates in the synthesis of molecules with biological interest (for example, see: [96–99]). Among the methodologies for generating stannylpyrazines, the Stille cross-coupling reaction of halogenopyrazines and a hexaalkyldistannane derivative is one of the most useful routes. For example, 2-amino-5-trimethystannylpyrazine **158** was successfully synthesized from 2-amino-5-bromopyrazine **157** using the Stille cross-coupling reaction (Scheme 60) [100].

This compound was described as an intermediate in the synthesis of terpyridine-derived ligands that contain amino-pyrazine moities and its zinc complex. This kind of ligand allowed the interesting formation of well-defined double hydrogen bonds by self-complementarity. As a result, co-crystallisation of two different metal complexes of these terpyridine-like ligands generated an alternating sequence of the two metals in infinite two-dimensional sheets in a chess-board-like manner (Fig. 12).

The second example showed that 2-chloropyrazine derivatives could be converted to the corresponding organostannes using a palladium catalyzed Stille reaction with hexamethyldistannane (Scheme 61) [101, 102]. These reactions were carried out in the presence of lithium chloride and 2,6-di-*tert*-butyl-4-methylphenol (BHT).

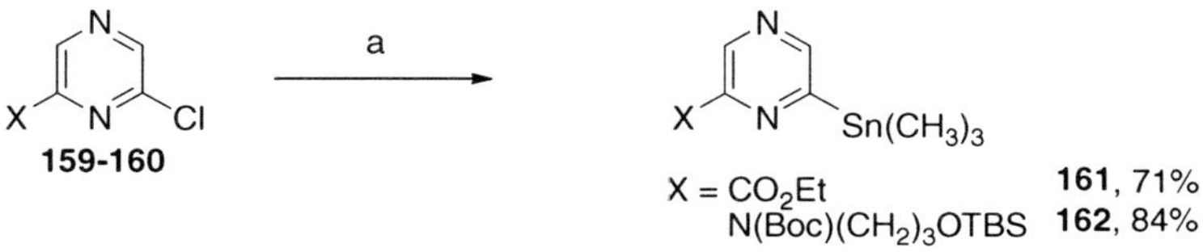

Scheme 61 (**a**) 1.1–2 equiv Me_6Sn_2, 5 mol% $Pd(PPh_3)_4$, 3 equiv LiCl, 4 mol% BHT, 1,4-dioxane, 90–100°C, 1.5–4 h [101, 102]

These two compounds were engaged in the synthesis of pyrazine-pyridine biheteroaryl scaffolds as a novel series of potent vascular endothelial growth factor receptor-2 inhibitor (VEGFR-2). The precise function of VEGFR-2 is to mediate the vascular endothelial cell mitogenesis and survival, as well as angiogenesis. Its activity warranted the development of potent inhibitors for anti-angiogenisis therapy [101, 102].

7 Conclusions

The methods described above, affording metalated pyrazines and quinoxalines, provided an efficient synthetic tool to access a great variety of building blocks and, therefore, to more elaborate structures such as natural products with a pyrazine or a quinoxaline ring in their backbone. That highlights the key role played by the functionalization of these heterocycles, using metalation and cross-coupling reactions. If the lithiated species, obtained with mono metallic lithium amides have been historically first developed and widely used, a new series of bimetallic alkali-metal complexes (known as "ate complexes") have recently emerged opening new perspectives in synthesis in this area due to their particular reactivity. Besides the methods of dehydrometalation mentioned and leading to metalated heterocycles, other methods such as dehalometalation, halogen–metal exchange, and oxidative additions have been reported. The transmetalation reactions combined with other reactions such as cross-couplings are also of great interest and constitute a powerful synthetic tool to access a wide range of structures. Undoubtedly, the formation and the use of metalated pyrazines and quinoxalines will keep an interesting challenge and will show a continuous flow of applications in the next years making them an indispensable synthetic tool.

References

1. Maga JA, Sizer CE (1973) J Agric Food Chem 21:22
2. Allen MS, Lacey MJ, Boyd S (1994) J Agric Food Chem 42:1734
3. Katritzky AR, Rees CW, Scriven EFV (1996) Comprehensive heterocyclic chemistry II. Pergamon, Oxford
4. Gribble G, Joule J (2007) Progress in heterocyclic chemistry, vol 18. Elsevier, Oxford

5. Weissleder R, Pittet MJ (2008) Nature 452:580
6. Vance E (2008) Nature 455:726
7. Duran R, Zubia E, Ortega MJ, Naranjo S, Salva J (1999) Tetrahedron 55:13225
8. Garg NK, Stolz BM (2005) Tetrahedron Lett 46:2423
9. Hirsh AJ, Molino BF, Zhang J, Astakhova N, Geiss WB, Sargent BJ, Swenson BD, Usyatinsky A, Wyle MJ, Boucher RC, Smith RT, Zamurs A, Johnson MR (2006) J Med Chem 49:4098
10. Dell A, William DH, Morris HR, Smith GA, Feeney J, Roberts GCK (1975) J Am Chem Soc 97:2497
11. Bailly C, Echepare S, Gago F, Waring MJ (1999) Anticancer Drug Des 14:291
12. Sato S, Shiratori O, Katagiri K (1967) J Antibiot 20:270
13. Waring MJ, Wakelin LPG (1974) Nature 252:653
14. Waring MJ, Wakelin LPG (1976) Biochem J 157:721
15. Addess KJ, Feigon J (1994) J Biochem 33:12397
16. Dailey S, Feast JW, Peace RJ, Sage IC, Till S, Wood ELJ (2001) Mater Chem 11:2238
17. O'Brien D, Weaver MS, Lidzey DG, Bradley DDC (1996) Appl Phys Lett 69:881
18. Patidar AK, Jeyakandan M, Mobiya AK, Selman G (2011) Int J Pharm Tech Res 3:386
19. Cheeseman GWH, Werstiuk ESG (1972) Advances in heterocyclic chemistry, vol 14. Academic, New York, p 99
20. Barlin B (1982) The chemistry of heterocyclic compounds, vol 41. Wiley, New York
21. Büchi G, Galindo J (1991) J Org Chem 56:2605
22. Porter AEA (1984) In: Katritzky AR, Rees CW (eds) Comprehensive heterocyclic chemistry. Pergamon, Oxford, pp 157–197
23. Woo GHC, Snyder JK, Wan ZK (2002) In: Gribble GW, Gilchrist TL (eds) Progress in heterocyclic chemistry, vol 14. Pergamon, Oxford, pp 279–284
24. Sato N (1996) In: Katritzky AR, Rees CW, Scriven EFV (eds) Comprehensive heterocyclic chemistry II, vol 6. Pergamon, Oxford, pp 233–278
25. Cheeseman GWH, Werstiuk ESG (1978) Adv Heterocycl Chem 22:367
26. Wiberg KB, Nakaji D, Breneman CM (1989) J Am Chem Soc 111:4178
27. Bird CW (1997) Tetrahedron 53:13111
28. Mandado M, Otero N, Mosquera RA (2006) Tetrahedron 62:12204
29. Wang Y, I-Chia Wu J, Li Q, von Ragué Schleyer P (2010) Org Lett 12:4824
30. D'Auria M (2005) Tetrahedron 46:6333
31. Kaim W (1983) Angew Chem Int Ed 22:171
32. Turck A, Plé N, Tallon V, Quéguiner G (1993) J Heterocycl Chem 30:1491
33. Turck A, Plé N, Mongin F, Quéguiner G (2001) Tetrahedron 57:4489
34. Chinchilla R, Nájera C, Yus M (2004) Chem Rev 104:2667
35. Shen K, Fu Y, Li JN, Gui QX (2007) Tetrahedron 63:1568
36. Beak P, Meyers AI (1986) Acc Chem Res 19:356
37. Breit B (2000) Chem Eur J 6:1519
38. Whisler MC, Mac Neil S, Snieckus V, Beak P (2004) Angew Chem Int Ed 43:2206
39. Quéguiner G, Marsais F, Snieckus V, Epsztajn J (1991) Adv Heterocycl Chem 52:187
40. Godard A, Turck A, Plé N, Marsais F, Quéguiner G (1993) Trends Heterocycl Chem 3:16
41. Turck A, Plé N, Quéguiner G (1994) Tetrahedron 37:2149
42. Plé N, Turck A, Couture K, Quéguiner G (1995) J Org Chem 60:3781
43. Quéguiner G (2000) J Heterocycl Chem 37:615
44. Seggio A, Chevallier F, Vaultier M, Mongin F (2007) J Org Chem 72:6602
45. Chevallier F, Mongin F (2008) Chem Soc Rev 37:595
46. Turck A, Mojovic L, Quéguiner G (1988) Synthesis 11:881
47. Turck A, Plé N, Quéguiner G (1994) Heterocyles 37:214
48. Decrane L, Plé N, Turck A (2005) J Heterocycl Chem 42:509
49. Plé N, Turck A, Heynderickx A, Quéguiner G (1998) Tetrahedron 54:4899
50. Chartoine A, Comoy C, Fort Y (2011) Org Biomol Chem 9:1839

51. Toudic F, Plé N, Turck A, Quéguiner G (2002) Tetrahedron 58:283
52. Toudic F, Heynderickx A, Plé N, Turck A, Quéguiner G (2003) Tetrahedron 59:6375
53. Fruit C, Turck A, Plé N, Mojovic L, Quéguiner G (2001) Tetrahedron 57:9429
54. Fruit C, Turck A, Plé N, Quéguiner G (1999) Heterocycles 51:2349
55. Fruit C, Turck A, Plé N, Quéguiner G (2002) J Heterocycl Chem 39:1077
56. Berghian C, Darabantu M, Turck A, Plé N (2005) Tetrahedron 61:9637
57. Zhang CY, Tour JM (1999) J Am Chem Soc 121:8783
58. Liu W, Wise DS, Townsend LB (2001) J Org Chem 66:4783
59. Toudic F, Turck A, Plé N, Quéguiner G, Darabantu M, Lequeux T, Pommelet JC (2003) J Heterocycl Chem 40:855
60. Ward JS, Merritt L (1991) J Heterocycl Chem 28:765
61. Le Fur N, Mojovic L, PLé N, Turck A, Marsais F (2005) Tetrahedron 61:8924
62. Gautheron-Chapoulaud V, Salliot I, Plé N, Turck A, Quéguiner G (1999) Tetrahedron 55:5389
63. Mulvey RE, Mongin F, Uchiyama A, Kondo Y (2007) Angew Chem Int Ed 46:3802
64. Mulvey RE (2009) Acc Chem Res 42:743
65. Krasovkiy A, Krasovkaya V, Knochel P (2006) Angew Chem Int Ed 45:2958
66. Garcia-Alvarez P, Graham DV, Hevia E, Kennedy AR, Klett J, Mulvey RE, O'Hara CT, Weatherstone S (2008) Angew Chem Int Ed 47:8079
67. Morin M, Bresser T, Knochel P (2009) Org Lett 11:3406
68. Jaric M, Haag BA, Unsinn A, Karaghiosoff K, Knochel P (2010) Angew Chem Int Ed 49:5451
69. Dong ZB, Clososki GC, Wunderlich SH, Unsinn A, Li J, Knochel P (2009) Chem Eur J 15:457
70. Dong ZB, Zhu WH, Zhang ZG, Li MZ (2010) J Organomet Chem 695:775
71. Rohbogner CJ, Wirth S, Knochel P (2010) Org Lett 12:1984
72. L'Helgoual'ch JM, Bentabed-Ababsa G, Chevallier F, Derdour A, Mongin F (2008) Synthesis 24:4033
73. Wunderlich SH, Knochel P (2007) Angew Chem Int Ed 46:7685
74. Mosrin M, Knochel P (2009) Org Lett 11:1837
75. Mosrin M, Monzon G, Bresser T, Knochel P (2009) Chem Commun 5615
76. Wunderlich SH, Rohbogner CJ, Unsinn A, Knochel P (2010) Org Process Res Dev 14:339
77. Baillie SE, Blair VL, Blakemore DC, Hay D, Kennedy AR, Pryde DC, Hevia E (2012) Chem Commun 48:1985
78. Blair VL, Blakemore DC, Hay D, Hevia E, Pryde DC (2011) Tetrahedron Lett 52:4590
79. Mongin F, Uchiyama M (2011) Curr Org Chem 15:2340
80. Snégaroff K, L'Helgoual'ch JM, Bentabed-Ababsa G, Nguyen TT, Chevallier F, Yonehara M, Uchiyama M, Derdour A, Mongin F (2009) Chem Eur J 15:10280
81. Bentabed-Ababsa G, Ely SCS, Hesse S, Nassar E, Chevallier F, Nguyen TT, Derdour A, Mongin F (2010) J Org Chem 75:839
82. L'Helgoual'ch JM, Bentabed-Ababsa G, Chevallier F, Yonehara M, Uchiyama M, Derdour A, Mongin F (2008) Chem Commun 5375
83. Jeganmohan M, Knochel P (2010) Angew Chem Int Ed 49:8520
84. Leprêtre A, Turck A, Plé N, Knochel P, Quéguiner G (2000) Tetrahedron 56:265
85. Buron F, Plé N, Turck A, Marsais F (2006) Synlett 10:1586
86. Schnürch M, Spina M, Khan AF, Mihovilovic MD, Stanetty P (2007) Chem Soc Rev 36:1046
87. Nora de Souza MV (2007) Curr Org Chem 11:637
88. Leprêtre A, Turck A, Plé N, Quéguiner G (2000) Tetrahedron 56:3709
89. Sugimoto O, Sudo M, Tanji K-I (2001) Tetrahedron 57:2133
90. Hirasawa N, Takahashi Y, Fukuda E, Sugimoto O, Tanji K-I (2008) Tetrahedron Lett 49:1492
91. Hebbar N, Ramondenc Y, Plé G, Dupas G, Plé N (2009) Tetrahedron 65:4190
92. Hebbar N, Fiol-Petit C, Ramondenc Y, Plé G, Plé N (2011) Tetrahedron 67:2287

93. Buron F, Plé N, Turck A, Quéguiner G (2005) J Org Chem 70:2616
94. Saito R, Tokita M, Uda K, Ishikawa C, Satoh M (2009) Tetrahedron 65:3019
95. Sugimoto O, Yamada S, Tanji K-I (2003) J Org Chem 68:2054
96. Gohlke H, Schwarz S, Gündisch D, Tilotta MC, Weber A, Wegge T, Seitz G (2005) J Med Chem 46:2031
97. Handratta VD, Vsaitis TS, Njar VCO, Gediya LK, Kataria R, Chopra P, Newman D Jr, Farquhar R, Guo Z, Qiu Y, Brodie AMH (2005) J Med Chem 48:2972
98. Hirschhäuser C, Haseler CA, Gallagher T (2011) Angew Chem Int Ed 50:5162
99. Maindron N, Poupart S, Hamon M, Langlois JB, Plé N, Jean L, Romieu A, Renard PY (2011) Org Biomol Chem 9:2357
100. Ziener U, Breuning E, Lehn JM, Wegelius E, Rissanen K, Baum G, Fenske D, Vaughan G (2000) Chem Eur J 6:4132
101. Kuo GH, DeAngelis A, Emanuel S, Wang A, Zhang Y, Connolly PJ, Chen X, Gruninger RH, Rugg C, Fuentes-Pesquera A, Middleton SA, Jolliffe L, Murray WV (2005) J Med Chem 48:4535
102. Kuo GH, Wang A, Emanuel S, DeAngelis A, Zhang R, Connolly PJ, Murray WV, Gruninger RH, Sechler J, Fuentes-Pesquera A, Johnson D, Middleton SA, Jolliffe L, Chen X (2005) J Med Chem 48:1886

Top Heterocycl Chem (2013) 31: 171–222
DOI: 10.1007/7081_2012_96

Published online: 16 March 2013

Reactions of Pyridines, Benzopyridines, and Azapyridines with Organomagnesiums and Organolithiums

Manfred Schlosser

Abstract A survey is given on how simple pyridines can be elaborated into substituted and functionalized structures based on organometallic chemistry. "Organometallics" in the given context means "polar" reagents, i.e. those containing organoalkali or organoalkaline-earth metals, in particular, organolithiums and organomagnesiums. For comparison, non-organometallic methods are summarized at the beginning and transition-metal chemistry at the very end. The review is critical. Emphasis is placed on the practicability and superiority of a method; novelty alone is not a sufficient criterion for coverage.

Keywords Additions · Eliminations · Isomerizations · Pyridines · Pyrimidines · Quinolines · Substitutions

Contents

M. Schlosser (✉)
Swiss Federal Institute of Technology, 1015 Lausanne, Switzerland
e-mail: manfred.schlosser@epfl.ch

Abbreviations and Notes

Ar	Argon
BOC	*tert*-Butoxycarbonyl
DABCO	1,8-Diazabicyclo[2.2.2]octane
DEE	Diethyl ether
El	Electrophilic part of a reagent *El*-X
El-X	Electrophilic reagent
HEX	Hexanes (or petroleum ether of bp ~65°C)
LIC-KOR	Superbasic 1:1 mixture of *n*-butyllithium and potassium *tert*-butoxide
LIDA	Lithium diisopropylamide
LIDMAE	Lithium 2-(dimethylamino)ethoxide
LIM-KOR	Superbasic 1:1 mixture of methyllithium and potassium *tert*-butoxide
LIPIP	Lithium piperid-1-ide (*N*-piperidyllithium)
LIS-KOR	Superbasic 1:1 mixture of *sec*-butyllithium and potassium *tert*-butoxide
LIT-KOR	Superbasic 1:1 mixture of *tert*-butyllithium and potassium *tert*-butoxide
LITMP	Lithium 2,2,6,6-tetramethylpiperid-1-ide (tetramethylpiperid-1-yllithium)
M-*Nu*	Nucleophilic reagent (M=metal, e.g., Li)
Nu	An amine or the nucleophilic part of a reagent
PMDTA	*N,N,N′,N″,N″*-pentamethyldiethylenetriamine
R	Alkyl or Aryl
spec.	Specifically (on this or that page)
Sv	Solvent
THF	Tetrahydrofuran
TMEDA	*N,N′*-tetramethylethylenediamine
+25°C	Ambient temperature (in general between 23°C and 26°C)
−75°C	Average temperature of dry ice/solvent baths
n-LiC_4H_9	The Old-German *n*-alkyl is perhaps the only unforgiveable sin one can commit against modern nomenclature. However, one does not find any butyllithium in the catalogs of chemicals suppliers, just *n*- or *sec*- or *tert*-butyllithium. To avoid confusion among our readers, we also specify *n*-butyllithium whenever appropriate. This violation of rules is of course not extended to any other compound, be it butyl chloride or hexyllithium.
inert gas	Reactions performed on a small scale (<25 mmol) may be run under an atmosphere of 99.999% pure nitrogen to minimize losses caused by moisture or oxygen. Such protection is not required for preparative scale reactions unless they last many hours.

1 Introduction

Not so many classes of organic compounds can rival with the prominence of pyridine and its congeners as structural motifs in natural and man-made products [1–6]. Unlike saturated six-membered (dihydropyridines, piperidine) or five-membered (pyrrolidine, furan) heterocycles, pyridines are 6π-aromatic [7] and hence are particularly stable molecular entities. Moreover, they display an unusually diversified reactivity profile.

The scope of this chapter restricts itself first of all to the parent compound pyridine and its substituted derivatives. Reactions of benzopyridines (quinolines, isoquinolines, acridines, phenanthridines) and azapyridines (pyrazidines, pyrimidines, pyrazines, cinnolines, quinazolines, quinoxalines, naphthyridines) will nevertheless be treated sporadically. However, "lateral" (exocyclic) deprotonations [8] of alkylpyridines (e.g., picolines) or even more remote ones (e.g., of trimethyl-2-pyridylsilane [9]) are not at all taken into account.

The reagents under inspection are organomagnesiums, organolithiums and, in addition, "mixed-metal" organoalkalis such as the superbasic 1:1 mixture of *n*-butyllithium and potassium *tert*-butoxide ("LIC-KOR" or Schlosser–Lím base). On the other hand, all transition element-mediated C–C coupling processes are disregarded.

2 Non-organometallic Reactions of Pyridines

Before turning to organometallic chemistry, the other pathways leading to pyridine congeners will be briefly reviewed. This knowledge will familiarize us with some reactivity patterns characteristic for this heterocyclic class and will prove indispensible for the preparation of the precursors to metal-bearing pyridines. Moreover, it provides a sound basis to compare the effectiveness of long established and more recent methods of synthesis and to select the most appropriate one in a given case.

Pyridines belong to the family of "electron-deficient" heterocycles. Their propensity for nucleophiles is therefore hardly surprising. But they also react with powerful electrophiles and even with carboradicals. Being rather rare, the latter kind of reactions will be examined first.

2.1 Radical and Radical-Anion Reactions

Carboradicals are known to add rather sluggishly to pyridines and other *N*-heterocycles, but quite readily—at least a thousandfold faster—to the 2- and 4-positions of pyridinium salts or other protonated heterocycles. Such reactions can be advantageously performed in aqueous media. The biphasic water/chlorobenzene mixture proves particularly suitable if the alkyl radical (e.g., methyl, butyl, cyclohexyl, *tert*-butyl) is generated from the corresponding carboxylic acid by oxidative decarboxylation) [10, 11]. Under such conditions, dialkylation can be minimized or completely avoided. γ-Picoline, 4-ethylpyridine, 4-cyanopyridine, pyrazine, quinoline, and quinoxaline were investigated as model substrates in great detail [10, 11].

Radicals are inevitably involved as transient species in the generation of organoalkali or alkali-earth intermediates by the insertion of the metal into a carbon–halogen (or also carbon–oxygen) bond [12]. It is advisable to wait until the reagent preparation is definitely finished before adding the heterocyclic reaction partner. Otherwise it may be difficult to tell whether the products originate from a radical or nucleophilic process. This ambiguity is illustrated by the reaction of 3-pyrid-3-ylpropyl chloride with magnesium. At ambient temperature and in the presence of a metal excess, in other words under radical favorable conditions, 3-propylpyridine, 3,4- and 2,3-cyclopentenopyridine are identified in a 7:2:1 ratio after neutralization and in a 3:2:5 ratio, if the organomagnesium solution is prepared first before being decanted from the unconsumed metal, diluted and heated to reflux [13]. The former conditions should be relatively favorable for radical pathways, the latter rather for organometallic reactivity. The preference for either one of the cyclization products depending on the experimental details suggests a coexistence of the two mechanistic options (Scheme 1).

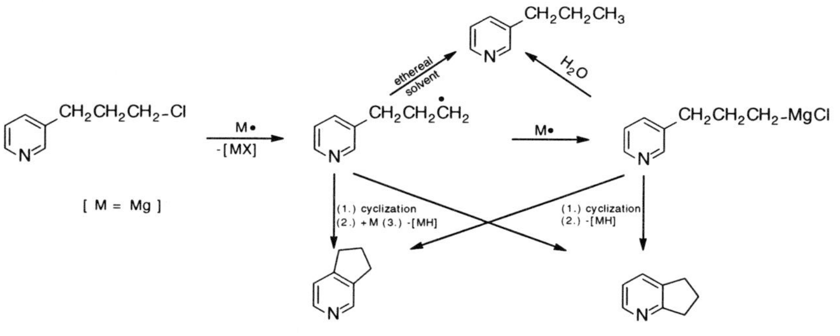

Scheme 1 Competing radical and organometallic pathways upon treatment of an organic halide with an alkali or alkali-earth metal

Pyridine undergoes dehydrogenative dimerization to afford 2,2′-bipyridyl when treated with lithium diisopropylamide (LIDA) in hexamethylphosphoric triamide (HMPA) [14]. Although one may feel tempted to invoke as the key steps of this transformation the lithiation of pyridine at its 2-position and the addition of this intermediate onto intact pyridine (see Sect. 3.1), a single electron-transfer (SET) mechanism [12] is far more plausible. At the beginning a lithium/pyridine 1:1 adduct is formed that dimerizes to give bis(1,2-dihydropyrid-2-yllithium). The latter spontaneously ejects lithium hydride, thus rearomatizing to 2,2′-bipyridyl (Scheme 2). The yellow sodium/pyridine 1:1 adduct is spectroscopically observable below 0°C. At +25°C, 4,4′-C,C-connection and sodium hydride elimination occur [15, 16]. The resulting 4,4′-bipyridyl combines with unconsumed sodium to produce a deep-blue "radical-anion" 1:1 adduct (Scheme 2) [15, 16].

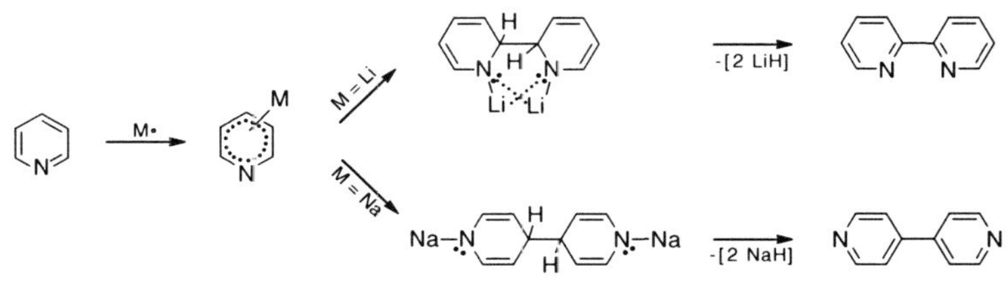

Scheme 2 Dimerization and subsequent rearomatization of lithium/pyridine and sodium/pyridine "radical anions"

2.2 *Electrophiles and Oxidants as Reactants*

Both electrophilic and oxidizing reagents are electron-deficient. Therefore it is reasonable to cover them in the same section (Sect. 2.2) in the same way as we shall practice this later with nucleophilic and reducing agents (Sect. 3.2).

2.2.1 Electrophilic Heteroaromatic Substitution

Electrophilic substitutions [17] require activation by acids and consequently act on pyridinium ion rather than on the neutral pyridine [18]. Its rate of reaction is hence inferior to that of the carba-analogous benzene by estimated 22 powers of ten [19] and hence even to that of nitrobenzene. To improve the practicability of pyridine nitration several modifications of the ordinary protocol have been advertized. They rely either on the assistance of removable electron-donor auxiliary groups [20] or on the replacement of the standard nitric and sulfuric acid mixtures by dinitrogen pentoxide [21–23].

Pyridine undergoes bromination with molecular bromine in fuming sulfuric acid at 130°C to give a mixture of 3-bromo- and 3,5-dibromopyridine in the course of 8 h [24, 25] and nitration with fuming nitric acid and oleum at 300°C [26, 27] mainly at the 3-position. In contrast, pyridine *N*-oxide incorporates the nitro functionality at the 4-position (Scheme 3) and under much milder conditions (90°C) [28]. Nitro groups can be readily reduced to amino groups. The latter are synthetically versatile entities, as they can be easily subjected to dediazotation–protonation or dediazotation–halogenations and can be smoothly oxidized to nitro compounds (using, e.g., hydrogen peroxide in acid medium), as exemplified with 2-aminopyridine (Scheme 3) [29–31]. 2-Aminopyridine is directly accessible from pyridine by the Tschitschibabin amination reaction.

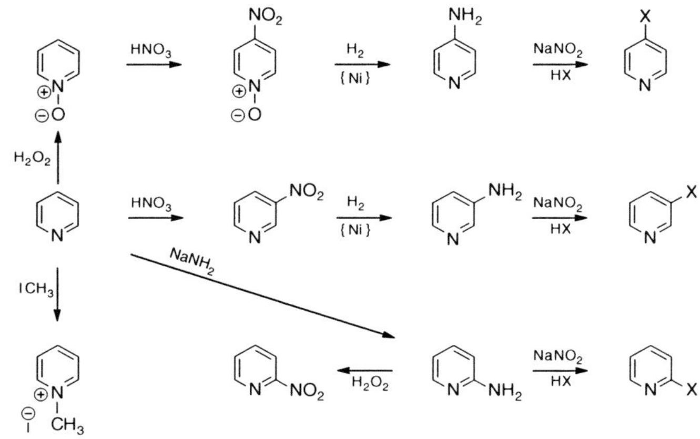

Scheme 3 Oxidation, quaternization, and nitration of pyridine, reduction of nitropyridines to aminopyridines and nucleophilic NH_2/X displacement [X = F, Cl, Br, I, OH, etc.]

Raising the bromination temperature to 300°C spoils both the regioselectivity and the product homogeneity by affording large amounts of 2,3-, 2,5-, 2,6-, 3,4-, and 3,5-dibromopyridines as well as 2,3,5- and 3,4,5-tribromopyridines [32]. At 500°C the 2- and 6-positions react almost exclusively [33]. The chlorination of pyridine gives rise to mainly 3-chloro- and 3,5-dichloropyridine at 170°C but mostly 2-chloro- and 2,6-chloropyridine are obtained in the range of 270–420°C [34]. When the nitration temperature is increased beyond 300°C more and more of the 2-nitropyridine isomer is formed [28]. At elevated temperatures electrophilic substitutions become reversible. Moreover, radical or radical-chain processes are launched at the expense of electrophilic ones. As we know meanwhile (Sect. 2.1) they deliver heterosubstituents preferentially to the nitrogen-neighboring positions (Scheme 4).

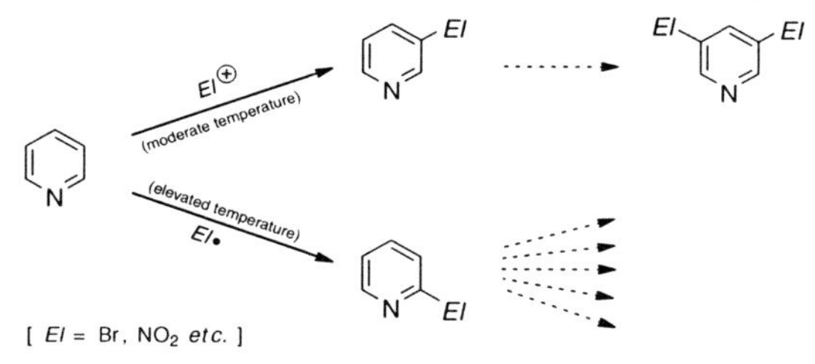

Scheme 4 Electrophilic and radical-mediated substitutions exhibit different regioselectivities

Electron-rich substituents facilitate electrophilic reactions of course. Potent donors generally are *para* directing. Thus 2-aminopyridine engages in bromination [35] and nitration [36] selectively at the 5-position (Scheme 5). Depending on the reaction conditions, small amounts of the 2-amino-3-nitropyridine isomer may nevertheless be formed [36, 37]. If this is the case, it can be easily removed by water vapor distillation due to its much higher volatility (caused by intramolecular rather than intermolecular hydrogen bonding). 4-Aminopyridine shows again selective bromination [38] and nitration [39–41], this time at the 3-position. In contrast, 3-aminopyridine reacts also readily but affords 5-amino-2-bromopyridine (6%) and 3-amino-2,6-dibromopyridine (20%) along with the main component 3-amino-2-bromopyridine (43%) [42]. Similarly 3-amino-2-nitropyridine is the major, though not sole product upon nitration [39].

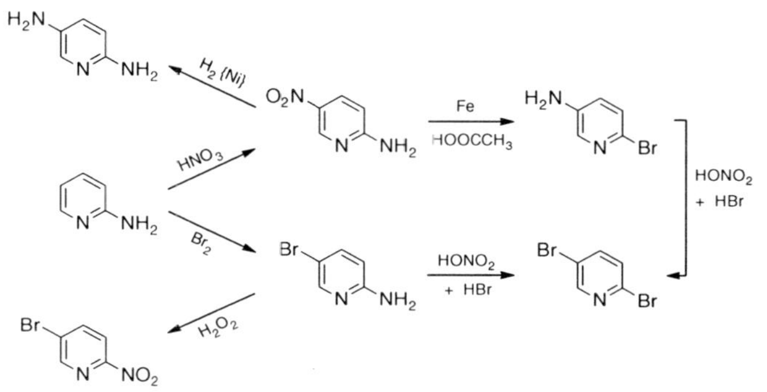

Scheme 5 The regioselective electrophilic substitution of 2-aminopyridine opens opportunities for further elaboration

The *N*-oxidation of pyridines is a well-known and widely exploited option (Scheme 3). The resulting pyridine 1-oxides have a characteristic reactivity pattern of their own (Sects. 3.3 and 4.2.6). More vigorous oxidants such as elemental fluorine in aqueous medium promote 2-hydroxylation (e.g., 51% of 6-hydroxy-2-picolinic acid and 62% of 2-hydroxy-4-picolinic acid from 2- and 4-picolinic acid [43]) and may even trigger under somewhat different conditions the oxidative removal of 2-carboxy groups from 2-pyridinecarboxylic acids 1-oxides (e.g., 81% of 6-acetoxypyridine-3-carboxylic acid by treating pyridine-2,5-dicarboxylic acid 1-oxide for 1 h at 30°C with acetic acid anhydride and triethylamine [44]). Similar decarboxylative chlorinations and aminations have been claimed too [45, 46]. As such reactions are carried out in the presence of base, generation of *N*-ylides by deprotonation of the 2-position is frequently assumed [47], although other intermediates deem more plausible (Scheme 6).

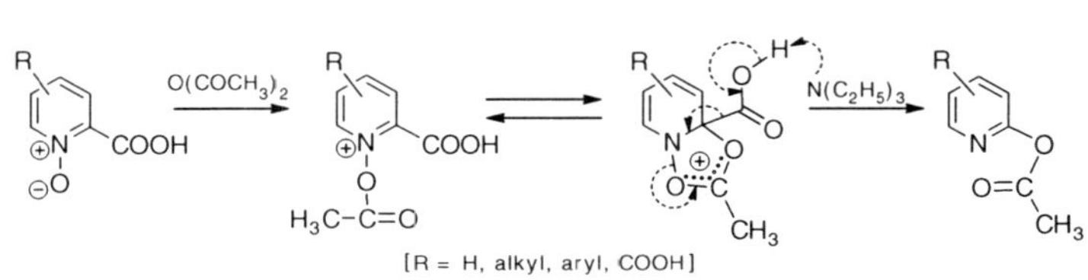

Scheme 6 Oxidative decarboxylation of 2-picolinic acids via their *N*-oxides

Oxidation of ring positions can be accomplished quite readily with potassium cyanoferrate(III) after quarternation of the pyridine nitrogen. Thus, the *N*-methylpyridinium ion can be smoothly converted into 1,2-dihydro-2-hydroxy-1-methylpyridine and subsequently into 1-methyl-2-pyridinone [48].

2.3 Nucleophiles and Reducing Agents as Reactands

The venerable Tschitschibabin amination reaction [49, 50] is already mentioned in a preceding section (Sect. 2.2.1, Scheme 3). Sodium amide adds nucleophilically onto the 2-position of pyridine heated to about 150°C in an inert hydrocarbon medium. Spontaneous elimination of sodium hydride produces 2-aminopyridine (which is deprotonated by the base prior to neutralization). Applying a large excess of the reagent and prolonged reaction times a second amino group can be introduced into the 6-position. Only when both nitrogen-adjacent positions are occupied the nucleophilic attack can be reoriented to the 4-position.

The harsh reaction conditions seriously limit the scope of the method. In particular, base-sensitive functional groups are not tolerated. On the other hand, the reaction can be extended to di-, tri-, and tetraazines if carried out in the presence of potassium permanganate [51]. The intermediacy of the amide adduct prior to hydride expulsion has been established in the case of quinoline [52, 53].

2.3.1 Nucleophilic Additions

The combined action of benzoyl chloride and potassium amide on pyridine *N*-oxide gives rise to a relatively stable 1,2-adduct that upon heating loses benzoic acid and rearomatizes to 2-cyanopyridine [54, 55]. When pyridine is simultaneously treated with ethyl chloroformate and trimethylsilyl cyanide [56] or diethylaluminum cyanide [57], respectively, 1-ethoxycarbonyl-2-cyano-1,2-dihydropyridine and 1-ethoxycarbonyl-4-cyano-1,4-dihydropyridine are formed (Scheme 7). Such so-called Reissert compounds [58] can serve as versatile intermediates for organic synthesis.

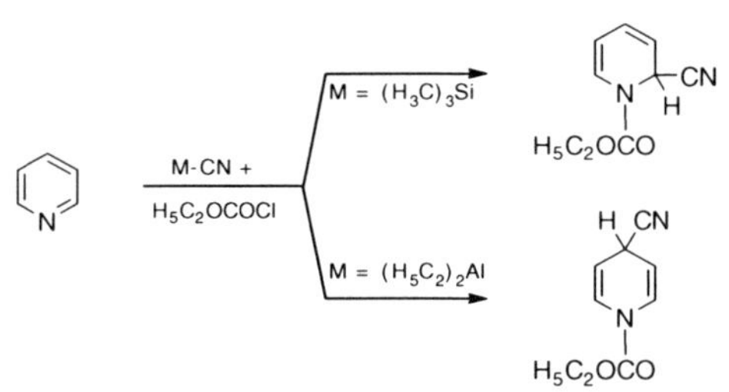

Scheme 7 Regioisomeric Reissert compounds by 1,2- or 1,4-addition of ethoxycarbonyl and cyano groups

2.3.2 Nucleophilic Substitutions

Good leaving groups located at the 2-, 4-, and 6-positions of pyridines can be easily displaced. In that way inexpensive chloropyridines can be converted into their bromo- or iodo analogs. Those offer the advantage to react with butyllithium under permutational halogen/metal interconversion. Bromo- and iodotrimethylsilane are the reagents of choice to perform such a halogen/halogen exchange on the laboratory scale (Scheme 8) [59]. *N*-Silylpyridinium salts are the presumed intermediates.

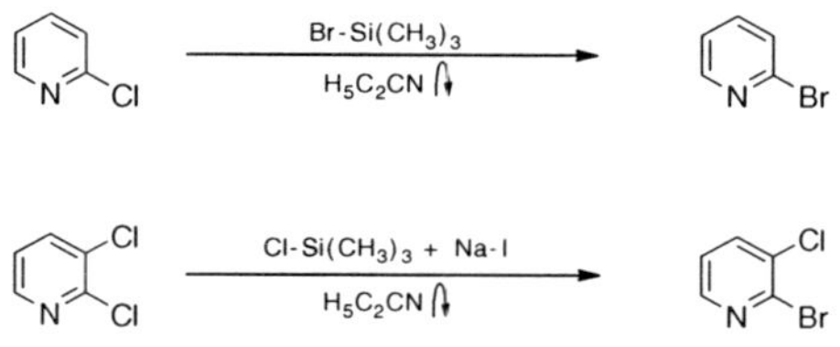

Scheme 8 Chlorine/heavy halogen exchange

When large-scale operations are projected, the use of hydrobromic or hydroiodic acid is more economical. Hydrofluoric acid is generally a poor exchange promoter. Optimized conditions are required to achieve satisfactory results (e.g., 76% of 2-fluoropyridine from 2-chloropyridine at 150°C and 97% of 2-fluoropyrimidine from 2-chloropyrimidine at 50°C [60]).

Fluoropyridines are starting materials much sought after in pharmaceutical and agricultural research. Treating 2- or 4-chloro- and -bromopyridines with spray-dried potassium fluoride in the presence of solubilizing solvents and auxiliaries provides the fluorinated products in yields averaging 70% (Scheme 9) [61]. Catalytic hydrogenation or treatment with zinc removes chlorine and heavier halogens in the presence of fluorine atoms without touching the latter [61].

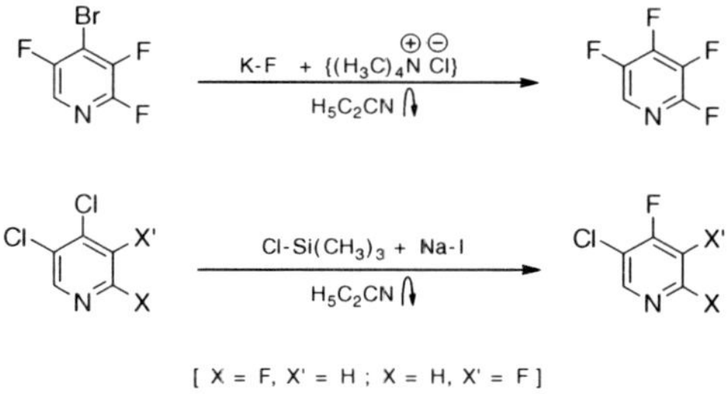

Scheme 9 Bromide/fluoride and chloride/fluoride displacement at the 4-position of pyridine

When two or more identical substituents are available for nucleophilic displacement, the 2-position often gains a lead over the 4-position (Scheme 10) [62]. Conversely, the 3-positions never can really enter into competition. In pyrimidines the 4- and 6-positions are generally distinctly more reactive than the 2-positions, let alone the 5-position [63].

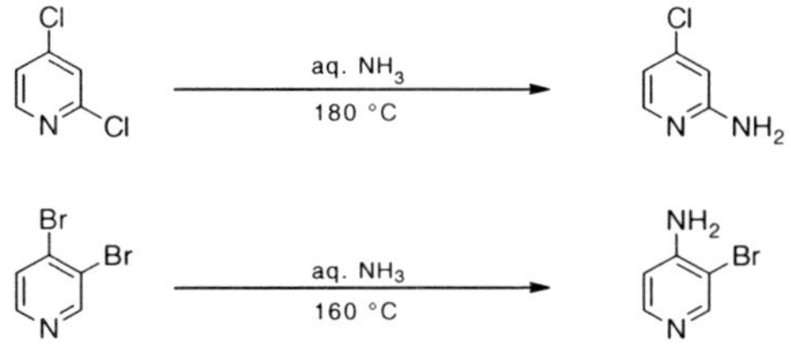

Scheme 10 Ammonia displacing halogen from the 2-position of 2,4-dichloropyridine and from the 4-position of 3,4-dibromopyridine

In this context the site indifference of pyridine *N*-oxides is noteworthy. Such substrates undergo nucleophilic substitution not only at the 2- and 4-positions but also at the 3-position (Scheme 11) [64].

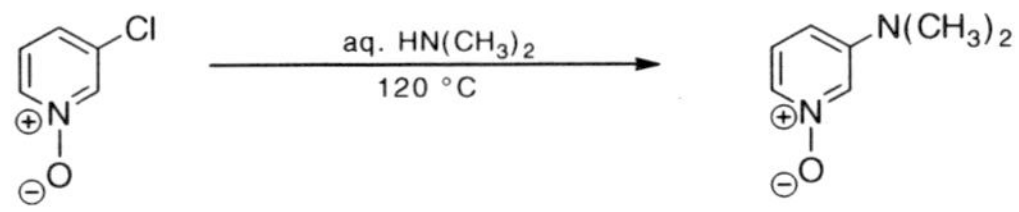

Scheme 11 Substitution of halogen by a dimethylamino group at the 3-position of 3-chloropyridine 1-oxide

When the nucleophilic displacement affords a product mixture or an undesired isomer, regioselectivity can be implemented by exploiting the screening effect of a trialkylsilyl group (the "silyl trick") [65, 66]. For example, 2,4-difluoropyridine reacts with dimethylamine, hydrazine, or other *N*-nucleophiles cleanly at the 4-position, whereas (2,4-difluoropyrid-5-yl)trimethylsilane is selectively substituted at the 2-position (Scheme 12). The silyl steering group can be easily removed to be replaced by hydrogen (desilicoprotonation) or bromine (desilicobromination). The hydrazino group can be replaced by hydrogen when treated with a mild oxidant such as copper (II) sulfate [61].

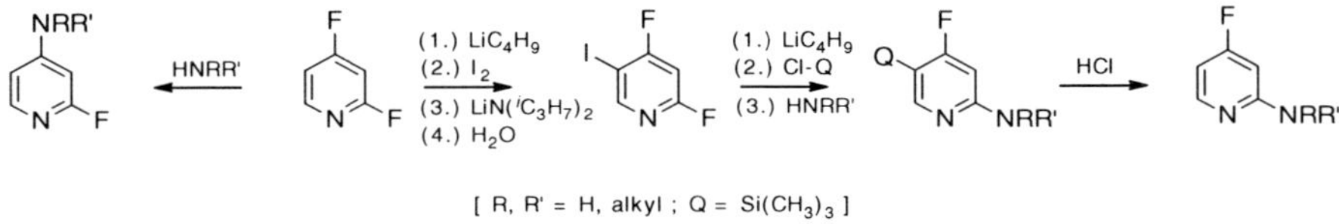

Scheme 12 Optional nucleophilic substitution of halogen from either the 2- or 4-position of 2,4-difluoropyridine

Dediazohalogenations are the most widely applied functional group transformations in the pyridine area. The aminopyridines are treated with hydrochloric, hydrobromic, or hydroiodic acid following closely the procedures elaborated in the arene series. Often Gattermann–Sandmeyer-like protocols are employed although it has not always been checked whether the addition of copper powder or cuprous salts really improves the yield. Representative examples (Scheme 13) are the

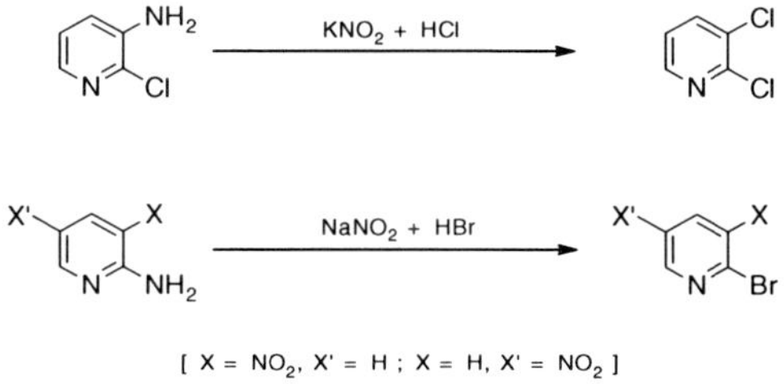

Scheme 13 Typical dediazohalogenations

preparation of 2-chloropyridine [67] and 2-bromopyridine [68–71], 3-chloropyridine [20], and 3-bromopyridine [20] from 2- or 3-aminopyridine, 2,3- [62], 2,4- [72], and 2,5-dichloropyridine [73] and 2,5-dibromopyridine [74] from the corresponding halo-2-aminopyridines or 2-bromo-3- [75] and -5-nitropyridine [76] from 2-amino-3- and -5-nitropyridine, respectively (Scheme 13).

Amino/fluorine displacements are performed according to Balz and Schiemann (i.e., by thermolysis of pyridinediazonium tetrafluoborates [77, 78]) or Osswald and Scherer (i.e., by diazotation in anhydrous hydrogen fluoride [79] or hydrogen fluoride/pyridine complexes [80]). The three monofluoropyridines (Scheme 14)

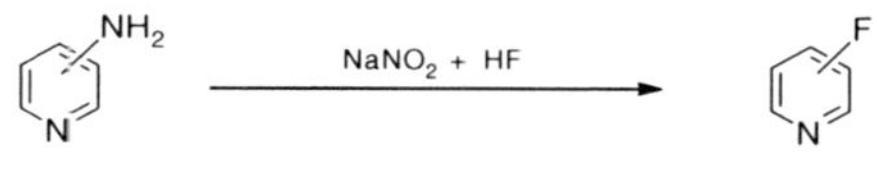

Scheme 14 2-, 3- and 4-Fluoropyridine by dediazotation/halogenation

and numerous other fluorinated heterocycles have been prepared in that way [80–83]. 4-Fluoropyridine is particularly reactive and hence labile upon storage.

In aqueous medium and in the absence of more powerful nucleophiles dediazotation produces phenols and, in the pyridine series, hydroxypyridines or the tautomeric pyrimidinones by *N*/*O* displacement [84, 85]. Thus, for example, 2-amino-5-nitropyridine is converted into 5-nitro-2-pyridinone [86, 87].

N/*N*-Displacements are scarce. Among the few cases known is the reaction of 5-bromo-2-nitropyridine with aqueous ammonia affording 2-amino-5-bromopyridine along with 5-amino-2-nitropyridine [39].

If located at the 2- or 4-position, hydroxy groups can be quite readily displaced by halogen from the pyridine ring. 2- and 4-hydroxypyridines exist predominantly in the tautomeric structure of 2- and 4-pyridinones. However, both the lactam and the azaphenol form can produce the same intermediate required to promote the *O*/*X*(oxygen/halogen)-displacement (Scheme 15). Suitable reagents are phosphorus pentachloride and pentabromide, phosphoryl chloride and bromide or *N*-chloro- and *N*-bromosuccinimide in the presence of triphenylphosphine [79–90]. Analogously 2,4-dihydroxypyridine can be converted into 2,4-dichloro- or 2,4-dibromopyridine [91], 5- and 6-chloro-2-pyridinone into, respectively, 2,5- [92] and 2,6- [93] pyridine and 5-nitro-2-pyridinone [94] into 2-bromo-5-nitropyridine.

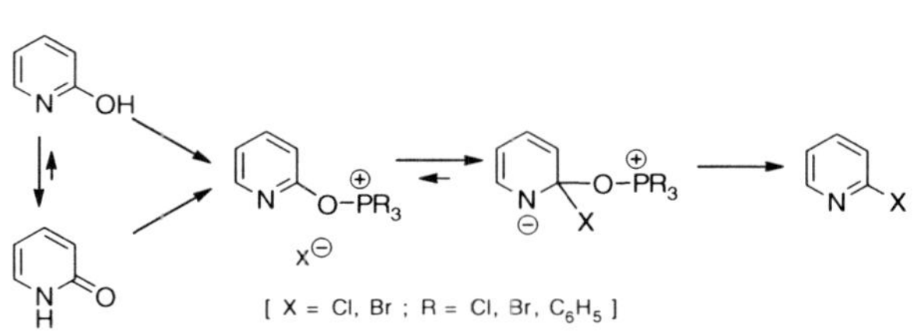

Scheme 15 Conversion of 2- and 4-pyridinones into 2- and 4-halopyridines

4-Nitropyridine 1-oxide reacts with hydrogen bromide and hydrogen chloride at, respectively, 120°C and 160°C to give 4-bromo- and 4-chloropyridine. This straightforward *N*/*X* (nitrogen/halogen)-displacement is an exception rather than the rule. Pyridine *N*-oxides in general undergo site-switching ("deflected") *O*/*X*-displacements when treated with halogenating reagents such as phosphoryl chloride or sulfuryl chloride (Scheme 16) [95, 96].

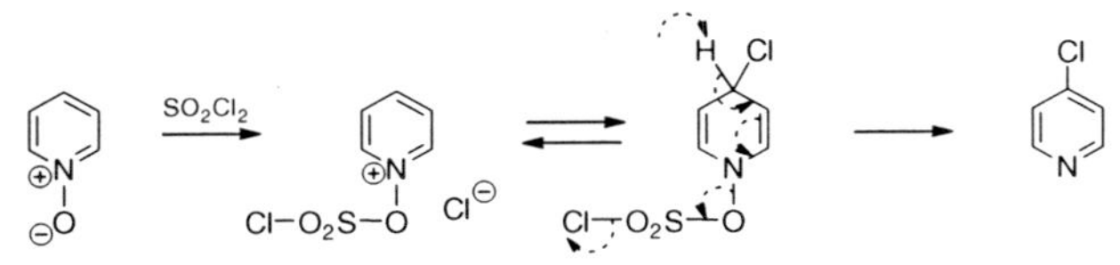

Scheme 16 "Deflected" *O*/*X* displacement upon reaction of pyridine *N*-oxide with sulfuryl chloride

In the same way 2,5-dichloropyridine is obtained with phosphoryl chloride at 110°C from 3-pyridine 1-oxide [97]. The conversion of 3-aminopyridine with a mixture of hydrogen chloride and hydrogen peroxide may well involve a transient *N*-oxide too [98]. However, the exact course taken by such transformations always depends on the exact conditions. 3-Chloro-4-hydroxypyridine 1-oxide, made from 4-hydroxypyridine 1-oxide with sulfuryl chloride [96], reacts with phosphoryl chloride to afford 3,4-dichloropyridine at 100°C and its *N*-oxide at 60°C [96].

The nitro entity is an excellent leaving group that can be displaced readily by halogen. Thus 4-nitropyridine and sulfuryl chloride at 110°C give 2,4-dichloropyridine [27], whereas 2-bromo-4-nitropyridine 1-oxide and acetyl bromide produce 2,4-dibromopyridine at 130°C and the corresponding *N*-oxide at 80°C [99].

The Tschitschibabin dehydroamination, an *H*/*N*-displacement, has already been mentioned at the beginning of this Sect. 2.3). Analogous *O*/*N*-displacements do exist but are hard to find. The conversion of 4-hydroxypyridine (4-pyridinone) with sodium hydroxide into 2,4-dihydroxy-pyridine represents one of these rare examples [100].

2.3.3 Reduction

As seen above, pyridine *N*-oxides often deoxygenize on their own. If the oxygen has to be removed in a discrete step, this can be conveniently done with iron powder in acetic acid [91]. If acid-sensitive substituents do not tolerate this medium, low-valent titanium (made from titanium tetrachloride and lithium aluminum hydride) may be used in tetrahydrofuran instead [101]. 4-Nitropyridine 1-oxide and congeners can be reduced to aminopyridines with iron powder [28, 91], or by catalytic hydrogenation [102] in a single operational step (Scheme 17).

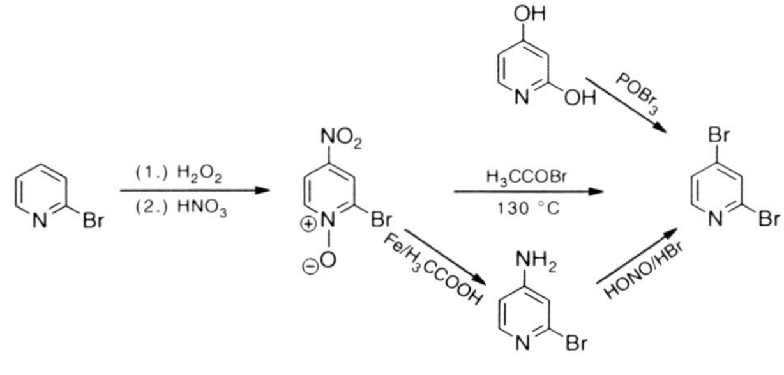

Scheme 17 Pathways leading from 2-bromo-4-nitropyridine 1-oxide to 2,4-disubstituted pyridines

Of course not only nitropyridine *N*-oxides but also nitropyridine themselves can be reduced to the corresponding aminopyridines using a metal (tin foil in hydrochloric acid, iron powder in acetic acid, or zinc in aqueous potassium hydroxide) or catalytic hydrogenation. Examples are the reduction of 2-amino-5-nitropyridine [103] and of 2-bromo-5-nitropyridine [94, 104].

To remove a hydroxy group from a pyridine ring is not as simple. To achieve the C–O bond cleavage the nucleofugality of the prospective leaving group has to be enhanced. The 4-(nonafluorobutane)sulfonyloxy group meets this condition. It is almost quantitatively cutoff from the 3-methoxy-2,6-di(2-thienyl)pyridine nucleus by 1,3-bis(diphenylphosphino)propane/palladium(II) diacetate catalyzed transfer hydrogenation (Scheme 18) [105].

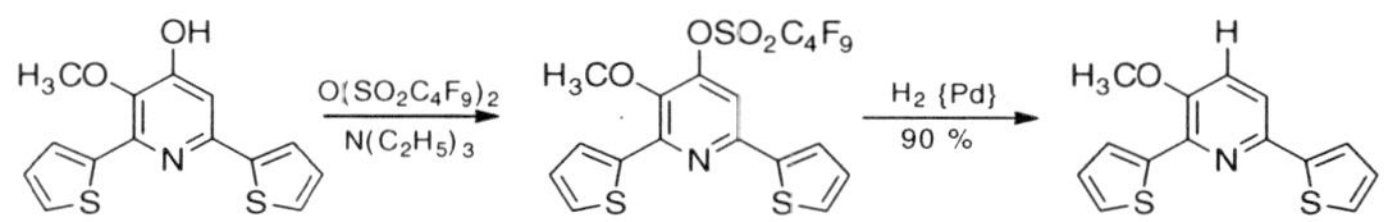

Scheme 18 Hydrogenolytic removal of an activated hydroxy group

A more challenging task than the manipulation of pyridine substituents or the total saturation of the ring to give piperidine (H_2, Raney-Ni, 120°C) is its controlled partial hydrogenation to obtain selectively 1,2- or 1,4-dihydropyridines and 1,2,3,4- or 1,2,3,6-tetrahydropyridines. Lithium aluminum hydride (or other complex hydrides) leads to mixtures of piperidines, dihydropyridines, and ring fission products [106]. However, 1,4-dihydropyridines are readily accessible by the treatment of pyridine with lithium (or sodium) in liquid ammonia and in the presence of ethanol (Scheme 19) [107]. To get rid of their more vulnerable NH-parts the di- and tetrahydropyridines are frequently converted into *N*-trimethylsilyl or *N*-methyl derivatives prior to isolation (Scheme 19) [107, 108].

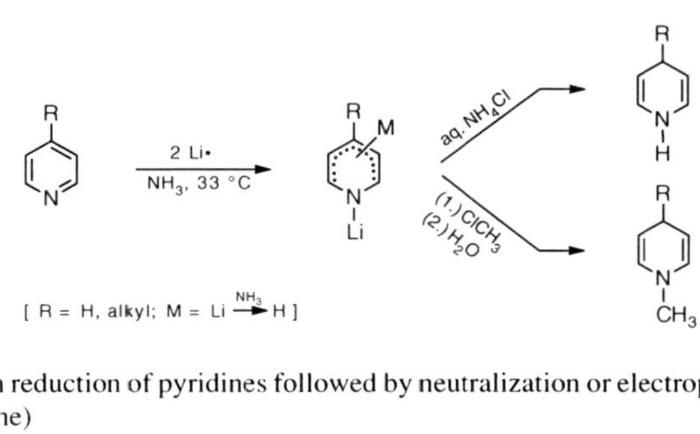

Scheme 19 Birch reduction of pyridines followed by neutralization or electrophilic trapping (e.g., with chloromethane)

The latter transformation suggests an inversion of the order of addition. *N*-Alkylation of the pyridine means to impose a positive charge and thus to boost the reactivity toward nucleophiles and reductands. In fact, 1,4-dimethylpyridinium iodide gives selectively 1,4-dimethyl-1,4-dihydropyridine under Birch conditions (Scheme 20) [109]. It further reacts with sodium borohydride to form 1,4-dimethyl-1,2,3,6-tetrahydropyridine although merely as one component in a product mixture

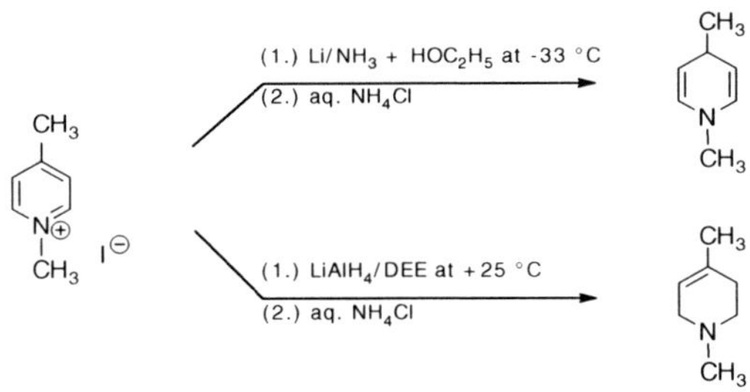

Scheme 20 Reduction of *N*-methylpyridinium salts to 1,4-dihydro- or 1,2,3,6-tetrahydropyridines

(Scheme 20) [110, 111]. 1-Methylpyridinium iodide affords 1-methyl-1,2,3,4-tetrahydropyridine along with some *N*-methylpyridine under the same conditions [112, 113]. Sodium dithionite suffices to bring about the reduction of *N*-benzylpyridinium iodides (carrying electron-withdrawing substituents) to the 1,4-dihydropyridines [114, 115].

3 Reaction of Pyridines with Polar Organometallics

One very characteristic reaction mode exhibited by organometallic reagents and intermediates toward pyridines, benzopyridines and azapyridines is Tschitschibabin-like substitution following the nucleophilic addition/nucleofugal elimination pattern. On the other hand, organoalkalis are often powerful enough to accomplish the addition step at temperatures low enough to avoid the normally ensuing elimination of metal hydride. In such a case we encounter nucleophilic addition rather than substitution.

3.1 Nucleophilic Substitution

When halopyridines serve as the substrates, metalation (hydrogen/metal permutation; Sect. 4.2.2) at a heteroatom-neighboring position usually outpaces the equally envisageable nucleophilic displacement of halogen. As this possibility is precluded for pentafluoropyridine it reacts with methyllithium [116], 1-propenyllithium [117] and phenyllithium [118] under fluorine/organyl substitution at the 4-position (Scheme 21). Tripyrid-2-ylmethyllithium is too weak a base to metalate 2-chloropyridine and, therefore, also condenses with it (Scheme 21) [119]. 2-Methyl- and 2-ethylsulfinylpyridines react with 2-pyridyllithiums to afford symmetrical or asymmetrical 2,2′-bipyridyls (Scheme 21) [120]. The high nucleofugal mobility of the sulfinyl entity is noteworthy.

Although N^9-lithiated chloropurine reacts quite readily with phenyllithium, the substitution product is formed in only poor yield (23%). A satisfactory result is assured when N^9-benzyl-6-iodopurine is treated with lithium diphenylcuprate [121].

Discovered by Karl Ziegler et al. [122, 123] in 1930, the reaction of alkyl- and aryllithiums with pyridine, quinoline, isoquinoline, acridine, and phenanthridine is

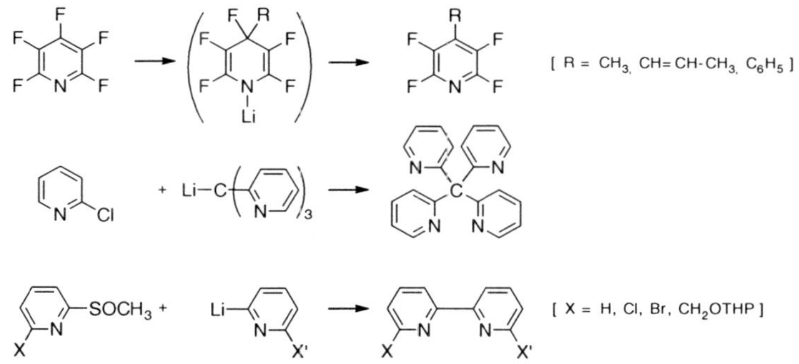

Scheme 21 Organometallic substitutions of halo- and sulfinylpyridines

still nowadays of utmost practical importance. If the 1,2-dihydro intermediate does not lose lithium hydride spontaneously, the rearomatization can be brought about thermally (in the temperature range between 40°C and 120°C) or oxidatively. Thus, 2,6-di-*tert*-butylpyridine can be readily made in a two-step procedure [124].

This method has been recently employed to access a series of 2′-*ortho*-substituted 2-arylpyridines (Scheme 22) [125]. Their torsional barriers, measured by dynamic NMR spectroscopy, were found to be substantially (by some 4 kcal mol^{-1}) lower than those of the carba-analogous biphenyls. In other words, the lone pair of a pyridine nitrogen atom is effectively smaller than that of an aromatic C–H bond [125].

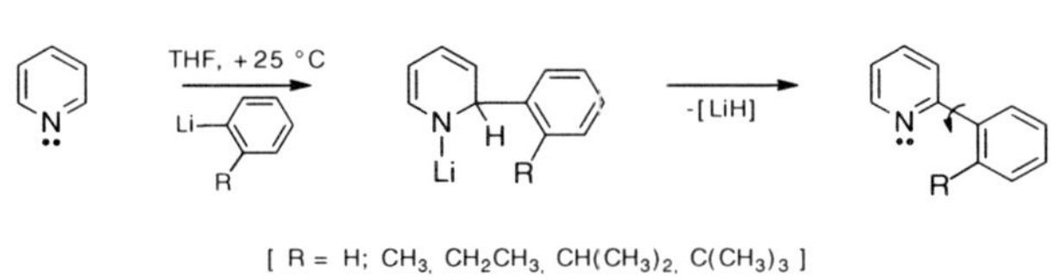

Scheme 22 Arylation at the 2-position of pyridine

As foreseeable, the nucleophilic addition onto diazene and triazenes is facilitated compared to pyridines. Conversely, the rearomatization of the transient nitrogen-rich dihydro intermediates usually requires oxidants such as 2,3-dichloro-5,6-dicyano-*p*-benzoquinone ("DDQ"). This procedure allows to introduce alkyl, aryl, pyridyl, thienyl, and thiazolyl groups into the 4-position of 2-chloropyrimidine and 2,6-dichloropyrimidine and into the 6-position of 2,4-dichloropyrimidine (Scheme 23) [126, 127]. Organolithiums add regioselectively onto the 3-position of 2-acetamido- and 2-pivaloylamidopiperazine (Scheme 23) [128].

Nucleophilic substitutions of pyridines can be accomplished under mild conditions with organomagnesiums if the latter are allylic or benzylic. Even then the yields remain poor. Just 8.5% of 4-benzylpyridine (note, attack on the 4-position!) were isolated after heating an ethereal solution of benzylmagnesium chloride and pyridine for 24 h under reflux [129]. Their feeble reactivity disqualifies Grignard reagents in the pyridine area. Otherwise they would be attractive reagents

Scheme 23 Nucleophilic additions onto pyrimidines and piperazines followed by oxidative rearomatization

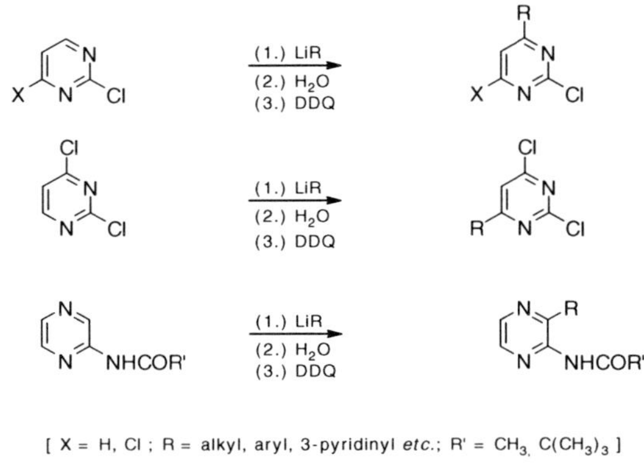

as magnesium salts are environmentally less harmful than lithium compounds and, as a consequence, do not need to be rigorously eliminated from waste waters. Of course, the Mase–Oshima–Knochel "turbo" magnesiates [130–132] do not help out of this situation as they contain both elements.

3.2 Nucleophilic Addition onto Pyridines

Unlike organomagnesiums, organolithiums add even at dry-ice temperatures onto the formal C=N bond of pyridines. At −75°C, however, no lithium hydride will be eliminated. This offers the unique opportunity to "fix" the 1,2-dihydropyridyl species by recurring to the entire inventory of electrophiles. The electron excess being concentrated at its 1-, 3-, and 5-centers, the regiochemical outcome of the quenching is uncertain a priori. Whereas protons [122], silyl [133] and acyl [134] groups dock preferentially at the nitrogen atom (i.e., the 1-position), bromine [135], alkyl halides [135–137], oxiranes [136], carbon dioxide [138], disulfides [139] and di-*tert*-butyl azodicarboxylate [140] are preferentially attached to the 5-position (Scheme 24).

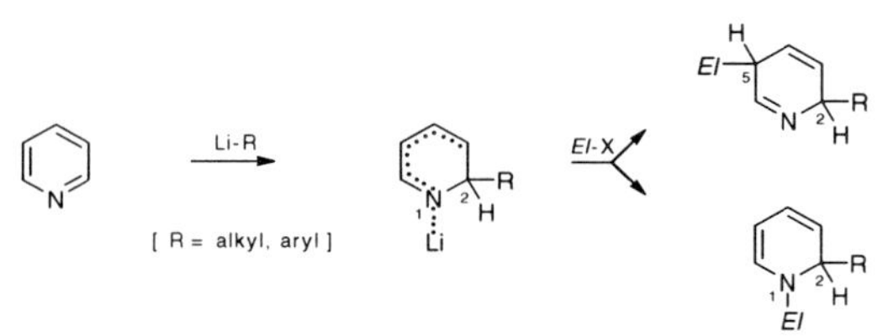

Scheme 24 Electrophilic trapping of the pyridine/organolithium adducts at either the nitrogen or the 5-position

In the same way adducts of diazenes and triazene can be obtained. Consecutive treatment of 1,3,5-triazene with an alkyl- or aryllithium in benzene and with water provides a 4-substituted 1,4-dihydro-1,3,5-triazene [141].

3.3 *Nucleophilic Addition to Pyridinium Salts and Pyridine Oxides*

The positively charged heteroatom makes pyridinium salts [142] and pyridine *N*-oxides very prone to nucleophilic attack. Thus Grignard reagents add smoothly onto 1,4-dimethylpyridinium iodide to produce 2-substituted 1,2-dihydro-1,4-dimethylpyridines [143]. *N*-activation is not restricted to alkylation. Acylation [144] and *N*-methylbenzenecarboximidoylation [145] have been proposed as further possibilities.

The reactions between pyridine *N*-oxides and organoalkalis and alkaline-earth compounds have been studied in great detail. Whereas alkyllithiums abstract a proton from a 2- and 6-position (Sect. 4.2.6) [146, 147], organomagnesiums mainly or exclusively add onto the ring if there is a vacant nitrogen-neighboring site. Phenylmagnesium bromide combines with pyridine *N*-oxide in diethyl ether [148] or, better, tetrahydrofuran [149]. High yields of 2- or 4-substituted pyridines are secured if the adduct is formed at −25°C or below before being treated consecutively with methanol and trifluoroacetic acid (Scheme 25) [150, 151].

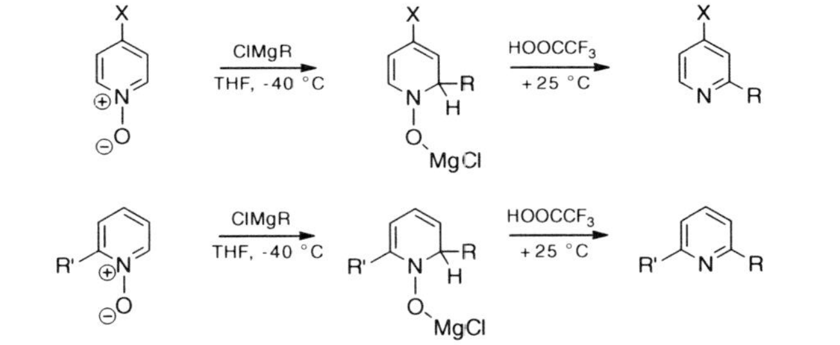

Scheme 25 2-Substituted pyridines by the addition of Grignard reagents onto pyridine *N*-oxides and subsequent dehydration

The *N*-oxides of pyridine and quinoline form with phenylmagnesium bromide a semistable 1,2-adduct [149, 152]. It collapses to 2-phenylquinoline and its *N*-oxide even if worked up under nitrogen (Scheme 26) [153]. When quinoline 1-oxide is treated with an alkyl- or (het)aryllithium in the presence of fluorenone as a hydride acceptor, the 2-substituted derivatives are isolated in moderate to good yields [153].

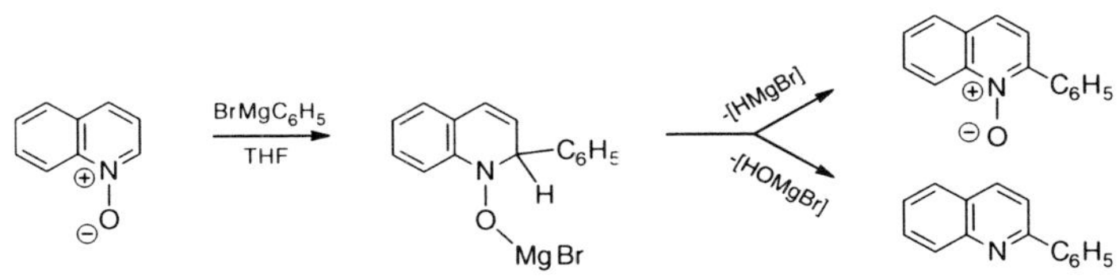

Scheme 26 Quinoline *N*-oxide and phenylmagnesium bromide

Scheme 27 Opening of pyridine *N*-oxide/organomagnesium adducts

Other complications results from the tendency of the adducts to undergo ring opening valence isomerization to 5-substituted (2*Z*,4*E*)-pentadiene aldoximes bromomagnesium salts (Scheme 27) [152, 154]. Analogous pericyclic processes have been reported for the sodium acetylide/pyridinium *N*-oxide adduct [155] and the reaction between aqueous sodium hydroxide and *N*-alkoxypyridinium salts [156], 1-[(isobutoxy-carbonyl)oxy]pyridinium chloride [157], *N*-pyridinium sulfonate [158, 159] or pyridinium *N*-nitrite [160].

4 Metalated Pyridines, Benzopyridines, and Azapyridines

A most effective way to elaborate aromatic six-membered nitrogen heterocycles is to introduce a metal atom at the targeted site and thus to activate the ring toward electrophiles. This approach opens a plenitude of possibilities for organic synthesis. We shall first systematize the known reaction patterns before addressing the crucial issue of how to prepare the specific metal derivatives of pyridines. To stay within the scope of this chapter, "metal" continues to mean a "polar" or "electropositive" one, that is first of all lithium and, to a lesser extent, also sodium, potassium, and magnesium.

4.1 Pyridylmetal Reactivity

There are no fundamental differences between the behavior of arylmetals and pyridylmetals. The same principal pathways are open to both of them.

The first reaction class to be considered is the substitutive displacement (Fig. 1) of the metal by hydrogen (including an isotope thereof), by carbon, by another tetravalent element Q (in particular, the metalloids, silicon and tin), by a trivalent pnictide Z (such as nitrogen or phosphorus), by a divalent chalcogenide (like oxygen or sulfur) and by one of the monovalent halogens X (fluorine, chlorine, bromine, or iodine).

The addition reactions onto carbon–carbon or carbon–heteroelement multiple bonds come next. They comprise the combination of pyridylmetals with simple or activated alkenes or alkynes, with azomethines (Schiff bases), cyanides, and isonitriles and, finally, aldehydes, ketones, carbon dioxide, and carbon monoxide (Fig. 2).

As the metal occupies a ring position by definition, the sole possibility to realize a β-elimination is to generate a 2,3- or 3,4-didehydropyridine, a so-called pyridyne

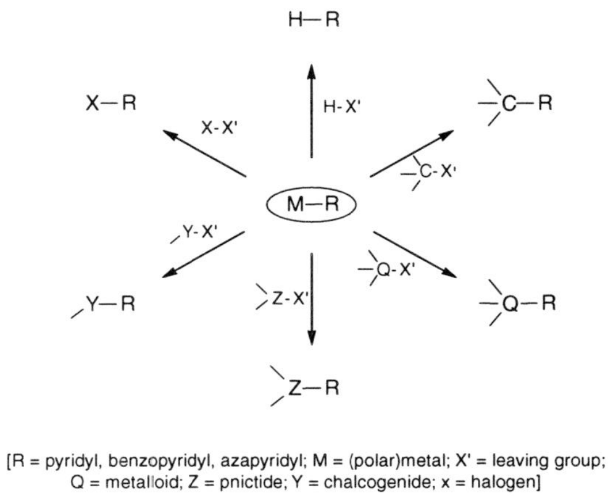

Fig. 1 Substitutive displacement of the metal by hydrogen, a carbon residue, by a metalloid Q, a pnictide Z, a chalcogenide Y, or a halogen X

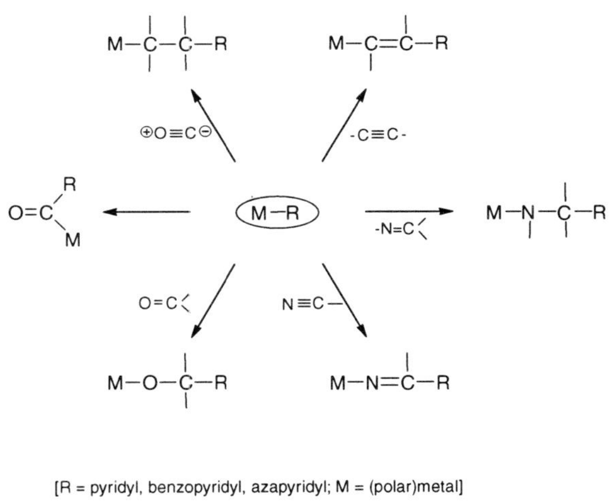

Fig. 2 Addition reactions of pyridylmetals

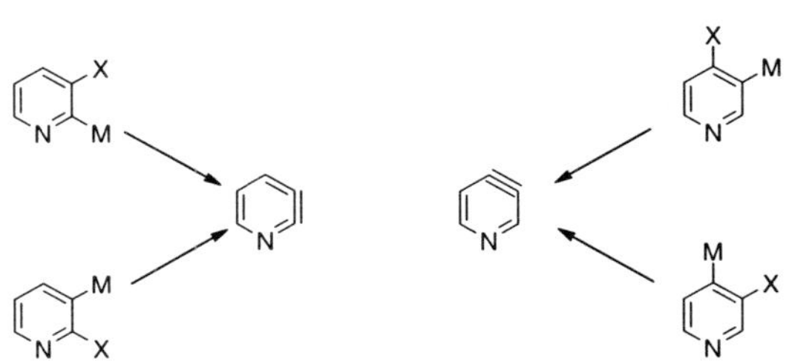

Fig. 3 2,3- and 3,4-didehydropyridine by β-elimination of metal halide or alkoxide from a vicinally counterpolarized pyridine

(Fig. 3). Tortured by an extreme angle deformation such cyclic alkynes or cumulenes are of course highly strained and consequently short-lived.

Pyridylmetals are not frequently involved in rearrangements or other isomerization processes. However, there is still one pericyclic reaction mode that is quite typical for them. The intermediates issued from nucleophilic addition to pyridinium

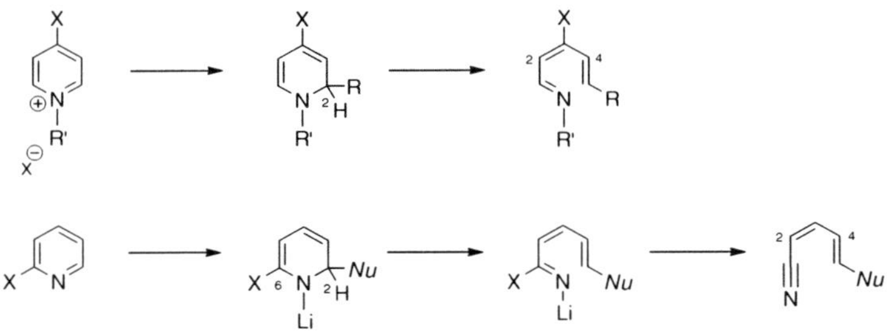

Fig. 4 Electrocyclic ring opening of the intermediates formed upon nucleophilic addition on pyridinium salts or 6-halopyridines

salts (as to pyridine *N*-oxides, see Sect. 3.3) are quite prone to electrocyclic ring opening giving rise to 2,4-pentadiene aldoximes (Fig. 4). If the substrate is a pyridine that carries a halogen or another good leaving group at a nitrogen-adjacent position, subsequent β-elimination will convert the immediate ring opening product into a 2,4-pentadiene cyanide (Fig. 4).

A few illustrations should help to concretize the schematic survey presented above. Though more or less randomly picked, those examples are, nevertheless, quite characteristic.

4.1.1 Substitution Reactions

Hydrolysis is one of the two key steps for the assessment of organometallic concentrations by means of the Wittig–Harborth [161] or Gilman–Haubein [162] "double titration" protocols. In addition, quenching the intermediate with heavy water or a deuterated acid enables very conveniently the introduction of an isotopic label at a predetermined position of the substrate (Scheme 28) [163, 164].

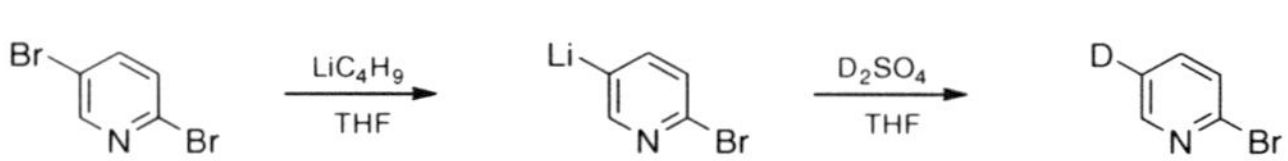

Scheme 28 Introducing an isotopic label at a specific position

Pyridyllithiums condense readily with methyl iodide and primary alkyl halides, in particular allylic and benzylic ones. Thus, 2-chloropyrid-3-yllithium combines with methyl iodide to give 2-chloro-3-methylpyridine in 80% yield [165]. 2-Fluoropyridine reacts with *n*-butyllithium in ethereal solvents almost exclusively under *X*/*C* (halide/alkyl) displacement providing 2-butylpyridine [166]. However, when treated consecutively or simultaneously [167] with LIDA and chlorotrimethylsilane, 2-fluoropyrid-3-yllithium is generated and instantaneously trapped as 2-fluoro-3-(trimethylsilyl)pyridine (Scheme 29) [168, 169].

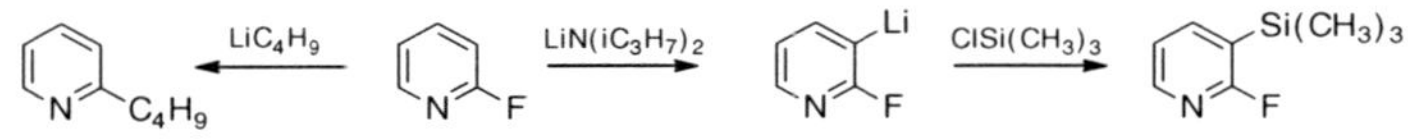

Scheme 29 Trapping of a pyridyllithium with chlorotrimethylsilane

Phosphine ligands are the indispensable modulators of metal complexes in homogeneous catalysis. The design and preparation of novel bidentate pyridylphosphines, such as tris(2,6-dimethoxypyrid-3-yl)phosphine [170], is therefore an area that currently attracts much activity.

The reaction of an organometallic intermediate with fluorodimethoxy borane [171], methyl borate or any other boric acid ester gives a boronate that can be oxidized with aqueous hydrogen peroxide in alkaline medium. This two-step procedure offers an effective and reliable access to alcohols and phenols. Trapping with dialkyl or diaryl disulfides opens a direct entry to sulfides and, after oxidation, sulfoxides and sulfones. In this way 3-bromo-4-(phenylthio)pyridine (61%) was made from 3-bromopyrid-4-yllithium (Scheme 30) [168, 169].

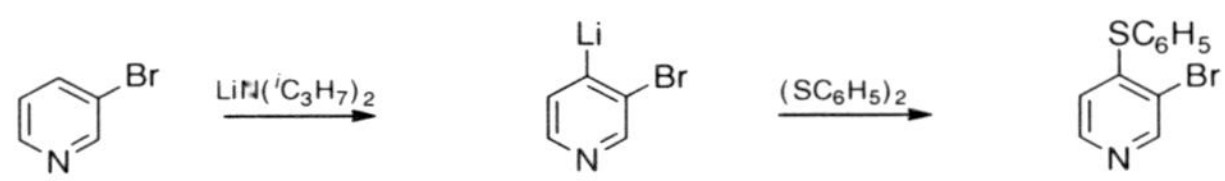

Scheme 30 Phenylthiylation of a pyridyllithium

Pyridyllithiums can be readily chlorinated, brominated, and iodinated using halogen sources such as hexachloroethane, *N*-chlorosuccinimide, 1,2-dibromo-1,1,2,2-tetrafluoroethane and molecular bromine or iodine. 4-Methoxypyrid-3-yllithium (generated with mesityllithium to avoid concomitant nucleophilic substitution) thus affords 3-iodo-4-methoxypyridine in 65% yield (Scheme 31) [172].

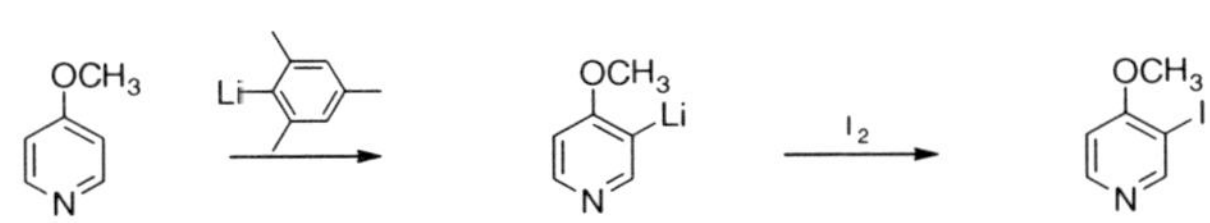

Scheme 31 Iodination of 4-methoxypyrid-3-yllithium

The pyridyllithiums shown in the last three schemes were generated by permutational hydrogen/metal interconversions, so-called metalations. Such kind of reactions will be covered systematically and in considerable detail in a later subsection (Sect. 4.2.3)

4.1.2 Addition and Addition/Elimination Reactions

A great variety of synthetically important carbofunctional entities can be combined with pyridylmetals. Three typical examples are shown below (Scheme 32): the α-hydroxylation by 2-furylaldehyde [173], the acylation by ethyl chloroformate [174] and the carboxylation by carbon dioxide [175].

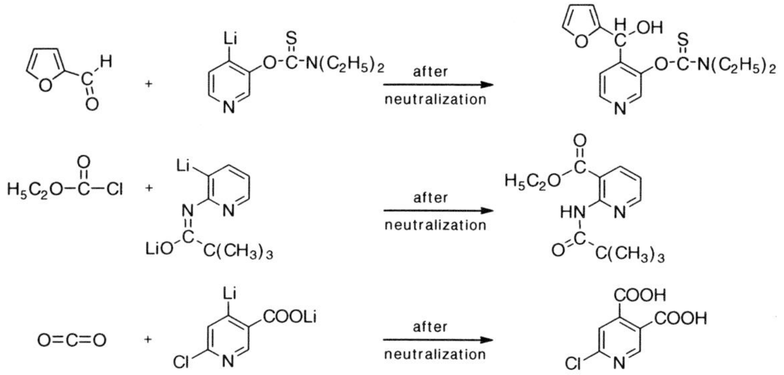

Scheme 32 Reaction of pyridyllithiums with carbofunctional units

The addition of pyridylmetals onto carboxamides takes place at −75°C. When the *O*-lithiated hemiaminal thus produced from 2,6-dichloropyrid-3-yllithium and dimethylformamide is neutralized without warming up the reaction mixture, the expected aldehyde can be isolated almost quantitatively [176]. However, if the hydrolysis is performed at ambient temperature, a 10:1 mixture of 2-chloro-6-dimethylamino- and 6-chloro-2-dimethylamino-3-pyridinecarbaldehyde is obtained instead [177]. The reason is the reversible decomposition of the intermediate into 2,6-dichloro-3-pyridinecarbaldehyde and the nucleophilically lurking lithium dimethylamide (Scheme 33).

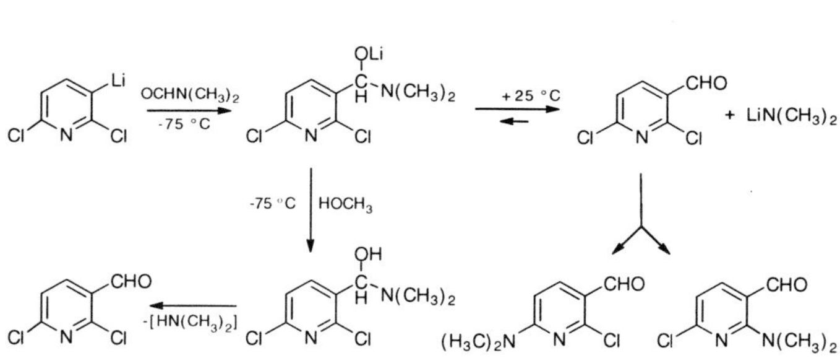

Scheme 33 Aldehyde formation compromised by the instability of the organolithium/dimethylformamide adduct at ambient temperature

O-Lithiated hemiaminals can provide neighboring group assistance to the metalation of ring positions. Thus, the species formed upon addition of 2-methoxypyrid-3-yllithium onto *N*-methyl-*N*-[2-(dimethylamino)ethyl]formamide is readily deprotonated at the 4-position (Scheme 34). Ensuing iodination and

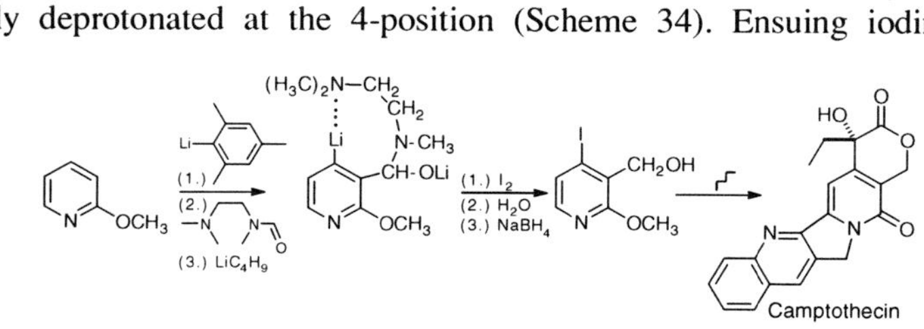

Scheme 34 Formylation as a key step on the route to Camptothecin

reduction of the aldehyde set free upon hydrolysis affords a key intermediate that can be elaborated to the DNA topomerase inhibitor (*S*)-Camptothecin (Scheme 34) [178].

Conjugate additions of pyridyllithiums onto α,β-unsaturated aldehydes, ketones, and esters are feasible. When treated with a pivaloyl-protected 2-aminopyrid-3-yllithium, β-(dimethyl-amino)acroleine undergoes an addition/elimination sequence that ultimately leads to 1,8-naphthyridine upon acid-promoted joining of the side chains (Scheme 35) [179].

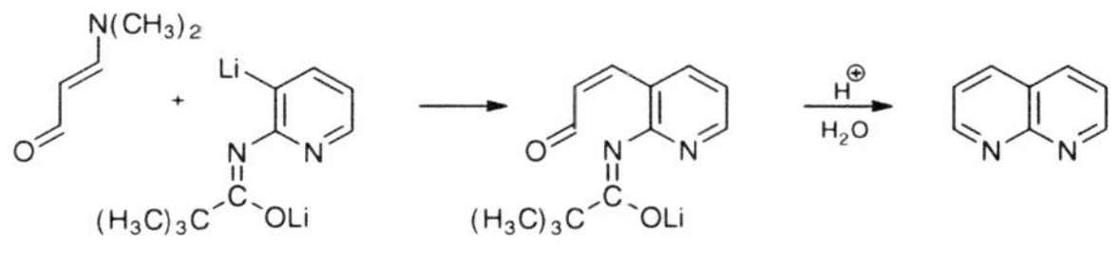

Scheme 35 Condensation of an amidopyridyllithium with β-(dimethylamino)acroleine and subsequent cyclization

In practice nucleophilic addition onto triple bonds narrows down to the use of nitriles as the substrates. The final products are ketones as the imine intermediates in general do not outlast hydrolysis (Scheme 36) [180].

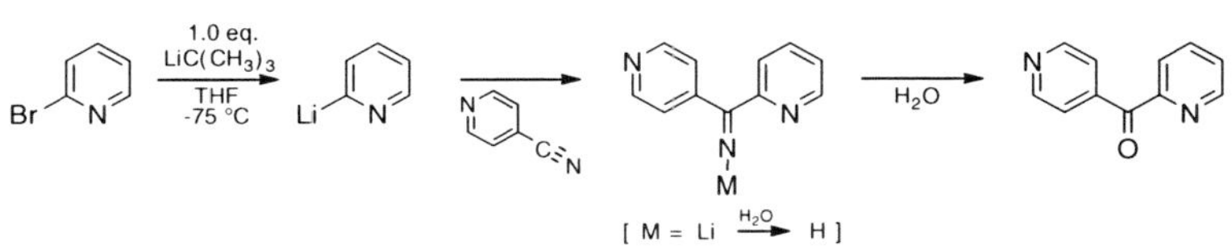

Scheme 36 Addition of pyrid-2-yllithium onto 4-cyanopyridine and hydrolysis to 2-pyridyl 4-pyridyl ketone

4.1.3 Elimination Reactions

If, as postulated, the metal is accommodated at one of the ring positions, β-elimination of a metal halide (or a metal organyloxide) will inevitably generate a dedihydropyridine ("pyridyne") [181, 182]. This will happen indeed whenever a halopyridine is exposed to a strong base at temperatures above that of dry ice. The emerging high-energy species seeks immediate stabilization by combining with any available *C*-, *O*-, or *N*-nucleophile and subsequent protonation. If such an addition/elimination (A/E) sequence is really operative, the isomeric precursors 4-halopyrid-3-ylmetals and 3-halopyrid-4-ylmetals must generate 3,4-dedihydropyridine as a common intermediate which has to exhibit its reactivity and selectivity profile without showing any memory of its origin. 3- and 4-Bromopyridine stand this test indeed as they produce with pyrrolidine in the presence of sodium amide and sodium *tert*-butoxide the same 1:1 mixture of 3- and 4-*N*-pyrrolylpyridine (Scheme 37) [183].

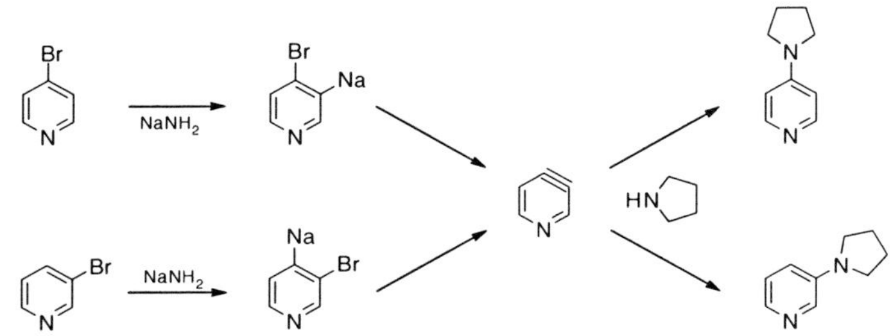

Scheme 37 3- and 4-Bromopyridine giving rise to the common intermediate 3,4-dedihydropyridine

The reaction of 2,3-dibromopyridine and potassium amide in liquid ammonia obviously passes through a bromopyridyne. However, the ensuing substitution to 2,4-diaminopyridine is more likely to be brought about by a different pathway, namely by following the nucleophilic addition/nucleofugal elimination (A/E) pattern. Coexistence of E/A and A/E processes can also be invoked for the transformation of 2,4-, 2,5-, 3,4- and 3,5-dibromopyridines [184].

4-Halopyrid-3-yllithiums can be readily generated from 4-fluoro-, 4-chloro- and 4-bromopyridine with *n*-butyllithium (if X = F or Cl) or LIDA in DEE or THF at dry ice temperatures. Upon warming, these species lose lithium halide. The 3,4-pyridyne set free from the fluoro compound enters into a Diels–Alder cycloaddition with furan if this diene is present in the solution (Scheme 38) [185].

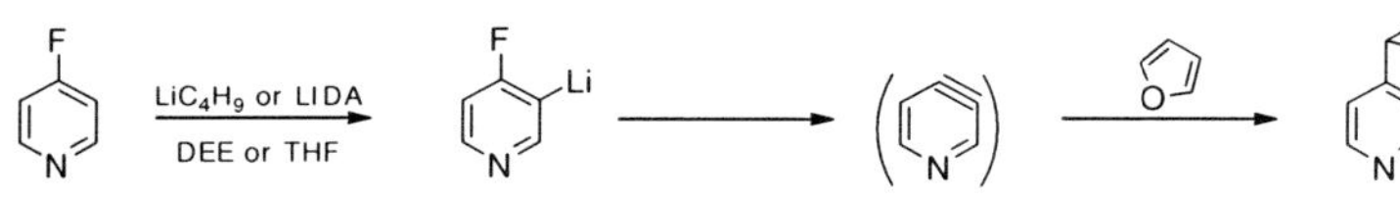

Scheme 38 In situ trapping of a pyridyne by Diels–Alder cycloaddition

Pyridynes combine with lithium enolates under [2+2] cycloaddition rather than by attack of the oxygen atom onto the formal triple bond. One of the two regioisomers produced undergoes spontaneous ring enlargement by the scission of the junction between the four- and six-membered ring (Scheme 39) [186].

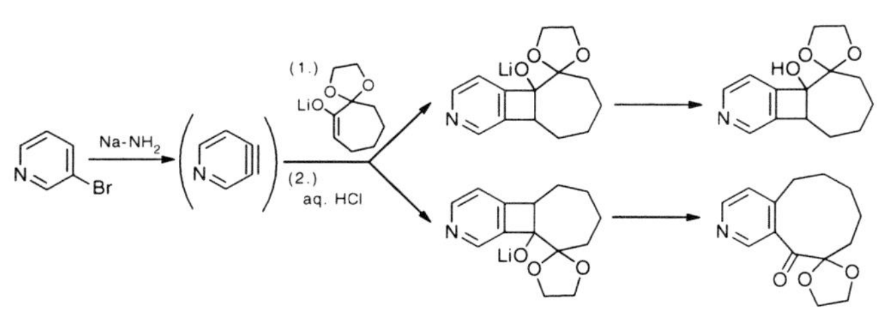

Scheme 39 [2+2] Cycloaddition between a pyridyne and a lithium enolate

4.1.4 Isomerization by Electrocyclic Ring Opening

As briefly reported above (Sect. 3.3, Scheme 27), the adducts formed between pyridine *N*-oxides and Grignard reagents tend to isomerize at ambient temperature to linear pentadiene aldoximes by electrocyclic ring opening. The smooth addition of organolithiums to the electron-deficient 1,3,5-triazine leading to 4-substituted 1,4-dihydro-1,3,5-triazines has also been mentioned (at the end of Sect. 3.2). Even the weakly basic and sterically hindered lithium bis(trimethylsilyl)amide is capable of forming such an adduct. However, this cannot be isolated as such because the lithium atom and one of the trimethylsilyl groups instantaneously swap places and thus trigger the cleavage of the single bond between the tetravalent carbon and its nitrogen neighbor, thus giving rise to the extensively delocalized 1,3,5,7-tetraaza-2,4,6-heptatrienyllithium (Scheme 40) [141].

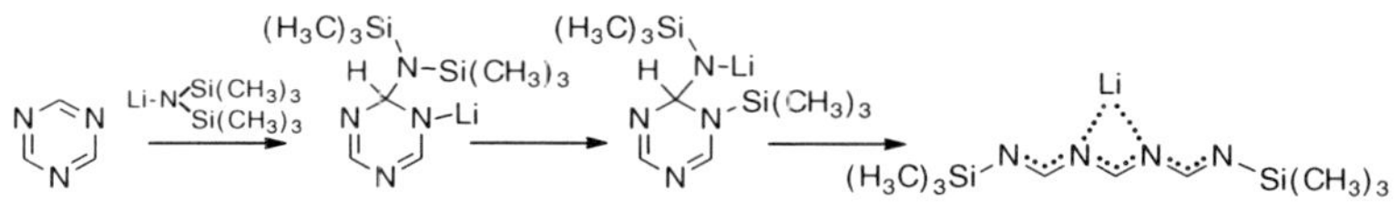

Scheme 40 Tetraazaheptatrienyllithium from 1,3,5-triazene

Similar structural reorganizations appear to occur when pyridines and pyrimidines bearing a good leaving group at their respective 2- or 6-positions are treated with a strong nucleophile. The adduct formed again opens the ring at the saturated C–N bond to produce an aza- or diazahexadiene that immediately eliminates metal halide to give an (aza)pentadienyl nitrile (Scheme 41) [187, 188].

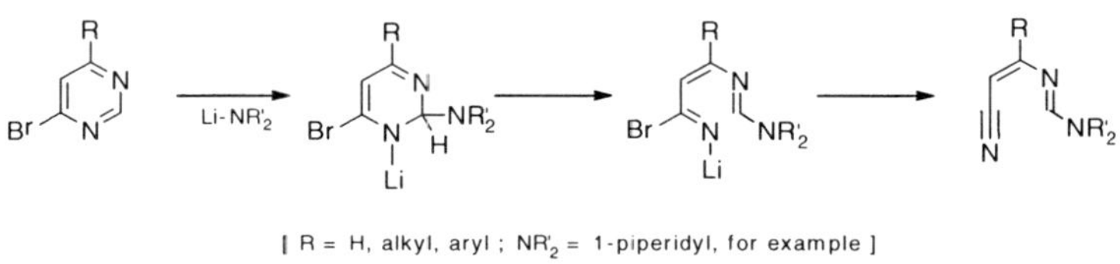

Scheme 41 Elimination-coupled electrocyclic pyrimidine ring cleavage

The stage for such a van der Plas–Utimoto–Nozaki scission of six-membered nitrogen heterocycles can also be set by the 1,2-migration of an alkyl group from a borate center to the 2-position of a 6-bromopyridine (Scheme 42) [189]. The β-elimination of lithium bromide acts of course as a splendid driving force that steers the reaction towards pentadienyl cyanide, its ultimate destination.

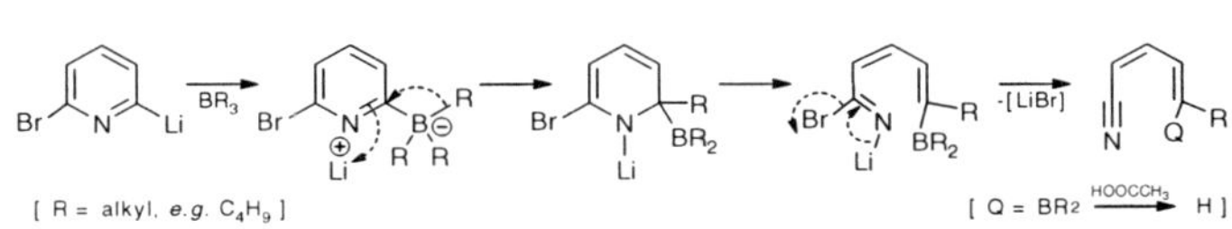

Scheme 42 Elimination-coupled electrocyclic pyridine ring cleavage

4.2 Generation of Pyridylmetals and Related Species

The preceding section has explored the synthetic potential of pyridylmetals without bothering about how one can get hold of such reactive intermediates. This is a crucial issue, all the more as some of them, in particular azapyridylmetals, tend to be fragile. The accessibility of pyridylmetals will now be examined in detail.

There are three universally applicable methods leading to pyridylmetals. The first and second of them rely on halopyridines as the starting materials. Those are treated either with an elemental metal or with an organometallic reagent. The third method subjects the pyridine to an acid/base reaction-like hydrogen/metal permutation accomplished with an organometallic reagent, in particular *n*-butyllithium.

4.2.1 Insertion of Elemental Metal into a Carbon–Halogen Bond

The preparation of pyridyl- and quinolylmagnesium halides by reductive insertion of magnesium – "reductive" regarding the fate of the organic backbone, not of the metal – into carbon–chlorine or carbon–bromine bonds of the corresponding halopyridines requires some experimental skills to avoid the menacing deactivation of the metal surface. All three pyridylmagnesium halide isomers [190–194] have been obtained in this way and also 2- and 8-quinolylmagnesium chloride (Scheme 43) [195]. 4-Bromo-2,3,5,6-tetrafluoropyridine [196], pentachloropyridine [197] and pentabromopyridine [198] accommodate the metal all in the 4-position (Scheme 43).

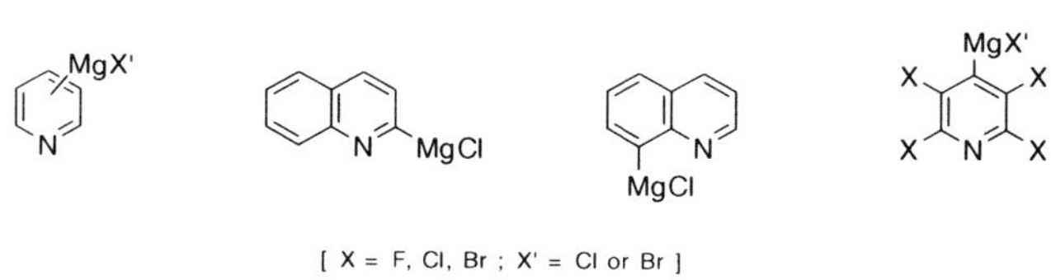

Scheme 43 Selected pyridyl- and quinolylmagnesium halides

The reductive insertion approach should not be attempted with alkali metals. The resulting organometallic species are too reactive and would cannibalize their own precursors.

4.2.2 Permutational Replacement of Halogen by Metal

The permutational interconversion of bromo- or iodopyridines or such quinolines with ethylmagnesium bromide or isopropylmagnesium chloride is well documented [199, 200]. The corresponding heterocyclic Grignard reagents are generally formed in acceptable or good yields. Recent work has capitalized on the coordinative power of magnesium by using a TADDOL [(4*R*,5*R*)-α^4, α^4, α^5, α^5-tetraphenyl-1,3-dioxolane-4,5-dimethanol] derived complex (Scheme 44) for enantioselective syntheses attaining er values up to 88:12 (ee 76%) [201].

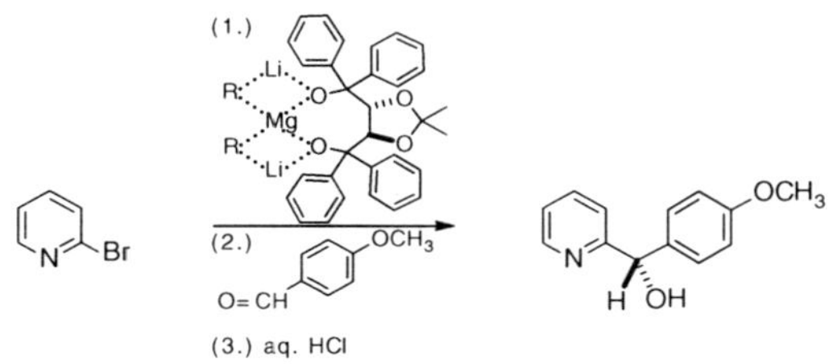

Scheme 44 Generation of a TADDOLate complexed pyrid-2-ylmagnesium and its addition onto *p*-anisaldehyde

n-Butyllithium, nevertheless, remains the exchange reagent of choice. It has been applied to 2-, 3-, or 4-bromopyridine [175, 180, 202, 203] and countless derivatives thereof. It converts 1- and 2- bromoisoquinoline into the lithiated species [204, 205]. It displaces the heavier and notoriously more mobile halogen at the 8-position of 4-bromo-8-iodo-2-(trifluoromethyl)quinoline but the bromine atom occupying the more acidic 4-position in its 4,8-dibromo analog (Scheme 45) [206].

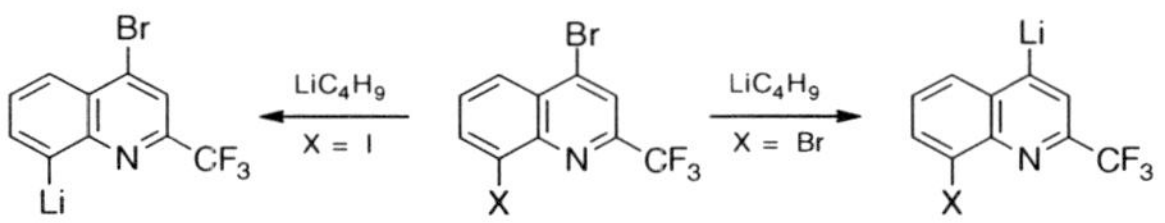

Scheme 45 Iodine is more exchange-performant than bromine

Two heavy halogen atoms present in the same molecule may cause problems. 2,6- or 3,5-dibromopyridine leaves of course no doubt about where the metal will show up. In contrast, 2,5-dibromopyridine does create an ambiguity. The intrinsic acidity of pyridine sites increases with the distance between the C–H bond and the ring nitrogen [207, 208]. If one conceives polar carbon–metal bonds as "partial carbanions," it becomes intelligible why magnesium and, a fortiori, lithium or potassium prefer to reside at a 3-(or 5-) rather than at a 2-position. Therefore, 2,5-dibromopyridine affords with *n*-butyllithium in tetrahydrofuran cleanly 6-bromopyrid-3-yllithium ("2-bromo-5-lithiopyridine"; Scheme 46) [163]. However, when the interconversion is performed in toluene, the reagent lacks the activation it usually owes to the coordination with ethers or other electron-donors. Under such circumstances the lithium does not despise the lone pair of the ring nitrogen, docks there and displaces the next-door bromine atom at the 2-position [164]. Analogously, 2,3,5-tribromopyridine reacts with

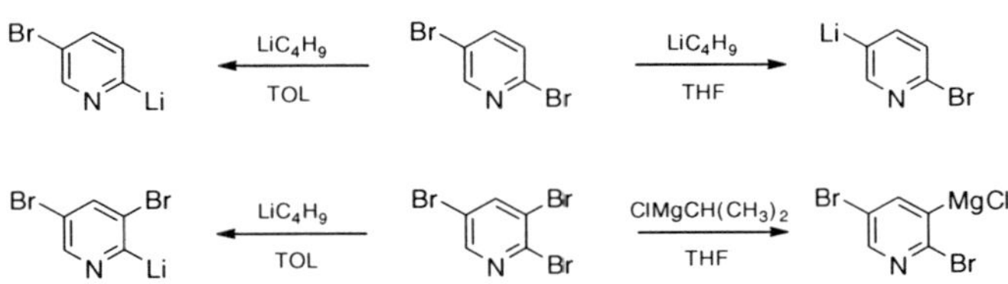

Scheme 46 Solvent control of halogen/metal permutation site

n-butyllithium in toluene at −75°C [209] and with isopropylmagnesium chloride in tetrahydrofuran at 0°C [210] selectively at the 2- and 3-positions, respectively (Scheme 46). These intermediates can be intercepted with molecular iodine to give the iododibromopyridines in 77% and 75% yield.

Pentachloropyridine [211] and pentabromopyridine [198] undergo the halogen/metal interconversion with *n*-butyllithium exclusively at the 4-position. The latter is statistically disfavored but more acidic than the 3-(5-), let alone the 2-(6-)position [207, 208].

(Trifluoromethyl)pyridyl substituents are not yet fully exploited and hence are quite attractive building blocks for pharmaceutical research. At each time four isomers of 2- and 3-(trifluoromethyl)pyridyllithiums have all been accessed by halogen/metal permutation using *n*-butyllithium, isopropylmagnesium chloride or lithium tributylmagnesiate in DEE, THF, or TOL (Scheme 47) [206].

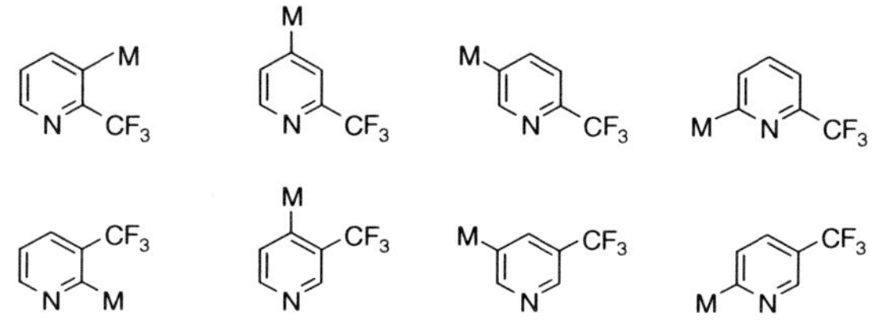

Scheme 47 Metal-bearing (trifluoromethyl)pyridines

Many halogen/metal permutations are documented in the diazene class too. Both pyrazyllithium [212, 213] and 3,6-dimethylpyrazyllithium [214] have been prepared from the iodo compounds. Acting on suitably substituted pyrimidines, *n*-butyllithium smoothly displaces tributylstannyl to afford pyrimidyl-2-yllithium [215] and bromine or iodine to give pyrimid-2-yllithium [215], 2,6-diethoxypyrid-4-yllithium [216], 2,4-dimethoxypyrimid-5-yllithium [217], 2,4-dibenzyloxypyrimid-5-yllithium [218, 219], 2-methylthio-pyrimid-4-yllithium [215], 4-(trifluoromethyl)pyrimid-2-yllithium [220], 2-chloro-4-(trifluoromethyl)pyrimid-5-yllithium [220], and 6-chloro-4-(trifluoromethyl)pyrimid-5-yllithium [221]. However, the permutational interconversion of 5,6-dibromo-4-(trifluoromethyl)pyrimidine has to be accomplished with isopropylmagnesium chloride as only this reagent discriminates strictly in favor of the 2-position and leaves the 6-bromo substituent completely intact (Scheme 48) [221].

4.2.3 Metalation of Pyridines: Regioselectivity

Halopyridines do not grow on trees. One has to make them. In general, the researcher resorts to one of the approaches outlined above (Sect. 2). For example, an aminopyridine is accessed by a nucleophilic or electrophilic substitution sequence to be eventually subjected to a dediazotation/halogenation. In other words, the halopyridine is obtained at the end of a fairly long reaction sequence.

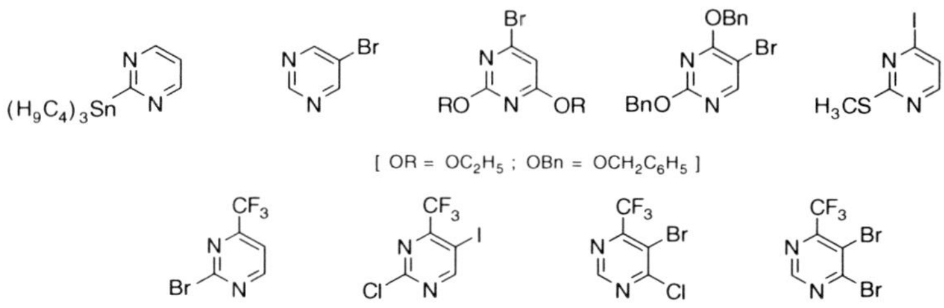

Scheme 48 Several pyrimidyllithiums and one pyrimidyl-magnesium from halogenated or stannylated precursors

It would of course be smarter to introduce the metal into the pyridine itself, the ultimate precursor to the halopyridine, by direct displacement of hydrogen. Obviously one has to solve the problem of regioselectivity. A variety of heterosubstituents provide an efficacious solution. They offer neighboring group assistance to the "metalation" (hydrogen/metal permutation) process and guide it to their immediate neighborhood. The most frequently solicited substituents are halogens, alkoxy or alkoxyalkoxy groups, and *N*-alkylated or *N*-acyclated amino functions. Once such a tool has been installed at the selected position, it can take over and control the reactivity and selectivity of the ensuing organometallic reactions.

Nitrogen-Assisted Metalation

Despite contradictory literature reports, it is not possible to metalate pyridine efficiently and reproducibly using our LIC-KOR superbase for this purpose. Single electron transfer compromises the outcome of such attempts.

Although far less basic, "Caubère's base" [186], the 1:1 mixture of *n*-butyllithium and lithium 2-(dimethylamino)ethoxide (LIDMAE), meets this challenge. Especially in apolar (HEX) or weakly polar (DEE) media the alkyllithium/alkoxide complex coordinates with the pyridine nitrogen and, reaching over to the neighboring 2-position, it grabs there a proton [222]. In the same way 4-methyl- and 3,5-dimethylpyrid-2-yllithium have been obtained from α-picoline and β,β-lutidine [223].

Of potential importance is the neighboring nitrogen assistance to the metalation of 2,4′-bipyridyl to afford 4-(2-pyridyl)pyrid-3-yllithium and of 2,2′-bipyridyl to afford a mixture of 2-(2-pyridyl)pyrid-3-yllithium and 2,2′-bipyrid-3,3′-diyldilithium (Scheme 49) [224]. LITMP may benefit even in THF from coordination to the

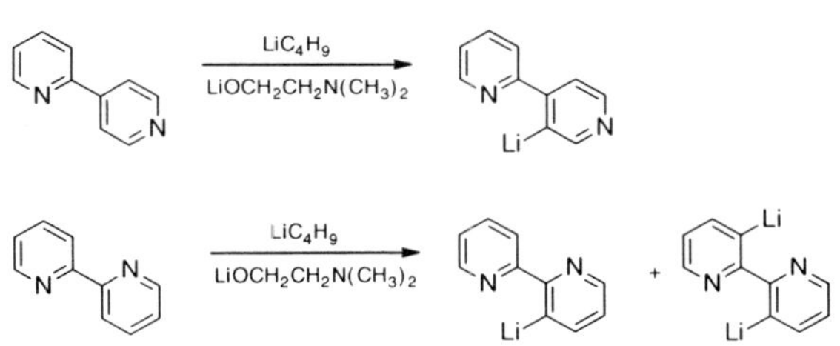

Scheme 49 LIDMAE-controlled metalation of bipyridyls

pyridine nitrogen. Using this base, a series of *ortho*(2′)-metalated 2-arylpyridines and 2-arylquinolines have been generated by Florence Mongin et al. [225].

As already shown above (Sect. 4.1.2, Scheme 35), lithiated NH-Piv (Piv= pivaloyl) groups promote effectively the metalation of pyridine rings. In the same way *N*-BOC-protected 2- and 3-aminopyridine can be metalated at, both times, the 3-position and subsequently condensed with phenyl cyanate (Scheme 50) [226].

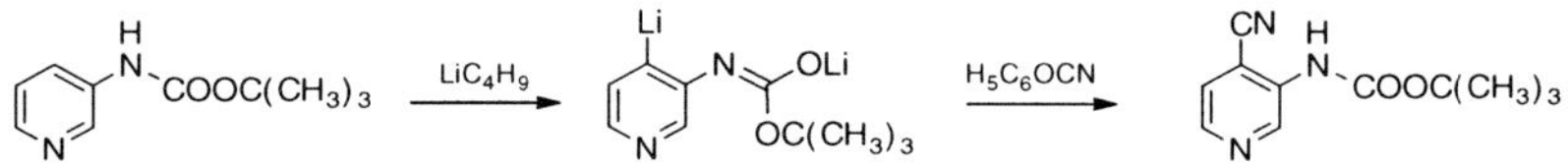

Scheme 50 Lithiation controlled by an *N*-BOC-amido group

The five-ring nitrogen directs the lithiation of 2-phenylimidazo[1,2-*a*]-pyrimidine (X=H) and its 7-chloro congener (X=Cl) to the 5-position (Scheme 51) [227]. Depending on the substitution pattern quinolines may be metalated at the 8-position, i.e. in the benzo part rather than in the heterocyclic ring (Scheme 51) [228, 229].

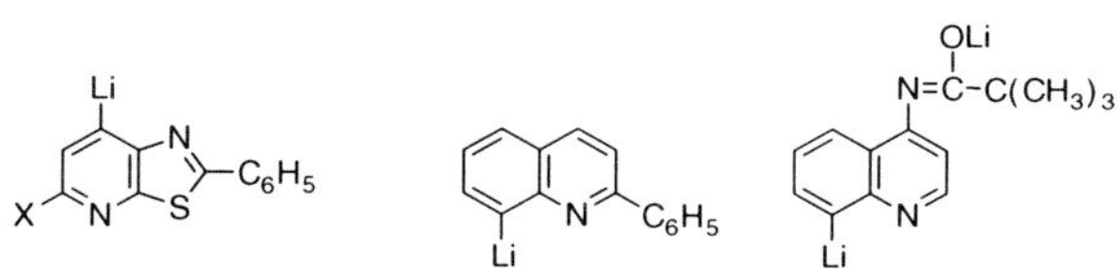

Scheme 51 Ring-nitrogen-assisted metalation of imidazopyrimidines (X = H or Cl) and quinolines

All three cyanopyridines are deprotonated by LITMP (Scheme 52) [230]. The nitrogen being no longer directly attached to the ring and pointing away from it is unavailable for coordination but exerts an inductive effect instead. Sterically less hindered lithium dialkylamides such as lithium diethylamide or lithium piperide nucleophilically displace the cyano substituent if located at the 2- or 4-position rather than to cause metalation [231].

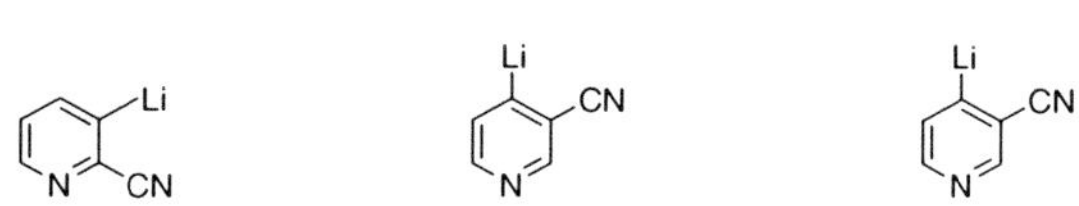

Scheme 52 Three readily accessible cyanopyridyllithiums

(Dialkylamino)alkyl substituents belong to the standard inventory of so-called *ortho*-directing neighboring groups. This structural feature can also be accessed indirectly by the addition of a lithium dialkylamide onto an aldehyde, for example of the chelating lithium *N*-methyl-*N*-2-(dimethylamino)ethylamide onto 6-methoxypyridine-2-carbaldehyde. Thus it provides protection of the formyl group against nucleophilic attack and activation of the adjacent 3-position toward metalation at the same time. Consecutive treatment of the adduct with *n*-butyllithium and iodomethane followed by neutralization affords 6-methoxy-3-methylpyridine-2-carbaldehyde and 6-methoxy-5-methylpyridine-2-carbaldehyde in a ratio of 97:3 and a yield of 77% (Scheme 53; see also Sect. 4.1.2, Scheme 34) [232]. If instead of lithium *N*-methyl-*N*-2-(dimethylamino)ethylamide lithium *N*-methylpiperazide is used for the hemiaminal formation, the isomeric ratio is inverted to 7:93 [232].

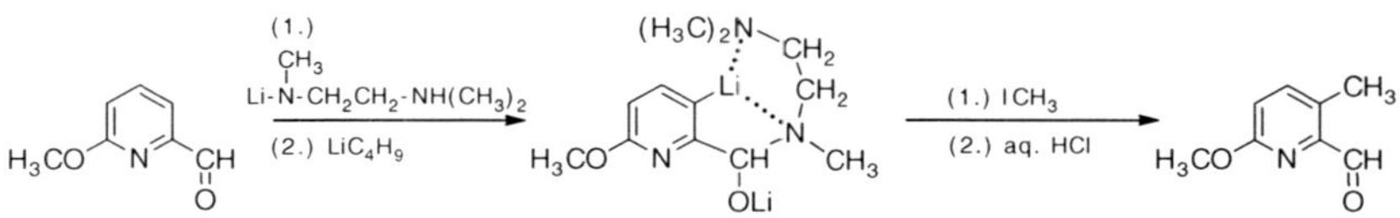

Scheme 53 Metalation of *O*-lithiated hemiaminals

Oxygen- and Sulfur-Assisted Metalation

2-Methoxypyridine is smoothly lithiated with LIDA at the 3-position [233] and with the Caubère base (*n*-butyllithium/LIDMAE) at the 6-position [234]. 3-Methoxypyridine is deprotonated by *n*-butyllithium in THF at the 2-position [235], whereas 3-(methoxymethoxy)pyridine reacts with *tert*-butyllithium in DEE at the 4-position [236, 237]. Clean metalation of 4-methoxypyridine at the 3-position can be accomplished with mesityllithium [172, 238].

LITMP attacks *N,N*-diisopropylpyridine-3-carboxamide at the 4-position [239]. It is presumably the electron-rich oxygen that acts as the coordinating center (Scheme 54).

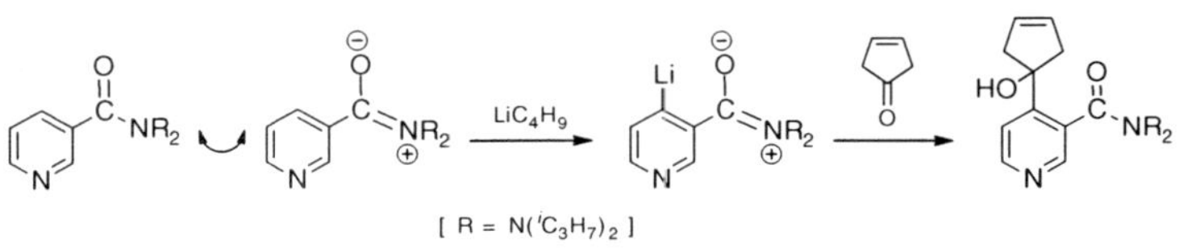

Scheme 54 Metalation of *N,N*-dialkylpyridinecarboxamides

Although less powerful, lithium carboxylates (O instead of NR_2 in Scheme 53) behave in a similar way. Upon consecutive treatment of each of the three pyridinecarboxylic acids with LITMP (2.0 eq.), carbon dioxide and mineral acid pyridine-2,3- and -3,4-dicarboxylic acid (the latter from the 3- and 4-isomeric precursors) were isolated in 65–85% yield [175].

Carbamoyl substituents may be used as metalation promoters too. 2-, 3- and 4-*N,N*-(diethylcarbamoyloxy)pyridines can be readily metalated with *sec*-butyllithium in THF at −75°C (the 3-isomer reacting at the 4-position) [240].

In the sulfur series the sulfinyl group revealed itself as an effective metalation promoter. Phenyl 2-, 3- and 4-pyridyl sulfoxide can be smoothly lithiated at the 3-, 4- and again 3-position when treated with LIDA in THF at −75°C [241].

Fluorine-Assisted Metalation

2-Fluoropyridine undergoes hydrogen/lithium permutation at the halogen-adjacent position 500–1,000 times faster than fluorobenzene does [242]. 2- and 4-fluorpyridine are both deprotonated at −75°C by *n*- and *sec*-butyllithium or LIDA in THF or DEE at the 3-position, respectively (Scheme 55) [168, 169, 242, 243]. TMEDA-complexed *n*-butyllithium in THF deprotonates 3-fluoropyridine cleanly at the 3-position [244], whereas *n*-butyllithium in the presence of DABCO (1,8-diazabicyclo[2.2.2]octane) exclusively attacks the 2-position (Scheme 55) [245].

Scheme 55 Four readily accessible fluoropyridines

Further electron-withdrawing substituents such as a second [65, 246, 247] or third [246, 248] fluorine atom or lithiooxy, alkoxy and alkoxyalkoxy groups [249] still enhance the reactivity.

The trifluoromethyl group exhibits some peculiar traits [250]. Although separated from the ring by one carbon atom, it acidifies the phenyl *ortho*-position in the gas phase approximately to the same extent as a single, directly attached fluorine atom does. Moreover, this acidifying effect of the CF_3 entity is only slightly diminished when monitored at *meta* and *para* positions whereas it rapidly levels off with distance in case of an individual fluorine atom [251]. Thus, behaving in an "altruistic" [249, 252] way, a remote trifluoromethyl group may prove instrumental for the successful metalation at a position neighboring another, less efficient heterosubstituent.

The trifluoromethyl group is relatively voluminous. Having a *B* value of 10.5 it is definitely more space-demanding than fluorine (*B* 4.3), trifluoromethoxy (*B* 5.5) or methyl (*B* 7.4) and is almost as bulky as isopropyl (*B* 11.1) [253–255].

The metalation of CF_3-bearing pyridines suffers from competing nucleophilic addition and SET processes. Those dominate the reaction between 3-(trifluoromethyl)pyridine and alkyllithiums to the extent of >98%; the claimed metalation and trapping in a yield of 75% was found to be irreproducible [206]. It proved, however, possible to lithiate at a nitrogen-neighboring position and subsequently carboxylate 2- and 4-(trifluoromethyl)pyridine in, respectively, 71% and 41% yield using Caubère's base. The same substrates can be deprotonated both at the 3-position with LITMP (73% and 84% trapped by carbon dioxide) [206]. Analogously, 2,6-bis(trifluoromethyl)pyridine was carboxylated at the 3-position [256] and 2,5- and 3,5-bis(trifluoromethyl)pyridine at the 2-position using *n*-butyllithium in THF at −75°C (Scheme 56) [257]. Trifluoromethylpyridines bearing further substitutents, in particular halogens, were also selectively metalated without any difficulty [258–260].

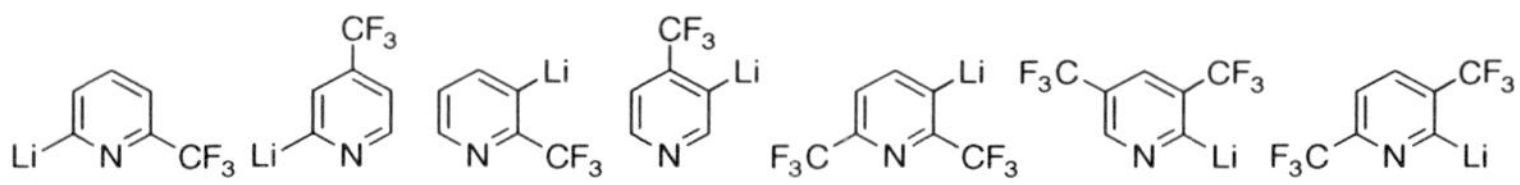

Scheme 56 Readily accessible (trifluoromethyl)pyridyllithiums

An even more peculiar unit is the trifluoromethoxy group, the most strongly acidifying one among the saturated substituents [251, 261–264]. It proved difficult to prepare (trifluoromethoxy)pyridines by extension of the existing techniques into the pyridine area. Standard methods such as Hiyama's oxidative desulfurization/fluorination [265] of pyridyldithiocarbamates failed [266]. The hetero-aromatic ring needs to be protected by at least one 2-chloro atom to survive the bromine/fluorine or chlorine/fluorine exchange in the alkoxy side chain. Once available, the chloro(trifluoromethoxy)pyridines were readily lithiated with LIDA in THF at −75°C (Scheme 57) [266]. Whenever there is an ambiguity about the site where the deprotonation is brought about, the metal seeks the vicinity of the trifluoromethoxy substituent at the expense of the chlorine atom (Scheme 57).

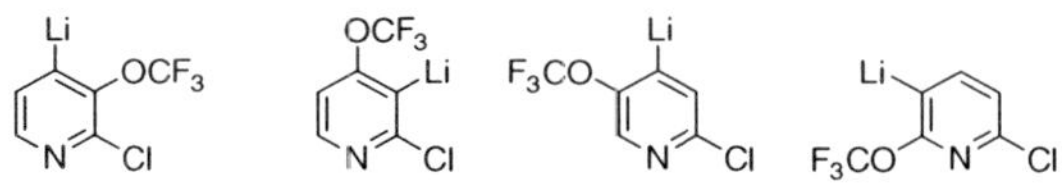

Scheme 57 Metalation of chloro(trifluoromethoxy)pyridines with LIDA

Chlorine-Assisted Metalation

The gas phase acidity of chlorobenzene exceeds that of fluorobenzene only minimally [267, 268]. This slight advantage is overcompensated by a steric effect due to the greater van der Waals radius of chlorine. As a consequence chlorobenzene is a little, exactly fivefold, less reactive than fluorobenzene toward *sec*-butyllithium in THF at −100°C [268].

LIDA suffices to metalate 2- and 4-chloropyridine, both at the 3-position [168, 269]. Lithium is alternatively introduced into the 6-position of 2-chloropyridine if the Caubère base (*n*-butyllithium/LIDMAE) is used [270]. 3-Chloropyridine

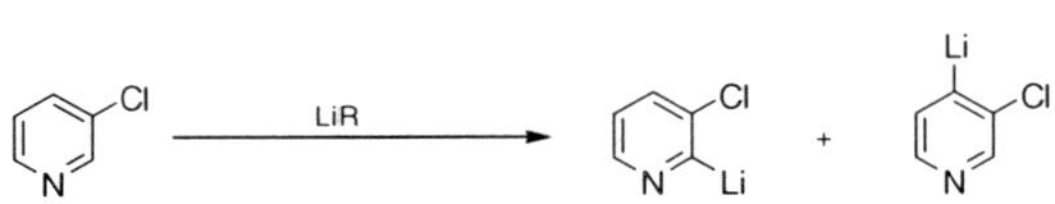

LiR = LiC_4H_9 + LIDMAE/DEE	85 %	- %
LiR = LiC_4H_9·TMEDA/DEE	60 %	9 %
LiR = $LiN(^iC_3H_7)_2$/THF	-	96 %

Scheme 58 Regioselectivities of 3-chloropyridine metalation

(Scheme 58) produces with LIDA, i.e. under reversible deprotonation, virtually neat 3-chloropyrid-4-yllithium whereas with TMEDA-complexed *n*-butyllithium 2- and 4-lithiation occur in a 7:1 ratio [271]. With *n*-butyllithium (or methyl- or phenyllithium) in THF, i.e. under irreversible deprotonation conditions, these two intermediates are obtained in a 1:1 ratio along with predominant nucleophilic addition onto the 2-position [271]. Clean metalation at the 2-position is achieved using Caubère's base (Scheme 58) [272].

2,4- [273] and 3,5- [274, 275] dichloropyridine react with particular ease at the positions flanked by the two halogen atoms. Depending on the reagent employed, *n*-butyllithium or LIDA, 2,6-dichloropyridine [276] affords a 1:4 or 10:1 mixture of 3- and 4-lithiated species. 2,5- And 3,4-dichloropyridine offer perfect site selectivity, the former undergoing metalation with *tert*-butyllithium at the 6-position and with PMDTA-activated *n*-butyllithium at the 4-position, the latter with Caubère's base at the 2- and with LIDA at the 4-position (Scheme 59) [277, 278].

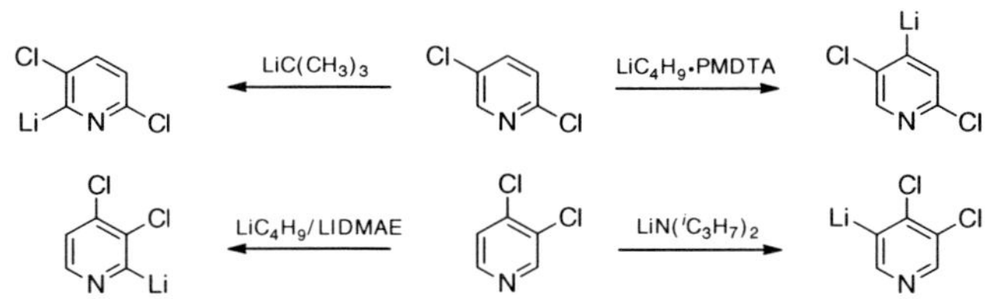

Scheme 59 Metalation of 2,5- and 3,4-dichloropyridine, both endowed with perfect optional site selectivity

Bromine- and Iodine-Assisted Metalation

Only metal amide bases such as LIDA and LITMP are suitable for the metalation of bromo- and iodopyridines as organolithiums would inevitably cause halogen/metal permutation. 2- [279, 280], 3- [168, 271] And 4-bromopyridine [281] can all be metalated with LIDA (Scheme 60). However, as a careful reexamination [282] has revealed, not all reactions proceed as regioselectively as claimed. 2-Bromopyrid-3-yllithium and 3-bromopyrid-4-yllithium are always contaminated with at least some (2–10%) 2-bromopyrid-4-yllithium and 3-bromopyrid-2-yllithium, respectively (Scheme 60) [282].

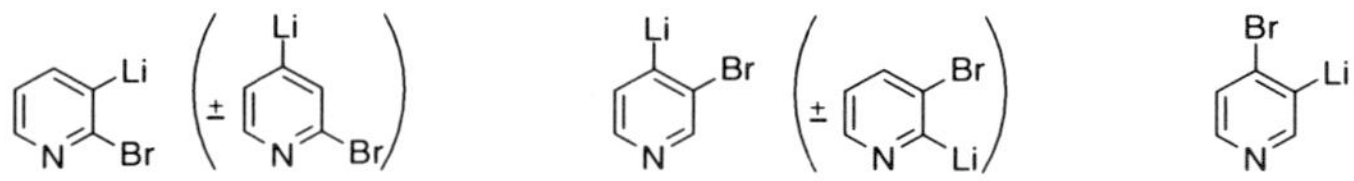

Scheme 60 More or less regioselective metalation of bromopyridines

4.2.4 Metalation of Pyridines: Regioflexibility

Let us assume a development project depends on the readily availability of 3-aminopyridines bearing specific functional groups at the 4-position. As specified (Sect. 4.2.3, Scheme 50) 3-(*N*-BOC-amino)pyridine would qualify as a suitable precursor because it can be cleanly metalated and subsequently substituted at the 4-position. This *Lucky Strike* approach would fail if the 2-position was targeted unless one blocks the more reactive 4-position by a protective group beforehand (see below).

Optional Site Selectivity

On the other hand, it does not cause any trouble to functionalize 3-fluoropyridine or 3-chloropyridine at either the 4- or the 2-position (see Sect. 4.2.3, Schemes 54 and 57). One would merely have to pick the right reagent to attack specifically the selected position. Such a *How to Kill Two Birds with One Stone* scenario is of course a blessing in practical terms.

There are numerous model cases of optional sites selectivities, for example the alternative metalation of 2,5-dichloropyridine at the 4- or 6-position and of 3,4-dichloropyridine at the 2- or 5-position (Scheme 58), or of 2-(trifluoromethyl) pyridine at the 2- or 3-position (Scheme 55) or the alternative halogen/metal permutation performed with 2,5-dibromopyridine at the 2- or 5-position and with 2,3,5-tribromopyridine at the 2- or 3-position (Scheme 46).

The same principle can be applied in the quinoline series. 4-(Trifluoromethyl)-quinoline reacts with LITMP in THF at the 3- and with Caubère's base in DEE at the 2-position (Scheme 61) [283]. LIDA in THF lithiates the 2-isomer at the 3- and LITMP in DEE lithiates it at the 8-position [283].

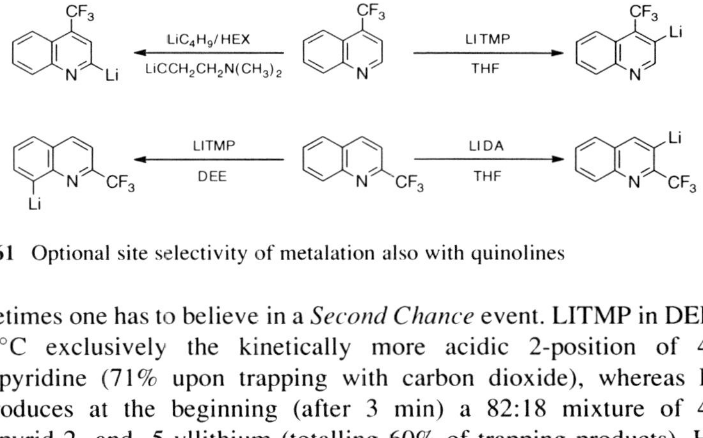

Scheme 61 Optional site selectivity of metalation also with quinolines

Sometimes one has to believe in a *Second Chance* event. LITMP in DEE attacks at −75°C exclusively the kinetically more acidic 2-position of 4-chloro-3-fluoropyridine (71% upon trapping with carbon dioxide), whereas LIDA in THF produces at the beginning (after 3 min) a 82:18 mixture of 4-chloro-3-fluoropyrid-2- and -5-yllithium (totalling 60% of trapping products). However, after 20 h and still at −75°C, the interplay between deprotonation and reprotonation converts the initially predominant 2-lithio species completely (91% of trapping product) into the thermodynamically more stable (i.e., less basic) 5-lithio isomer (Scheme 62) [284].

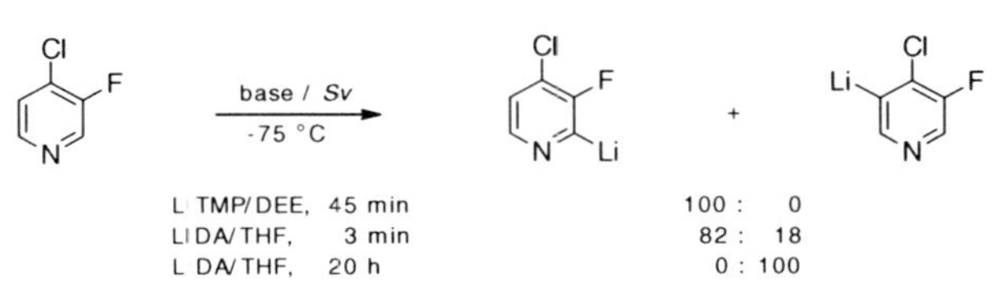

Scheme 62 Lithiation of 4-chloro-3-fluoropyridine: evolution of regioisomers depending on the base and the reaction time

A similar though less clear-cut situation is encountered with 2,6-dichloropyridine upon its treatment with LIDA in THF at −75°C. Interception with molecular iodine demonstrated the presence of 2,6-dichloropyrid-3- and -4-yllithium in a 1:1 ratio after 15 min, but in a 13:1 ratio after 2 h (72% of iodo compound) [285].

Transmetalations as those covered in the two preceding paragraphs are notoriously slow because relatively weak bases are converted into somewhat still weaker bases. The gain in free energy being consequently small the activation barriers are relatively high. However, halogen/metal (or metalloid/metal) interconversion processes have generally much lower activation barriers to overcome. In other words, bromine and iodine can swap places with lithium at much less expense and hence much faster than hydrogen does.

Basicity-diminishing migrations of bromine countercurrent to lithium were first discovered in the arene [286] and thiophene [287] field. Later extensions focuses on furans [288], pyridines [289], and quinolines [290]. Thus, 3-bromo-2-halopyrid-4-yllithiums were isomerized to 4-bromo-2-halopyrid-3-yllithiums [291], 3-fluoro-4-halopyrid-2-yllithiums or 3-fluoro-4-iodoquinol-2-yllithiums to 3-fluoro-2-halopyrid-4-yllithiums or 3-fluoro-2-iodoquinol-4-yllithiums [290], 4-bromo-3,5- dichloropyrid- 2-yllithium to 2-bromo-3,5-dichloropyrid-4-yllithium [277], and 2,6-dichloro-3-iodopyrid-4-yllithium to 2,6-dichloro-4-iodopyrid-3-yllithium (Scheme 63) [277].

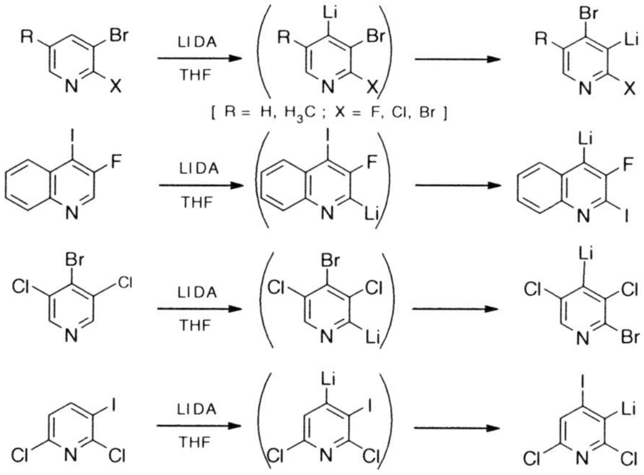

Scheme 63 Heavy halogen migration in exchange against lithium

When the 2-, 3-, and 4-positions are simultaneously occupied, the migration is unleashed by the deprotonation of the 5-position where eventually the bromine or iodine ends up. This has been demonstrated with 2-chloro-3-fluoro-4-iodopyridine [266] and with 2,4-dichloro-3-iodopyridine [271a] (Scheme 64). The course of the reaction always follows the basicity gradient, the isomerization products being less basic than their precursors.

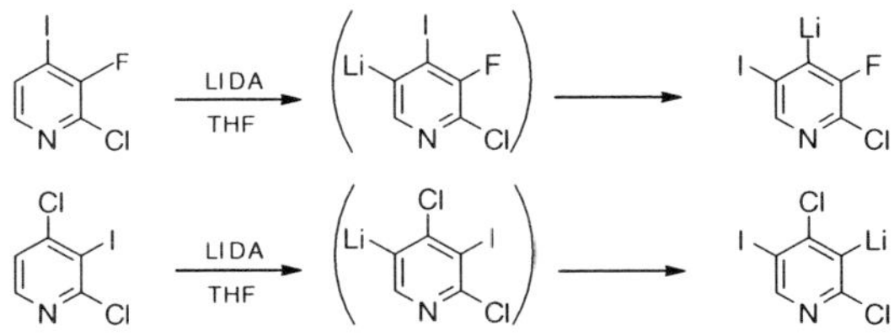

Scheme 64 Further basicity gradient-driven halogen migrations

The base employed generally affects the rate of isomerization but not its outcome. There is one exception, however. Consecutive treatment of 2-chloro-5-(trifluoromethyl)pyridine with LIDA and iodine affords a 1:1 mixture of the 3- and 4-substitution products. When lithium piperidide (LIPIP) is added to it, the 3-iodo component is lithiated at the relatively less congested 4-position to give 2-chloro-3-iodo-5-(trifluoromethyl)pyrid-4-yllithium (33% of pyridine recovered after neutralization) while the 4-iodo compound presumably becomes the prey of nucleophilic addition and degradation. Conversely, with TMEDA-activated LITMP as the base the 3- and 4-iodo components are apparently engaged in a dynamic equilibrium which channels all the material through the 3-iodo isomer to the 6-chloro-5-iodo-3-(trifluoromethyl)pyrid-2-yllithium intermediate. Neutralization affords 62% of 2-chloro-3-iodo-5-(trifluoromethyl)pyridine (Scheme 65) along with the dismutation products 2-chloro-5-(trifluoromethyl)pyridine (10%) and 6-chloro-2,4-iodo-3-(trifluoromethyl)pyridine (8%) [206].

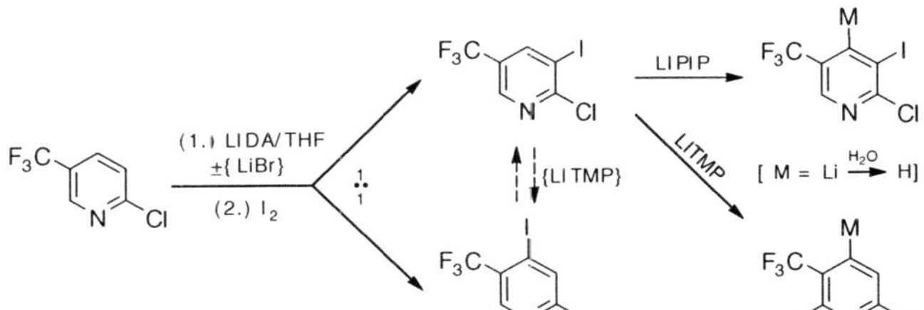

Scheme 65 Different products when using different bases

All such halogen migrations are brought about in a chain process as postulated for the isomerization of 2-bromo-3-(trifluoromethyl)phenyllithium. The isomerization is mediated by a catalytic amount of accidentally formed 1,2-dibromo-3-(trifluoromethyl)benzene that acts at −75°C as a self-restoring turntable (Scheme 66) [292]. At −100°C, however, the initially formed aryllithium remains intact and can be trapped without by-products (Scheme 66) [292].

It is of course most gratifying to identify conditions that enable the interception of both transient species, the isomerized one and its more basic precursor. If such a stop-and-go option cannot be implemented by simple temperature variation (as achieved above), a change of the migratory halogen [293] or of the solvent [293] may provide the solution.

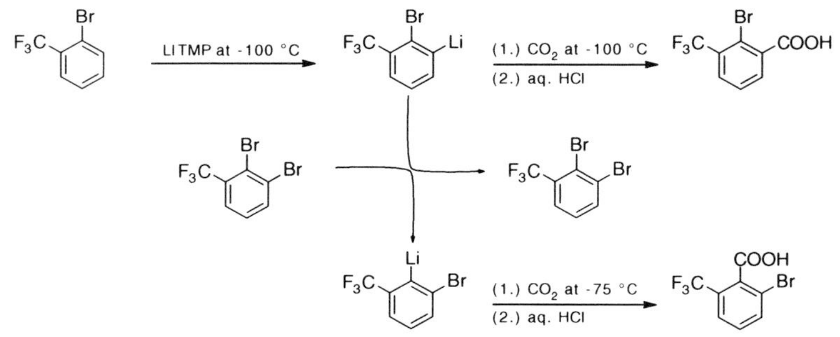

Scheme 66 Basicity-lowering isomerization of bromo- or iodo(het)aryllithiums by continuous halogen/metal permutation

Deploying Protective Groups

When a metal needs to be introduced at a position remote from an assisting heterosubstituent one generally resorts to protecting groups in order to modify local reactivities. The choice is essentially restricted to two tools. A chlorine atom precludes of course any deprotonation of the site it occupies itself and, at the same time, acidifies the neighboring positions thus making them prone to the attack of the reagent. For example, 3-chloro-2-fluoropyridine is readily metalated at the 4-position (Scheme 67) [294]. In contrast, trialkylsilyl groups exert no activating but merely a steric effect that makes inaccessible not only to their own binding site but also to its immediate vicinity. Both 2,3- and 2,5-difluoro-4-(trimethylsilyl)pyridine smoothly undergo deprotonation at the 6-position (Scheme 67) [247].

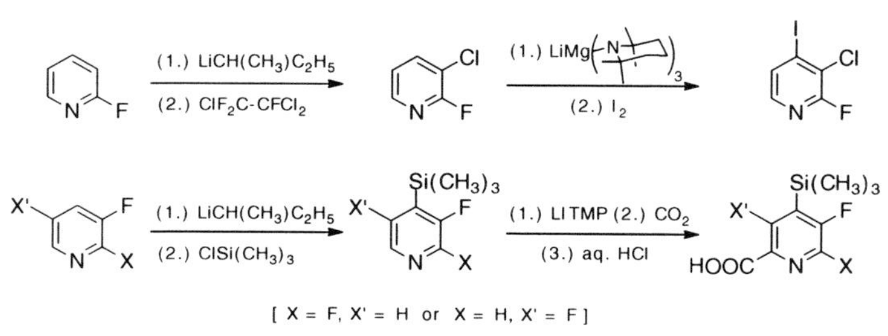

Scheme 67 Chlorine and trimethylsilyl, the two complementary metalation-orienting groups

Once the protective group has fulfilled its task it can be removed without any trouble. Treatment with zinc powder or catalytic hydrogenation suffices to replace chlorine by hydrogen, and silyl entities give way to hydrogen, bromine or iodine upon exposure to mineral acids, molecular bromine or iodine chloride.

6-Fluoroquinoline is selectively metalated at the 5-position when treated with an alkyllithium. However, 4-bromo-5-fluoro-2-trifluoromethyl-3-(trimethylsilyl)-quinoline (prepared with LIDA and chlorotrimethylsilane from the silyl-free

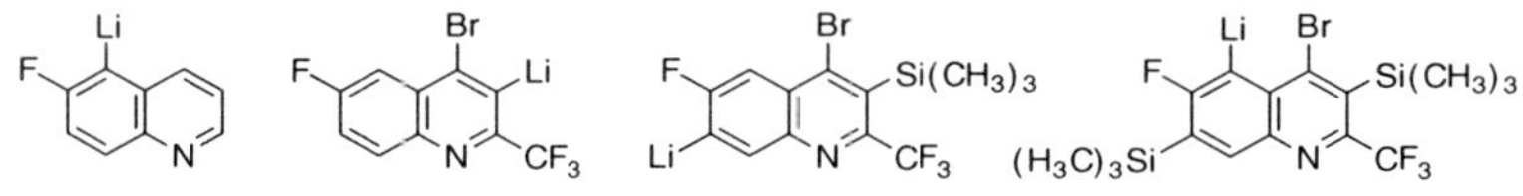

Scheme 68 Lithiation of 6-fluoroquinoline at the 5- or 7-position depending on the presence of other substituents

precursor) reacts at the 7-position, the 5-position being obviously congested by the silyl-buttressed bromine atom. To bring the 5-position back into the game, one has to introduce a second trimethylsilyl entity at the 7-position (Scheme 68) [295].

Regiochemical Exhaustiveness

A most stringent test for methodical flexibility is to introduce the same substituent or functional group into each of the vacant positions of a given substrate. The conversion of 3-fluoropyridine into all four 3-fluoropyridine-carboxylic acids represents a typical example of this kind of "regiochemical exhaustiveness" (Scheme 69) [284].

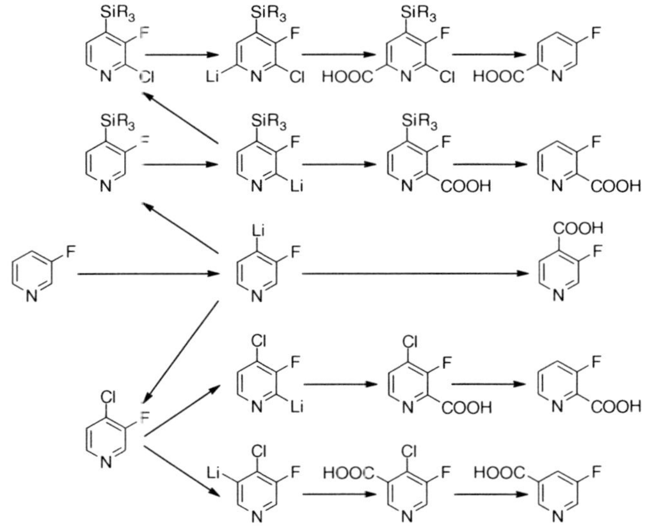

Scheme 69 "Regioexhaustive" carboxylation of 3-fluoropyridine

4.2.5 Metalation of Pyrimidines

A separate subsection is devoted to the metalation of pyrimidines because its outcome is far less predictable than that of pyridines. The hydrogen/metal interconversion mode permanently struggles against side reactions such as nucleophilic additions and SET processes. Success is often based on a trial-and-error approach rather than on guiding principles.

When treated with LITMP and chlorotrimethylsilane, 4-(*N*-BOC-amino)pyrimidine gives only 11% of the trapping product (Scheme 70) [296]. 5-Methoxypyrimidine undergoes lithiation at the 4-position [297], and 2,4- or 4,6-dimethoxypyrimidine [298] at the 5-position (Scheme 70).

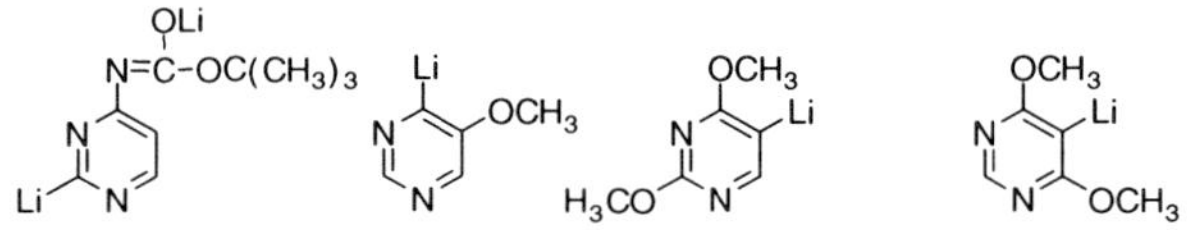

Scheme 70 Metalated pyrimidines bearing nitrogen or methoxy substituents

The combination of methoxy or methylthio groups with other heterosubstituents facilitates the metalation process and stabilizes the resulting intermediates. 2-Chloro-4-methoxypyrimid-5-yllithium [299], 4-chloro-2,6-dimethoxypyrimid-5-yllithium [297], 6-iodo-2-(methylthio)pyrimid-4-yllithium [300], and 2-methylthio-6-(trifluoromethyl) pyrimid-4-yllithium [301] are synthetically versatile species (Scheme 71).

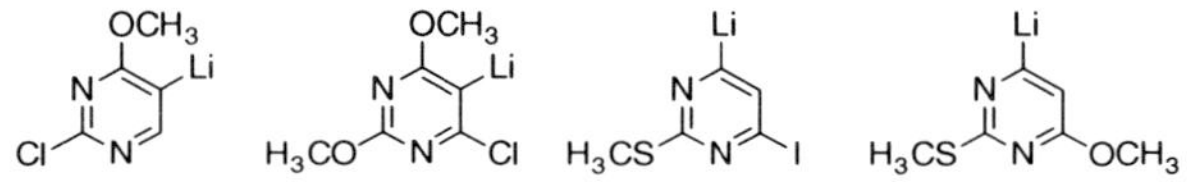

Scheme 71 Metalated pyrimidines bearing halogen and methoxy or methylthio substituents

Pyrimidyllithiums carrying halogens as the sole substituents are easy to generate and to transform further. Unfortunately, 2,4-dichloropyrimidine [273, 302] reacts simultaneously at the 5- and 6-position and thus gives rise to a product mixture. In contrast, 4,6-dichloropyrimidine [273] undergoes clean metalation at the 5-position. The substrates 2,4,6-trichloropyridine [273], 2,4-dichloro-6-(trifluoromethyl)-pyrimidine [221] and 2,4-dibromo-6-(trifluoromethyl)pyrimidine [221] leave no ambiguity anyway (Scheme 72).

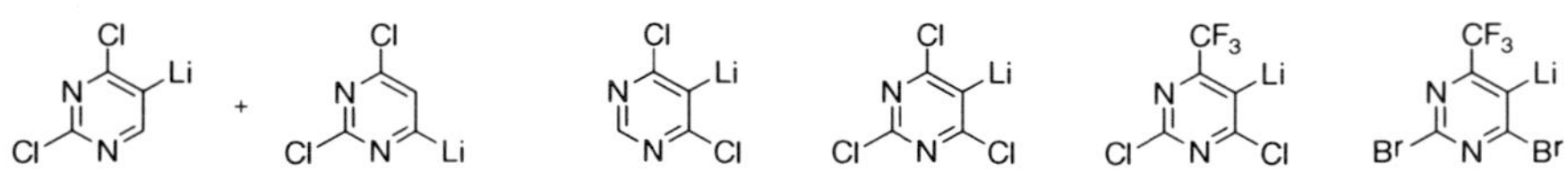

Scheme 72 Metalated pyrimidines bearing only halogen substituents

4.2.6 Metalation of Pyridine *N*-Oxides and Related Zwitterions

Pyridine *N*-oxides generally react with organomagnesiums under addition (Sect. 3.3) but with organolithiums (if in THF or DEE) under hydrogen/metal permutation at a nitrogen-adjacent α-position. The transformations are unclean however, the monometalation being in particular compromised by concomitant α,α-dimetalation (Scheme 73). Moreover, despite numerous efforts aiming at improvements the yields remain moderate to poor (5–66%) [146, 147, 303].

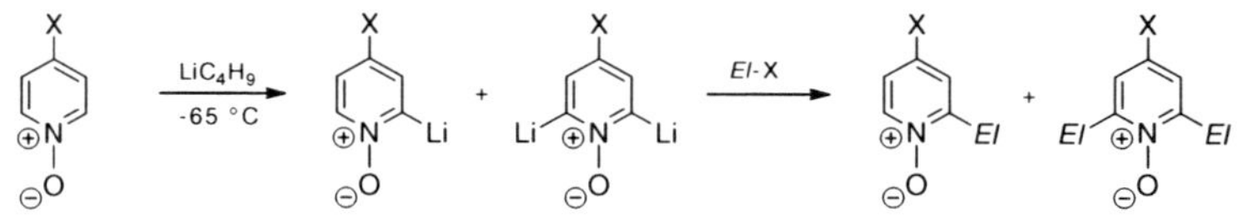

[X = H, Cl, OC_2H_5, CH_3 ; *El*-X = Cl_2, Br_2, H_3CCHO, $(H_2C)_5CO$, CO_2]

Scheme 73 Concomitant mono- and dimetalation of pyridine 1-oxides

More promising might be the activation of the pyridine nucleus with boron trifluoride although so far only preliminary results have been reported. Pyridine [304, 305], 4-methylpyridine [304, 305] and 4-(dimethylamino)pyridine [306] are smoothly α-lithiated by LITMP in DEE at −75°C to afford trapping products in up to 85% yield (Scheme 74). In the absence of boron trifluoride γ-picoline is deprotonated at the methyl group whereas the two other substrates stay inert.

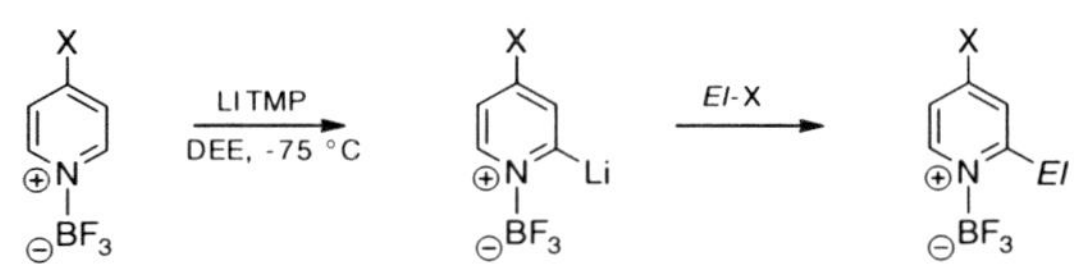

[X = H, CH_3, $N(CH_3)_2$; *El*-X = Cl_2, Br_2, H_3CCHO, $(H_2C)_5CO$, CO_2]

Scheme 74 Metalation and derivatization of pyridine/boron trifluoride adducts

The metalation of *N*-betaines composed of pyridines and hexafluoroacetone must be considered as a curiosity for the time being. Very low reaction temperatures are mandatory. The substrates employed so far were pyridine and 4-*tert*-butylpyridine, the electrophiles methanol-*O*-*d* and chlorotrimethylsilane (Scheme 75) [307].

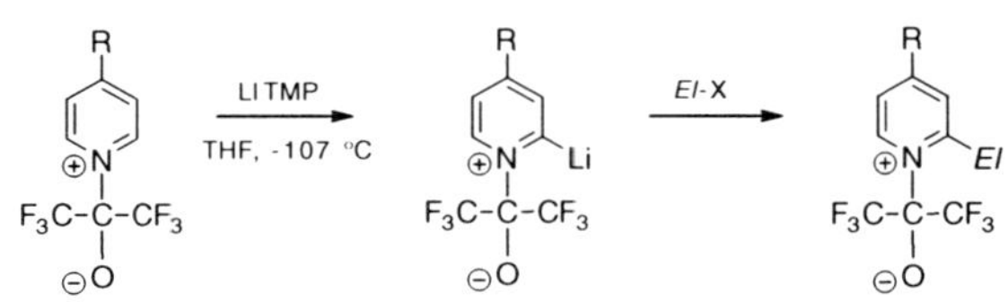

[X = H, $C(CH_3)_3$; *El*-X = $DOCH_3$, $ClSi(CH_3)_3$]

Scheme 75 Metalation of pyridine/hexafluoroacetone adducts

4.2.7 Metalation of Di- and Tetrahydropyridines

Dihydro- and tetrahydropyridines (Sect. 2.3.3) are fragile structures. Their metalated intermediates play therefore a rather marginal role in synthesis-oriented organic chemistry.

When treated with (trimethylsilyl)methylpotassium in THF at −75°C, 1,4-dihydro-1-methylpyridine generates a delocalized 8π-electron species (Scheme 76). This pentadienylpotassium (Scheme 76) condenses with iodomethane to give a 1:1 mixture of 1,2-dihydro-1,2-dimethylpyridine and 1,4-dihydro-1,4-dimethylpyridine in 55% total yield. Analogously 1,4-dihydro-1-methyl- and 1,4-dimethylquinoline are deprotonated at the 4-position. In contrast, both 1,4-dihydro-1-methyl- and −1,4-dimethylpyridine react with the LIC-KOR base at the olefinic 2-position (Scheme 76) [308].

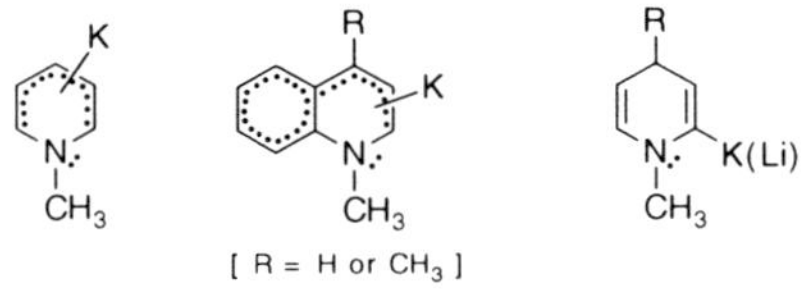

Scheme 76 Deprotonation of allylic (benzylic) vs. olefinic positions

The dilithiation of 1,4-dihydro-1-phenylpyridine has been claimed without providing any experimental details [106, 309]. 1,2-Dihydro-3-cyano-1-methylpyridine and its 1,4-dihydro isomer are readily deprotonated by LIDA at the 4- and 2-positions, respectively (Scheme 77) [310]. 1-*tert*-Butoxycarbonyl-1,4-dihydro-pyridines [311] and 1-methyl-4-pyridone [312] react smoothly at −40°C and −75°C, respectively, with *n*-butyllithium under hydrogen/metal interconversion at the 2-position (Scheme 77).

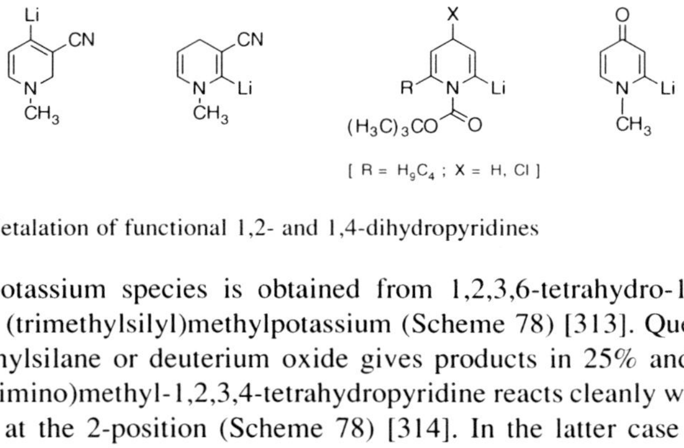

Scheme 77 Metalation of functional 1,2- and 1,4-dihydropyridines

An allylpotassium species is obtained from 1,2,3,6-tetrahydro-1,4-dimethylpyridine and (trimethylsilyl)methylpotassium (Scheme 78) [313]. Quenching with chlorotrimethylsilane or deuterium oxide gives products in 25% and 50% yield. 1-(*tert*-Butylimino)methyl-1,2,3,4-tetrahydropyridine reacts cleanly with *n*- or *tert*-butyllithium at the 2-position (Scheme 78) [314]. In the latter case an electron-withdrawing functional group facilitates the metalation step and provides dipole stabilization of the intermediate.

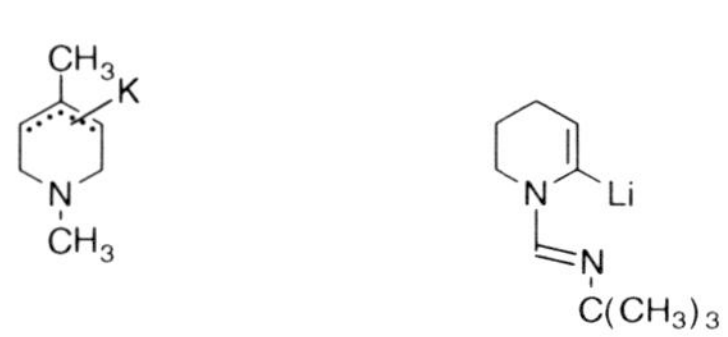

Scheme 78 Once more allylic vs. olefinic metalation

5 Outlook

Our survey began with a general account of familiar and long-established ("classical"?, "conventional"?) methods of pyridine elaboration in order to juxtapose them with more recent skills owed to the advent of polar organometallic chemistry. This dual view will be outdated before long. Stoichiometric or catalytic transition elements are demanding their share of the market.

Transition metal chemistry [315] means in the heterocyclic domain first of all the embracement of the most popular aryl-aryl connecting methods. Although some problems have been encountered with the standard Suzuki–Miyaura protocol [316], many impressive condensations have yet been achieved between bromopyridines or bromoquinolines and (het)arylboronic acids [317] or between halo(het)arenes and pyridylboronic acids [317, 318]. 2,6-Dichloropyridines have even been enticed to perform double substitution [319] with boronic acids and 2-chloro-3,4-iodopyridine was made to react sequentially first at the 4-, next at the 3- and finally at the 2-position [320]. Pyridine *N*-oxides were found to undergo arylation by arylboronic acids at the 2-position when silver nitrate was used as a catalyst and potassium persulfate as an oxidant [321].

An effective protocol has been designed for the Kumada–Tamao–Corriu-like cross-coupling of pyridylmagnesium bromides and five- or six-membered bromoheterocycles [322]. Negishi-type reactions have been realized between pyridyl- and quinolylzinc halides [323] or 1-pyrrylzinc chloride [324] and chloro- or bromo(het)arenes, the catalyst being phosphine-coordinated Pd(II).

Stille coupling between halopyridines, -quinolines or -pyrazines works generally very well [325, 326]. Although not yet extensively explored in the heterocyclic series, Hiyama couplings between chloro-2-(trimethylsilyl)pyridines and bromopyridines or bromothiophenes equally post already a strong record [327, 328].

The challenging Heck–Mizoroki linking of halopyridines, halodiazenes, and haloquinolines with five- or six-membered heterocycles has been intensively explored. The halogen-free component may be an imidazole [329], indole [330], imidazopyrimidine [331], purine [332], adenine [333], furan [334], thiophene [334], benzothiophene [335], oxazole [336], and (benzo)thiazole [334, 337, 338] or a pyridine and quinoline *N*-oxide [339, 340], a pyridine *N*-benzoylimide [341] or pyridine *N*-phenacylide [342].

Almost all such catalytic methods rely on palladium or palladium/copper complexes. Only occasionally other elements may take over. The condensation between 3-iodothiophene and pyrazine is mediated by potassium *tert*-butoxide [343]. The arylation at the 2-position of quinoline is best brought about with arylethylzinc at 130°C (for 20 h) using catalytic amounts of $Ni(COD)_2$ [di(1,5-cyclooctadiene)nickel] and tricyclohexylphosphine [344]. Tetracarbonyldi-μ-chlorodirhodium (i.e., the dimeric dicarbonylchlororhodium) is the catalyst of choice for the substitution of 2-alkylpyridines (pyridine itself fails!) by bromoarenes at 190°C (in 24 h) [345]. Gold catalysis promotes the condensation of

2-bromopyridine with pyrazine [346]. The attachment of 2-chloropyridine to the 10-position of benzo[*h*]quinoline requires a ruthenium/cerium dioxide complex [347]. The latter reaction reminds of course of the *ortho*-palladation and subsequent arylation [348–351] or functionalization [351] of 2-phenylpyridine.

Although still lagging behind arylations, transition element-mediated alkylations of pyridines gain more and more attention [352]. 2-Alkyl substituted pyridines react with ethylene, propene, and other 1-alkenes under zirconium hydride catalysis at 25–50°C [353, 354] or in the presence of a diorganylscandium/tris(pentafluorophenyl)borane complex at 70°C [355] to afford 2-ethyl-, 2-isopropyl- and branched 2-alkylpyridines, respectively. Styrene-type olefines are attached straight-chain, again selectively to the 2-position, if the scandium is replaced by an yttrium complex [355]. 3,3-Dimethyl-1-butene (*tert*-butylethylene) can be attached, again straight-chain, onto the 6-position of 2-alkylpyridines using a Rh(I) catalyst at 165°C [356].

Jointly operating high oxidation-state nickel and palladium complexes are capable of picking up a perfluoroalkyl chain from an adequate bromo compound or carboxylic acid. This chain is eventually delivered to the *ortho* position of 2-phenylpyridine [357].

The regioselectivity can be completely switched from the 2- to the 4-position of pyridine (or quinoline) when $Ni(COD)_2$ [di(1,5-cyclooctadiene)nickel] is used as the catalyst in combination with "MAD" [bis(di-2,6-*tert*-butyl-4-methylphenoxy)-methylaluminum]. Depending on whether a 1-alkene or (trimethylsilyl)ethylene is the coupling partner, the yields range from 48 to 91% [358]. The ruthenium hydride-catalyzed reaction between 3- or 4-acetylpyridine and (triethoxysilyl)ethylene benefits from the neighboring group assistance offered by the carbonyl oxygen and hence proceeds regioselectively. 3-Acetyl-4-[(triethoxysilyl)ethyl]pyridine and its 4,3-isomer are obtained in 56% and 60% yield, respectively [359].

Pyridine, pyrimidine, or quinoline in the presence of palladium(II) acetate, 1,10-phenanthroline, silver carbonate and oxygen attach at 140°C acrylic acid esters or amides to the 3-position. Hydride removal leads to the final products 3-pyrid-3-yl-acrylic acid esters or amides [360].

When a mixture of pyridine and 1-hexene in the presence of nonacarbonyltri-ruthenium and 10 atm of carbon monooxide is heated to 150°C for 16 h, the heterocycle is acylated under CO insertion rather than alkylated. 2-Pentanoylpyridine is formed along with minor amounts (<10%) of branched isomers [361].

The dehydrogenative cross-coupling of arenes or hetarenes with five-membered heterocycles [361] and of electron-rich with electron-poor heterocycles [362–364] belongs to the most fascinating achievements of recent times. This novel reaction pattern is up to now still dominated by five-membered ring heterocycles as their structural and electronic variability offers better chances to prevent the menacing homocoupling. But extensions to the field of six-membered heterarenes will not take long to manifest themselves. The oxidative coupling [365] of pyridine and quinoline *N*-oxides with furans and thiophenes is just a first step and more success stories will come before long.

Why this *tour-de-force* excursion into the area of transition element chemistry? It reveals trend lines of current research activities and allows us to anticipate future evolutions. The pyridine universe is becoming tripartite. Polar organometallic and transition element chemistry enter into competition with the older first-generation methods of pyridine elaboration and functionalization. Competition is good for the trade.

References

1. Abramovitch RA (ed) (1974–1975) The chemistry of heterocyclic compounds: pyridines and its derivatives, vol. 14, supplements part 1–4. Wiley, New York
2. Newkome GR (ed) (1984) The chemistry of heterocyclic compounds: pyridines and its derivatives, vol. 14, supplement part 5. Wiley, New York
3. Joule JA, Mills K, Smith GF (1998) Heterocyclic chemistry, 3rd edn. Stanley Thornes, Cheltenham
4. Joule JA, Mills K (2000) Heterocyclic chemistry, 4th edn. Blackwell, Oxford
5. Spitzner D (1991) Houben-Weyl: methods of organic synthesis, vol E7. Thieme, Stuttgart
6. Spitzner D (2005) Science of synthesis, vol 15. Thieme, Stuttgart
7. Wang Y, Wu JI-C, Li Q, Schleyer PVR (2010) Org Lett 12:4824–4827
8. Clark RD, Jahangir A (1995) Org React 47:1–314
9. Itami K, Kamei T, Mitsudo K, Nokami T, Yoshida J-i (2001) J Org Chem 66:3970–3976
10. Fontana F, Minisci F, Nogueira Barbosa MC, Vismara E (1990) Tetrahedron 46:2525–2538
11. Minisci FV, Vismara E, Fontana F (1989) Heterocycles 28:489–519
12. Schlosser M (2002) in: Schlosser M (ed) Organometallics in synthesis: 2nd manual. Wiley, Chichester, pp 1–352 (spec. 73–81)
13. Eisch JJ, Russo DA (1968) J Organomet Chem 14:P13–P16
14. Clarke AJ, McNamara S, Meth-Cohn O (1974) Tetrahedron Lett 15:2373–2376
15. Ward RL (1961) J Am Chem Soc 83:3623–3626
16. Schmulbach CD, Hinckley CC, Wasmund D (1968) J Am Chem Soc 90:6600–6602
17. Comins DL, O'Connor S (1988) Adv Heterocycl Chem 44:199–267
18. Olah GA, Olah JA, Overchuk NA (1965) J Org Chem 30:3373–3376
19. Katritzky AR, Fan W-Q (1992) Heterocycles 34:2179–2229
20. Räth C (1931) Liebigs Ann Chem 486:95–106
21. Bakke JM, Hegbom I, Øvereeide E, Aaby K (1994) Acta Chem Scand 48:1001–1006
22. Bakke JM (2003) Pure Appl Chem 75:1403–1415
23. Katritzky AR, Scriven EFV, Majumder S, Akhmedova RG, Vakulenko AV, Akhmedov NG, Murugan R, Abboud KA (2005) Org Biomol Chem 3:538–541
24. den Hertog HJ, Wibaut JP (1932) Recl Trav Chim Pays-Bas 51:381–388
25. den Hertog HJ, van der Does L, Landheer CA (1962) Recl Trav Chim Pays-Bas 81:864–870
26. Kirpal A, Reiter E (1925) Ber Dtsch Chem Ges 58:699–701
27. den Hertog HJ, Overhoff J (1930) Recl Trav Chim Pays-Bas 49:552–556
28. den Hertog HJ, Overhoff J (1950) Recl Trav Chim Pays-Bas 69:468–473
29. Kirpal A, Böhm W (1932) Ber Dtsch Chem Ges 65:680–682
30. Fox GJ, Hepworth JD, Hallas G (1973) J Chem Soc Perkin Trans 1:68–69
31. Dolci L (1999) Bioorg Med Chem 7:467–480
32. den Hertog HJ (1945) Recl Trav Chim Pays-Bas 64:85–101
33. Wibaut JP, den Hertog HJ (1945) Recl Trav Chim Pays-Bas 64:55–64
34. Wibaut JP, Nicolai JR (1939) Recl Trav Chim Pays-Bas 58:709–721
35. Cañibano V, Rodríguez JF, Santos M, Sanz-Tejedor MA, Careño MC, Gonzáles G, García-Ruano JL (2001) Synthesis 2175–2179
36. Tschitschibabin A (1915) Chem Zbl 86I:1066

37. Płaźek E, Sucharda E (1928) Ber Dtsch Chem Ges 61
38. Paudler WW, Jovanovic MC (1983) J Org Chem 48:1064–1069
39. den Hertog HJ, Jouwersma C (1953) Recl Trav Chim Pays-Bas 72:125–134
40. Bakke JM, Riha J (1999) J Heterocycl Chem 36:1143–1146
41. Antoine M, Czech M, Gerlach M, Günther E, Schuster T, Marchand P (2011) Synthesis 794–806
42. Cañibano V, Rodriguez JF, Santos M, Sanz-Tejedor MA, Ascensión M, Carreño MC, Gonzáles G, Garcia-Ruano JL (2001) Synthesis 2175–2179
43. Van Der Puy M, Nalwajek D, Wicks GE (1988) Tetrahedron Lett 29:4389–4392
44. Quarroz D (1984) Swiss Patent 644 847 A5 19840831 (to Lonza)
45. Quarroz D (1983) Eur Patent Appl EP 84118 A1 19830727 (to Lonza)
46. Quarroz D (1983) Eur Pat Appl EP 90173 A1 19831005 (to Lonza)
47. Umemoto T, Tomizawa G (1987) Tetrahedron Lett 28:2705–2708
48. Prill EA, McElvain SM (1943) Org Synth Coll 2:419–421
49. Tschitschibabin A, Seide O (1915) Chem Zbl 86I:1064–1065
50. Vorbrüggen H (1990) Adv Heterocycl Chem 49:117–192
51. Rykowski A, van der Plas HC (1985) Synthesis 884–886
52. Zoltewicz JA, Helmick LS, Oestreich TM, King RW, Kandetzki PE (1973) J Org Chem 38:1947–1949
53. Woźniak M, Baránski A, Nowak K, van der Plas HC (1987) J Org Chem 52:5643–5646
54. Fife WK, Scriven EFV (1984) Heterocycles 22:2375–2394
55. Popp FD, Takeuchi I, Kant J, Hamada Y (1987) Chem Commun 1765–1766
56. Kant J, Popp FD (1985) Chem Ind (Lond) 125–126
57. Duarte FF, Popp FD, Holder AJ (1993) J Heterocycl Chem 30:893–896
58. Popp FD, Uff BC (1985) Heterocycles 23:731–740
59. Schlosser M, Cottet F (2002) Eur J Org Chem 4181–4184
60. Fukuhara T, Yoneda N (1993) Chem Lett 509–512
61. Bobbio C, Rausis T, Schlosser M (2005) Chem Eur J 11:1903–1910
62. den Hertog HJ, Schogt JCM, de Bruyn J, de Klerk A (1950) Recl Trav Chim Pays-Bas 69:673–699
63. Parks EL, Sandford G, Yufit DS, Howard JAK, Christopher JA, Miller DD (2010) Tetrahedron 66:6195–6204
64. Sakamoto T, Kaneda S-I, Nishimura S, Yamanaka H (1985) Chem Pharm Bull 33:565–571
65. Schlosser M, Bobbio C, Rausis T (2005) J Org Chem 70:2494–2502
66. Schlosser M, Ruzziconi R (2010) Synthesis 2111–2123
67. Marckwald W (1894) Ber Dtsch Chem Ges 27:1317–1346 (spec. 1323)
68. Tschitschibabin AE, Tyazhelova VS (1923) Chem Zbl 94III:1021
69. Craig LC (1934) J Am Chem Soc 56:231–232
70. Allen CFH, Thirtle JR (1955) Org Synth Coll 3:136–138
71. Kondo S, Nakanishi M, Tsuda K (1984) J Heterocycl Chem 21:1243–1244
72. Graf R, Lederek-Ponzer E, Freiberg L (1931) Ber Dtsch Chem Ges 64:21–26
73. Tschitschibabin AE, Jegorow A (1928) Chem Zbl 99II:1670–1671
74. Case FH (1946) J Am Chem Soc 68:2574–2577
75. Baik W, Luan W, Lee HJ, Yoon CH, Sangho K, Kim BH (2005) Can J Chem 83:213–219
76. Talik T, Talik Z (1964) Tetrahedron (suppl 6):81–88
77. Balz G, Schiemann G (1927) Ber Dtsch Chem Ges 60:1186–1190
78. Hand ES, Baker DC (1989) Synthesis 905–908
79. Osswald P, Scherer O (1933) *DR Pat* 600 706 (to IG Farben)
80. Yoneda N, Fukuhara T (1996) Tetrahedron 52:23–26
81. Wibaut J-P, Holmes-Kamminga WJ (1958) Bull Soc Chim Fr 424–428
82. Roe A (1949) Org React 5:193–228
83. Rutherford KG, Redmond W (1973) Org Synth Coll 5:133–136
84. Camps R (1902) Arch Pharm (Weinheim) 240:345–365 (spec. 358)

85. Zwart C, Wibaut JP (1955) Recl Trav Chim Pays-Bas 74:1062–1069
86. Schmidt-Thome J, Göbel H (1951) Hoppe-Seylers Z Physiol Chem 288:237–243
87. Kijima A, Sekiguchi S (1987) Bull Chem Soc Jpn 60:3597–3602
88. Dimroth P, Radtke V (1979) Liebigs Ann Chem 769–775
89. Sugimoto O, Mori M, Moriya K, Tanji K-i (2001) Helv Chim Acta 84:1112–1118
90. Sun Z, Wang H, Wen K, Fan E (2011) J Org Chem 76:4149–4153
91. Garcia-Lago R, Alonso-Gómez JL, Sicre C, Cid M-M (2008) Heterocycles 75:57–64
92. Iddon B, Mack AG, Suschitzky H, Taylor JA, Wakefield BJ (1980) J Chem Soc Perkin Trans 1:1370–1380
93. den Hertog HJ, De Bruyn J (1951) Recl Trav Chim Pays-Bas 70:182–190
94. Binz A, Schickh O (1935) Ber Dtsch Chem Ges 68:315–324
95. Bobrański B, Kochańska L, Kowalewska A (1938) Ber Dtsch Chem Ges 71:2385–2388
96. den Hertog HJ, Maas J, Kolder CR, Combe WP (1955) Recl Trav Chim Pays-Bas 74:59–68
97. Yamanaka H, Araki T, Sakamoto T (1988) Chem Pharm Bull 36:2244–2247
98. Dehmlow EV, Schulz H-J (1985) Tetrahedron Lett 26:4903–4906
99. Kolder CR, Den Hertog HJ (1953) Recl Trav Chim Pays-Bas 72:285–295
100. Schickh v O, Binz A, Schulz A (1936) Ber Dtsch Chem Ges 69:2593–2605
101. Malinowski M, Kaszmarek L (1987) Synthesis 1013–1015
102. Tschitschibabin AE, Kirsonow AW (1927) Ber Dtsch Chem Ges 60:766–776 (spec. 775)
103. Balicki R, Maciejewski C (2002) Synth Commun 32:1681–1683
104. Kasparian AJ, Sacvarin C, Allgeier AM, Walker SD (2011) J Org Chem 76:9841–9844
105. Bera MK, Hommes P, Reissig H-U (2011) Chem Eur J 17:11838–11843
106. Stout DM, Meyers AI (1982) Chem Rev 82:223–243
107. Birch AJ, Karakhanov EA (1975) Chem Commun 480–481
108. Schneider P (1979) Doctoral Dissertation. Ecole Polytechnique Fédérale Lausanne, pp 55–56
109. de Koning AJ, Budzelaar PHM, Brandsma L, de Bie MJA, Boersma J (1980) Tetrahedron Lett 21:2105–2108
110. Ferles M (1958) Collect Czech Chem Commun 23:479–484
111. Ferles M, Attia A, Šilhánková A (1973) Collect Czech Chem Commun 38:615–619
112. Anderson PS, Kruger WE, Lyle RE (1965) Tetrahedron Lett 6:4011–4017
113. Holik M, Ferles M (1965) Collect Czech Chem Commun 30:3067–3070
114. Biellmann JF, Callot HJ (1968) Bull Soc Chim Fr 1154–1159
115. Biellmann JF, Callot HJ (1968) Bull Soc Chim Fr 1159–1165
116. Chambers RD, Iddon B, Musgrave WKR (1964) J Chem Soc 3736–3739
117. Banks RE, Burgess JE, Cheng WM, Haszeldine RN (1968) J Chem Soc 575–581
118. Chambers RD, Iddon B, Musgrave WKR, Chadwick L (1968) Tetrahedron 24:877–885
119. Matsumoto K, Kannami M, Oda M (2003) Tetrahedron Lett 44:2861–2864
120. Uenishi J, Tanaka T, Wakabayashi S, Oae S, Tsukube H (1990) Tetrahedron Lett 31:4625–4628
121. McKenzie TC, Glass D (1987) J Heterocycl Chem 24:1551–1553
122. Ziegler K, Zeiser K (1930) Ber Dtsch Chem Ges 63:1847–1851
123. Ziegler K, Zeiser K (1931) Liebigs Ann Chem 485:174–192
124. Francis RF, Davis W, Wisener JT (1974) J Org Chem 39:59–62
125. Mazzanti A, Lunazzi L, Lepri S, Ruzziconi R, Schlosser M (2011) Eur J Org Chem 6725–6731
126. Harden DB, Mokrosz MJ, Strekowski L (1988) J Org Chem 53:4137–4140
127. Bursavitch MG, Lombardi S, Gilbert AM (2005) Org Lett 7:4113–4116
128. Nakamura H, Aizawa M, Murai A (1996) Synlett 1015–1016
129. Veer WLC, Goldschmidt S (1946) Recl Trav Chim Pays-Bas 65:793–795
130. Iida T, Wada T, Tomimoto K, Mase T (2001) Tetrahedron Lett 42:4841–4844
131. Inoue A, Kitagawa K, Shinokubo H, Oshima K (2001) J Org Chem 66:4333–4339
132. Ackermann L (2009) Chem Uns Zeit 43:74–83
133. Cook NC, Lyons JE (1966) J Am Chem Soc 88:3396–3403

134. Giam CS, Knaus EE, Pasutto FM (1974) J Org Chem 39:3565–3568
135. Giam CS, Stout JL (1970) Chem Commun 478–478
136. Giam CS, Knaus EE, Lockhart RA, Keener IG (1975) Can J Chem 53:2305–2310
137. Klemm LH, Burel W, Purtzer D (1992) J Heterocycl Chem 29:1673–1675
138. Doyle P, Yates RRJ (1970) Tetrahedron Lett 11:3371–3374
139. Finch N, Gemenden CW (1975) J Org Chem 40:569–574
140. Zhang L-H, Tan Z (2000) Tetrahedron Lett 41:3025–3028
141. Boesvel WM, Hitchcock PB, Lappert MF (1997) Chem Commun 2091–2092
142. Sliwa W (1992) N-substituted salts of pyridines. Springer, Berlin
143. Thiessen LM, Lepoivre JA, Alderweireldt FC (1974) Tetrahedron Lett 15:59–62
144. Wang X, Kauppi AM, Olsson R, Almquist F (2003) Eur J Org Chem 4586–4592
145. Charette AB, Grenon M, Lemire A, Pourashraf M, Martel J (2001) J Am Chem Soc 123:11829–11830
146. Abramovitch RA, Smith EM, Knaus EE, Saha M (1972) J Org Chem 37:1690–1696
147. Abramovitch RA, Coutts RT, Smith EM (1972) J Org Chem 37:3584–3587
148. Colonna M, Risaliti A (1953) Gazz Chim Ital 83:58–61
149. Kato T, Yamanaka H (1965) J Org Chem 30:910–930
150. Andersson H, Banchelin TS-L, Das S, Olsson R, Almquist F (2010) Chem Commun 3384–3386
151. Andersson H, Olsson R, Almquist F (2011) Org Biomol Chem 9:337–346
152. Kellogg RM, van Bergen TJ (1971) J Org Chem 36:1705–1708
153. Tagawa Y, Nomura M, Yamashita H, Goto Y, Hamana M (1999) Heterocycles 51:2385–2398
154. Schiess P, Ringele P (1972) Tetrahedron Lett 13:311–312
155. Fritzsche U, Hünig S (1974) Liebigs Ann Chem 1407–1414
156. Eisenthal R, Katritzky AR (1965) Tetrahedron 21:2205–2213
157. Al-Arnaoult A, Courtois G, Miginiac L (1987) J Organomet Chem 333:139–153
158. Becher J (1980) Synthesis 589–612
159. Becher J (1988) Org Synth Coll 6:640–644
160. Bakke JM, Ranes E, Riha J (1998) Tetrahedron Lett 39:911–912
161. Wittig G, Harborth G (1944) Ber Dtsch Chem Ges 77:315–325 (spec. 322)
162. Gilman H, Haubein AH (1944) J Am Chem Soc 66:1515–1516
163. Parham WE, Piccirilli RM (1977) J Org Chem 42:257–260
164. Wang X, Rabbat P, O'Shea P, Tillyer R, Grabowski EJJ, Reider PJ (2000) Tetrahedron Lett 41:4335–4338
165. Mallet M, Quéguiner G (1979) Tetrahedron 35:1625–1631
166. Bobbio C, Schlosser M (2002), unpublished results
167. Schlosser M, Guio L, Leroux F (2001) J Am Chem Soc 123:3822–3823
168. Gribble GW, Saulnier MG (1980) Tetrahedron Lett 21:4137–4140
169. Gribble GW, Saulnier MG (1993) Heterocycles 35:151–169
170. Chan ASC, Chen C-C, Cao R, Lee M-R, Peng S-M, Lee GH (1997) Organometallics 16:3469–3473
171. Rauchschwalbe G, Schlosser M (1975) Helv Chim Acta 58:1094–1099
172. Comins DL, Morgan LA (1991) Tetrahedron Lett 32:5919–5922
173. Beaulieu F, Snieckus V (1992) Synthesis 112–118
174. Turner JA (1983) J Org Chem 49:3401–3408
175. Mongin F, Trécourt F, Quéguiner G (1999) Tetrahedron Lett 40:5483–5486
176. Schlosser M, Rausis T (2004) Eur J Org Chem 1018–1024
177. Shutske GM, Roehr JE (1997) J Heterocycl Chem 34:789–795
178. Comins DL, Nolan JM (2001) Org Lett 3:4255–4257
179. Zhichkin P, Beer CMC, Rennels WM, Fairfax DJ (2006) Synlett 379–382
180. Wibaut JP, de Jonge AP, van der Voort HGP, Otto PPHL (1951) Recl Trav Chim Pays-Bas 70:1054–1066
181. den Hertog HJ, van der Plas HC (1965) Adv Heterocycl Chem 4:121–144

182. Reinecke MG (1982) Tetrahedron 38:427–498
183. Jamart-Gregoire B, Leger C, Caubère P (1990) Tetrahedron Lett 31:7599–7602
184. Streef JW, den Hertog HJ (1966) Recl Trav Chim Pays-Bas 85:803–816
185. Marsais F, Trécourt F, Bréant P, Quéguiner G (1988) J Heterocycl Chem 25:81–87
186. Caubère P (1993) Chem Rev 93:2317–2334
187. van der Plas HC (1978) Acc Chem Res 31:462–468
188. Streef JW, Den Hertog HJ, van der Plas HC (1985) J Heterocycl Chem 22:985–991
189. Utimoto K, Sakai N, Nozaki H (1974) J Am Chem Soc 96:5601–5602
190. Normant H (1957) Bull Soc Chim Fr 728–733
191. Overhoff J, Proost W (1938) Recl Trav Chim Pays-Bas 57:179–184
192. Wibaut JP, van der Voort HGP (1952) Recl Trav Chim Pays-Bas 71:798–804
193. Wibaut JP, Huls R (1952) Recl Trav Chim Pays-Bas 71:1021–1026
194. Wibaut JP, Heeringa LG (1955) Recl Trav Chim Pays-Bas 74:1003–1020
195. Kochetkov NK, Sokolov SD, Vagurtova NM, Nifantev EE (1960) Dokl Akad Nauk SSSR 133:598–601
196. Chambers RD, Hutchinson J, Musgrave WKR (1965) J Chem Soc 5040–5045
197. Dua SS, Gilman H (1968) J Organomet Chem 12:234–236
198. Berry DJ, Wakefield BJ (1969) J Chem Soc [C] 2342–2346
199. Furukawa N, Shibutani T, Fujihara H (1987) Tetrahedron Lett 28:5845–5848
200. Trécourt F, Breton G, Bonnet V, Mongin F, Marsais F, Quéguiner G (1999) Tetrahedron Lett 40:4339–4342
201. Catel D, Payen O, Chevallier F, Mongin F, Gros PC (2012) Tetrahedron 68:4018–4028
202. Murray A, Foreman WW, Langham W (1948) J Am Chem Soc 70:1037–1039
203. Gilman H, Spatz SM (1951) J Org Chem 16:1485–1494
204. Gilman H, Soddy TS (1957) J Org Chem 22:1715–1716
205. Wommack JB, Barbee TG, Thoeness DJ, Mcdonald MA, Pearson DE (1969) J Heterocycl Chem 6:243–245
206. Cottet F, Marull M, Lefebvre O, Schlosser M (2003) Eur J Org Chem 1559–1568
207. Tupitsyn IF, Zatspeina NN, Kirova AV, Kaminsky YL, Ivanenko AG (1973) Reakts Sposobn Org Soedin 10:143–162 [Chem Abstr 79:114928]
208. Schafman BS, Wenthold PG (2007) J Org Chem 72:1645–1651
209. Cottet F (2003) Doctoral Dissertation. Ecole Polytechnique Fédérale, Lausanne, pp 76–77
210. Cottet F (2002) Internal report, 3 March 2003
211. Cook JD, Wakefield BJ (1968) J Organomet Chem 13:15–23
212. Hertz HS, Kabacinski FF, Spoerri PE (1969) J Heterocycl Chem 6:239–241
213. Rho T, Abuh YF (1994) Synth Commun 24:253–256
214. Hirschberg A, Peterkoski A, Spoerri PE (1962) J Heterocycl Chem 2:209–210
215. Sandosham J, Undheim K (1994) Tetrahedron 50:275–284
216. Langley BW (1956) J Am Chem Soc 78:2136–2141
217. Ulbricht TLV (1959) Tetrahedron 6:225–231
218. Goudgaon NM, Naguib FNM, el Kouni MH, Schinazi RF (1993) J Med Chem 36:4250–4254
219. Asbun W, Binkley SB (1966) J Org Chem 31:2215–2219
220. Ondi L, Lefebvre O, Schlosser M (2004) Eur J Org Chem 3714–3718
221. Schlosser M, Lefebvre O, Ondi L (2006) Eur J Org Chem 1593–1598
222. Gros P, Fort Y, Caubère PJ (1997) Chem Soc Perkin Trans 1:3597–3600
223. Gros P, Viney C, Fort Y (2002) Synlett 628–630
224. Zoltewicz JA, Dill CD (1996) Tetrahedron 52:14469–14474
225. Mongin F, Rebstock A-S, Trécourt F, Quéguiner G, Marsais F (2004) J Org Chem 69:6766–6771
226. Sato N, Yue Q (2003) Tetrahedron 59:5831–5836
227. Couture A, Huguerre E, Grandclaudon P (1989) Tetrahedron Lett 30:183–184
228. Gilman H, Morton JW (1954) Org React 8:258–304
229. Godard A, Jacquelin J-M, Quéguiner G (1988) J Organomet Chem 354:273–285
230. Cailly T, Fabis F, Lemaître S, Bouillon A, Rault S (2005) Tetrahedron Lett 46:135–137
231. Penney JM (2004) Tetrahedron Lett 45:2667–2669

232. Comins DL, Killpack MO (1990) J Org Chem 55:69–73
233. Trécourt F, Mallet M, Marsais F, Quéguiner G (1988) J Org Chem 53:1367–1371
234. Gros P, Fort Y, Caubère P (1998) J Chem Soc Perkin Trans 1:1685–1690
235. Marsais F, Le Nard G, Quéguiner G (1982) Synthesis 235–237
236. Winkle MR, Ronald RC (1982) J Org Chem 47:2101–2108
237. Winkle MR, Ronald RC (1983) Tetrahedron 39:2031–2042
238. Trécourt F, Mallet M, Mongin O, Gervais B, Quéguiner G (1993) Tetrahedron 49:8373–8380
239. Iwao M, Kuraishi T (1983) Tetrahedron Lett 24:2649–2652
240. Miah MAJ, Snieckus V (1985) J Org Chem 50:5438–5440
241. Furukawa N, Shibutani T, Fujihara H (1989) Tetrahedron Lett 30:7091–7094
242. Schlosser M, Mongin F (2007) Chem Soc Rev 36:1161–1172
243. Estel L, Marsais F, Quéguiner G (1988) J Org Chem 53:2740–2744
244. Marsais F, Quéguiner G (1983) Tetrahedron 39:2009–2021 (spec. 2014–2015)
245. Barrière J-C, Bacqué E, Puchault G, Quenet Y, Molherat C, Cassayre J, Paris J-M (1998) Tetrahedron 54:12859–12886
246. Marzi E, Gorecka J, Schlosser M (2004) Synthesis 1609–1618
247. Bobbio C, Schlosser M (2005) J Org Chem 70:3039–3045
248. Marzi E, Schlosser M (2005) Tetrahedron 61:3393–3401
249. Marzi E, Mongin F, Spitaleri A, Schlosser M (2001) Eur J Org Chem 2911–2915
250. Schlosser M (2006) Angew Chem 118:5558–5572; Int Ed 45:5432–5446
251. Gorecka-Kobylinska J, Schlosser M (2009) J Org Chem 74:222–229
252. Schlosser M, Mongin F, Porwisiak J, Dmowski W, Büker HH, Nibbering NMM (1998) Chem Eur J 4:1281–1286
253. Ruzziconi R, Spizzichino S, Lunazzi L, Mazzanti A, Schlosser M (2009) Chem Eur J 15:2645–2652
254. Mazzanti A, Lunazzi L, Ruzziconi R, Spizzichino S, Schlosser M (2010) Chem Eur J 16:9186–9192
255. Ruzziconi R, Spizzichino S, Mazzanti A, Lunazzi L, Schlosser M (2010) Org Bioorg Chem 8:4463–4471
256. Cottet F, Schlosser M. (2004), unpublished results
257. Porwisiak J, Dmowski W (1994) Tetrahedron 50:12259–12266
258. Cottet F, Marull M, Mongin F, Espinosa D, Schlosser M (2004) Synthesis 1619–1624
259. Cottet F, Schlosser M (2004) Tetrahedron 60:11869–11874
260. Cottet F, Schlosser M (2004) Eur J Org Chem 3793–3798
261. Castagnetti E, Schlosser M (2001) Eur J Org Chem 691–695
262. Schlosser M, Castagnetti E (2001) Eur J Org Chem 3991–3997
263. Castagnetti E, Schlosser M (2002) Chem Eur J 8:799–804
264. Leroux F, Castagnetti E, Schlosser M (2003) J Org Chem 68:4693–4699
265. Shimizu M, Hiyama T (2006) Angew Chem 118:5558–5572; Int Ed 45:5432–5446
266. Manteau B, Genix P, Brelot L, Voss J-P, Pazenok S, Giornal F, Leuenberger C, Leroux FR (2010) Eur J Org Chem 6043–6066
267. Büker HH, Nibbering NMM, Espinosa D, Mongin F, Schlosser M (1997) Tetrahedron Lett 39:8519–8522
268. Schlosser M, Marzi E, Cottet F, Büker HH, Nibbering NMM (2001) Chem Eur J 7:3511–3516
269. Mallet M (1991) J Organomet Chem 406:49–56
270. Gros P, Fort Y, Quéguiner G, Caubère P (1995) Tetrahedron Lett 36:4791–4794
271. Marsais F, Bréant P, Ginguène A, Quéguiner G (1981) J Organomet Chem 216:139–147
272. Choppin S, Gros P, Fort Y (2001) Eur J Org Chem 603–606
273. Radinov R, Chanev C, Haimova M (1991) J Org Chem 56:4793–4796
274. Cale AD, Gero TW, Walker KR, Lo YS, Welstead WJ, Jaques LW, Johnson AF, Leonard CA, Nolan JC, Johnson DN (1989) J Med Chem 32:2178–2199 (spec. 2198)
275. Murata N, Sugihara T, Kondo Y, Sakamoto T (1997) Synlett 298–300

276. Güngör T, Marsais F, Quéguiner G (1981) J Organomet Chem 215:139–150
277. Marzi E, Bigi A, Schlosser M (2001) Eur J Org Chem 1371–1376
278. Marzi E (2001) Doctoral Dissertation. Université de Lausanne, pp 43–46, 91–96
279. Marsais F, Laperdix B, Güngör T, Mallet M, Quéguiner G (1982) J Chem Res (S) 278–279
280. Menzel K, Fisher EL, DiMichele L, Frantz DE, Nelson TD, Kress MH (2006) J Org Chem 71:2188–2199
281. Mallet M, Quéguiner G (1982) Tetrahedron 39:3035–3042
282. Effenberger F, Daub W (1991) Chem Ber 124:2119–2125
283. Schlosser M, Marull M (2003) Eur J Org Chem 1569–1575
284. Marzi E, Bobbio C, Cottet F, Schlosser M (2005) Eur J Org Chem 2116–2123
285. Marzi E (2001) Doctoral Dissertation. Université de Lausanne, pp 50–51, 101–102
286. Wotiz JH, Huba F (1959) J Org Chem 24:595–598
287. Moses P, Gronowitz S (1961) Arkiv Kemi 19:119–132
288. Fröhlich J, Hametner C, Kalt W (1996) Monatsh Chem 325–330
289. Rocca P, Cochennec C, Marsais F, Thomas-dit-Dumont L, Mallet M, Godard A, Quéguiner G (1993) J Org Chem 59:7832–7838
290. Arzel E, Rocca P, Marsais F, Godard A, Quéguiner G (1999) Tetrahedron 55:12149–12156
291. Mallet M, Branger G, Marsais F, Quéguiner G (1990) J Organomet Chem 382:319–332
292. Mongin F, Desponds O, Schlosser M (1996) Tetrahedron Lett 37:2767–2770
293. Heiss C, Rausis T, Schlosser M (2005) Synthesis 617–621
294. Awad H, Mongin F, Trécourt F, Quéguiner G, Marsais F (2004) Tetrahedron Lett 45:7873–7877
295. Marull M, Schlosser M (2004) Eur J Org Chem 1008–1013
296. Wada A, Yamamoto J, Hamaoka Y, Ohki K, Nagai S, Kanatomo S (1990) J Heterocycl Chem 27:1831–1835
297. Plé N, Turck A, Couture K, Quéquiner G (1996) Synthesis 838–842
298. Lee JI, Cho SH (1996) Bull Korean Chem Soc 17:868–869
299. Plé N, Turck A, Bardin F, Quéquiner G (1992) J Heterocycl Chem 29:467–470
300. Schlosser M, Wei H-x (1997) Tetrahedron 53:1735–1742
301. Plé N, Turck A, Heynderickx A, Quéquiner G (1997) J Heterocycl Chem 34:551–556
302. Plé N, Turck A, Martin P, Barbey S, Quéquiner G (1993) Tetrahedron Lett 34:1605–1608
303. Abramovitch RA, Knaus EE (1975) J Heterocycl Chem 12:683–690
304. Kessar SV, Singh P, Singh KN, Dutt M (1991) J Chem Soc Chem Commun 570–571
305. Kessar SV, Singh P (1997) Chem Rev 97:721–737
306. Vedejs E, Chen X (1996) J Am Chem Soc 118:1809–1810
307. Taylor SL, Lee DY, Martin JC (1983) J Org Chem 48:4156–4158
308. Schlosser M, Schneider P (1979) Angew Chem 91:515–517; Int Ed (1966) 18:489–490
309. Stout DM, Takaya T, Meyers AI (1975) J Org Chem 40:563–569
310. Schmidt RR, Berger G (1976) Chem Ber 109:2936–2947
311. Comins DL, Weglarz MA, O'Connor S (1988) Tetrahedron Lett 29:1751–1754
312. Meghani P, Joule JA (1988) J Chem Soc Perkin Trans 1:1–8
313. Moret E, Schneider P, Margot C, Stähle M, Schlosser M (1985) Chimia 39:231–232
314. Meyers AI, Edwards PD, Bailey TR, Jagdmann GE (1985) J Org Chem 50:1019–1026
315. Zhao D, You J, Hu C (2011) Chem Eur J 17:5466–5492
316. Kowada T, Matsuyama Y, Ohe K (2008) Synlett 1902–1906
317. Kondolff I, Doucet H, Santelli M (2008) J Heterocycl Chem 45:109–118
318. Billingsley K, Buchwald SL (2007) J Am Chem Soc 129:3358–3366
319. Jäger M, Ericksson L, Bergquist J, Johansson O (2007) J Org Chem 72:10227–10230
320. Daykin LM, Siddle JS, Ankers AL, Batsanov AS, Bryce MR (2010) Tetrahedron 66:668–675
321. Mai W, Yuan J, Li Z, Sun G, Qu L (2012) Synlett 145–149
322. Ackermann L, Potukuchi HK, Kapdi AR, Schulzke C (2010) Chem Eur J 16:3300–3303
323. Çalimsiz S, Sayah M, Mallik D, Organ MG (2010) Angew Chem 122:2058–2061; Int Ed 49:2014–2017

324. Rieth RD, Mankad NP, Calimano E, Sadighi JP (2004) Org Lett 6:3981–3983
325. Choppin S, Gros P, Fort Y (2000) Org Lett 2:803–805
326. Kuo G-H, Wang A, Emanuel S, DeAngelis A, Zhang R, Connolly PJ, Gruninger RH, Sechler J, Fuentes-Pesquera A, Johnson D, Middleton SA, Jolliffe L, Chen X (2005) J Med Chem 48:1886–1900
327. Pierrat P, Gros PC, Fort Y (2005) Org Lett 7:697–700
328. Loüerat F, Gros PC (2010) Tetrahedron Lett 51:3558–3560
329. Jiang W, Guan J, Macielag MJ, Zhang S, Qiu Y, Kraft P, Battacharjee S, John TM, Haynes-Johnson D, Lundeen S, Sui Z (2005) J Med Chem 48
330. Akita Y, Inoue A, Yamamoto K, Ohta A, Kurihara T, Shimizu M (1985) Heterocycles 23:2327–2333
331. Li W, Nelson DP, Jensen MS, Hoerrner RS, Javadi GJ, Cai D, Larsen RD (2003) Org Lett 5:4835–4837
332. Čerňa I, Pohl R, Klepetářová B, Hocek M (2006) Org Lett 8:5389–5392
333. Sahnoun S, Messaoudi S, Brion J-D, Alami M (2009) Org Biomol Chem 7:4271–4278
334. Derridj F, Roger J, Geneste F, Djebbar S, Doucet H (2009) J Organomet Chem 694:455–465
335. Fournier dit Chabert J, Joucla L, David E, Lemaire M (2004) Tetrahedron 60:3221–3230
336. Strotman NA, Chobanian HR, Guo Y, He J, Wilson JE (2010) Org Lett 12:3578–3581
337. Huang J, Chan J, Chen YX, Borths CJ, Baucom KD, Larsen RD, Faul MM (2010) J Am Chem Soc 132:3674–3675
338. Do H-Q, Khan RMK, Daugulis O (2008) J Am Chem Soc 130:15185–15192
339. Campeau L-C, Rousseaux S, Fagnou K (2005) J Am Chem Soc 127:18020–18021
340. Leclerc J-P, Fagnou K (2006) Angew Chem 118:7945–7950; Int Ed 45:7781–7786
341. Larivée A, Mousseau JJ, Charette AB (2008) J Am Chem Soc 130:52–54
342. Xu J, Cheng G, Su D, Liu Y, Wang X, Hu Y (2009) Chem Eur J 15:13105–13110
343. Yanagisawa S, Ueda K, Taniguchi T, Itami K (2008) Org Lett 10:4673–4676
344. Tobisu M, Hyodo N, Chatani N (2009) J Am Chem Soc 131:12070–12071
345. Berman AM, Lewis JC, Bergman RG, Ellman JA (2008) J Am Chem Soc 130:14926–14927
346. Li M-Y, Hua R (2009) Tetrahedron Lett 50:1478–1481
347. Miura H, Wada K, Hosokawa S, Inoue M (2010) Chem Eur J 16:4186–4189
348. Kasahara A (1968) Bull Chem Soc Jpn 41:1272
349. Hartwell GE, Lawrence RV, Smas MJ (1970) Chem Commun 912–912
350. Li W, Yin Z, Jiang X, Sun P (2011) J Org Chem 76:8543–8548
351. Dick AR, Hull KL, Sanford MS (2004) J Am Chem Soc 126:2300–2301
352. Nakao Y (2011) Synthesis 1951–1955
353. Jordan RF, Taylor DF (1989) J Am Chem Soc 111:778–779
354. Rodewald S, Jordan RF (1994) J Am Chem Soc 116:4491–4492
355. Guan B-T, Hou Z (2011) J Am Chem Soc 133:18086–18089
356. Lewis JC, Bergman RG, Ellman JA (2007) J Am Chem Soc 129:5332–5333
357. Dudkina YB, Mikhaylov DY, Gryaznova TV, Sinyashin OG, Vicic DA, Budnikova YH (2012) Eur J Org Chem 2114–2117
358. Nakao Y, Yamada Y, Kashihara N, Hiyama T (2010) J Am Chem Soc 132:13666–13668
359. Grigg R, Savic V (1997) Tetrahedron Lett 38:5737–5740
360. Ye M, Gao G-L, Yu J-Q (2011) J Am Chem Soc 133:6964–6967
361. Moore EJ, Pretzer WR, O'Connell TJ, Harris J, LaBounty L, Chou L, Grimmer SS (1992) J Am Chem Soc 114:5888–5890
362. Stuart DR, Villemure E, Fagnou K (2007) J Am Chem Soc 129:12072–12073
363. Potavathri S, Pereira KC, Govelsky SI, Pike A, LeBris AP, DeBoef B (2010) J Am Chem Soc 132:14676–14681
364. Han W, Ofial AR (2011) Synlett
365. Xi P, Yang F, Qin S, Zhao D, Lan J, Gao G, Hu C (2010) J Am Chem Soc 132:1822–1824

Top Heterocycl Chem (2013) 31: 223–268
DOI: 10.1007/7081_2012_97

Published online: 27 March 2013

Metalation of Pyridazine, Cinnoline, and Phthalazine

Ernst Horkel

Abstract This chapter reviews the metalation of pyridazine, cinnoline, and phthalazine derivatives as well as the usage of the hereby formed organometallic compounds in subsequent reactions. The main topics under discussion are lithiation, magnesation, and zincation, the focus being on deprotonative metalation by properly chosen strong bases, e.g. lithiumdiisopropylamide or lithium 2,2,6,6-tetramethylpiperidide. A short discussion upon boronyl and stannyl derivatives of the three diazines under investigation is enclosed at the end of the chapter.

Keywords Deprotonation of diazines · Lithiation · Magnesation · Zincation

Contents

Abbreviations

Bn	Benzyl
Bu	Butyl

E. Horkel (✉)
Institute of Applied Synthetic Chemistry, Vienna University of Technology, Getreidemarkt 9, 1060 Vienna, Austria
e-mail: ernst.horkel@tuwien.ac.at

Bz	Benzoyl
dba	Dibenzylideneacetone
DEE	Diethyl ether
DMF	Dimethylformamide
DMSO	Dimethyl sulfoxide
equiv	Equivalent(s)
h	Hour(s)
i-Pr	Isopropyl
LDA	Lithium diisopropylamide
LTMP	Lithium 2,2,6,6-tetramethylpiperidide
min	Minute(s)
rt	Room temperature
s-Bu	*sec*-Butyl
TBDMS	*tert*-Butyldimethylsilyl
t-Bu	*tert*-Butyl
THF	Tetrahydrofuran
TIPS	Triisopropylsilyl
TMEDA	*N,N,N′,N′*-tetramethyl-1,2-ethylenediamine
TMP	2,2,6,6-Tetramethylpiperidyl
TMS	Trimethylsilyl

1 Introduction

Among the large number of nitrogen-containing heterocyclic compounds, diazines can be considered as one of the classes gaining highest interest. A lot of natural compounds contain diazine moieties, the most prominent probably being the nucleoside bases thymine, cytosine, and uracil, which are all pyrimidine derivatives. In contrast to the ubiquitous appearance of pyrimidine derivatives, only a small number of natural compounds contain the pyradizine core. A selection is shown in Scheme 1, including the first pyridazine derivative isolated from a natural source, namely pyridazomycin [1, 2]. This antifungal and antibiotic compound was obtained from *Streptomyces violaceoniger sp. griseofucus*. The dihydropyridazines antrimycin [3] and cirratiomycin A were isolated from *Streptomyces xanthocidicus* MG-125-CF1 and *Streptomyces cirratus* 248-Sq2, respectively. Antrimycin shows activity in vitro against *Mycobacterium smegmatis* and *Mycobacterium tuberculosis*, whereas cirratiomycin A is active in vitro against a small band of *Lactobacilli* and some strains of *Streptococcus* [4–8].

However, in the field of pharmaceutical, medicinal, and agro-chemistry, pyridazine derivatives have gained much interest in the last decades, this being reflected by numerous publications and patents contributing to the chemistry and applications of these diazine scaffolds. Among pharmaceutically active pyridazines, two classes have to be highlighted. Amino- and hydrazino-pyridazins are often elements of dopaminergic, serotoninergic, cholinergic, and GABA-ergic ligands [9, 10], whereas derivatives

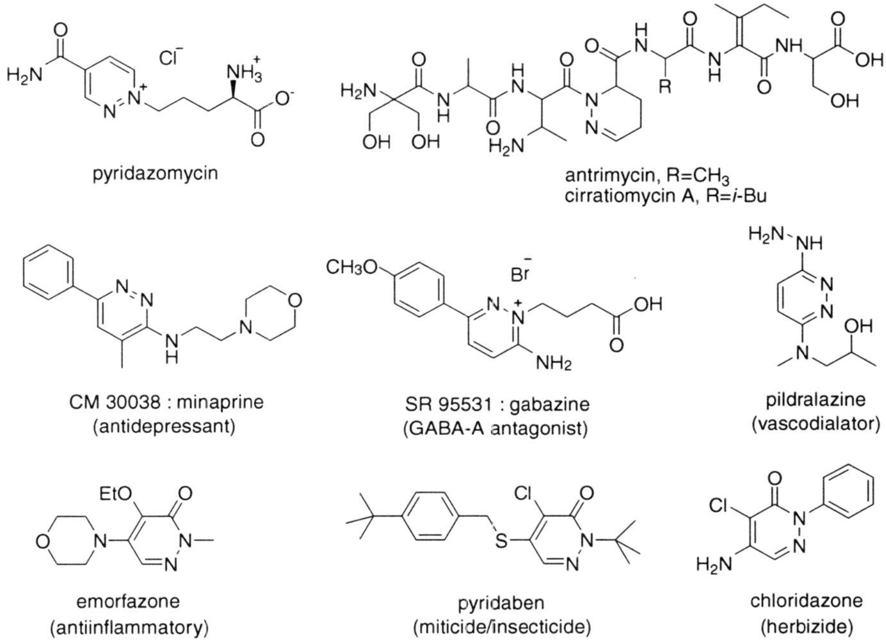

Scheme 1 Naturally occurring and biologically active pyridazine derivatives

containing the 3-(2H)-pyridazinone system show e.g. cardiovascular [11–14] and inflammatory activity [15–23]. Pyridazine derivatives are also widely used in agrochemical applications. Known for decades, chloridazone is still used for weed control. Evidences for the ongoing research activity in the field of pyridazine derivatives are, e.g., the recently commercialization of a selective contact miticide/insecticide, namely pyridaben [24].

Considering all these important applications of various pyridazine derivatives it becomes obvious that there is a strong demand for reliable and flexible synthetic methods that allow both synthesis from scratch and modification of complex structures containing the pyridazine motif. For building up the pyridazine core, numerous methods have been established, most of them being classical heterocycle formation reactions, incorporating cyclocondensations, intramolecular condensations, and cycloadditions [25, 26]. However, for the synthesis of more sophisticated motifs, neither the preparation of the according synthetic precursors is simple nor the methodology is of modular fashion. Hence, the post-cyclization decoration of simple pyridazine derivatives with appropriate substituents is much more appealing and therefore highly desirable. To fulfill this task, metalation and subsequent reaction of the hereby formed organometallic species is often the method of choice. To achieve metalation of (hetero)aryl compounds, two different approaches are generally considered: (a) metal–halogen exchange and (b) deprotonative metalation. Metal–halogen exchange is commonly used in heterocyclic chemistry. The halogen atom of an (hetero)aromatic compound is replaced by the metal

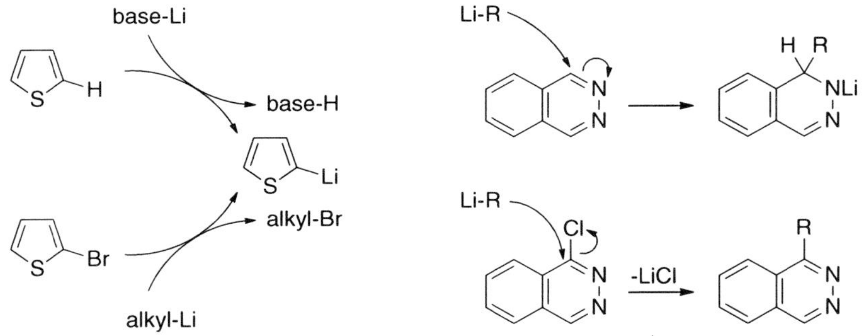

Scheme 2 Commonly used deprotonative lithiation and metal–halogen exchange reaction illustrated on thiophene derivatives (*left*); unwanted addition of alkyl-lithium and nucleophilic substitution (*right*)

Scheme 3 Calculated pK_a values of various *N*-heterocycles

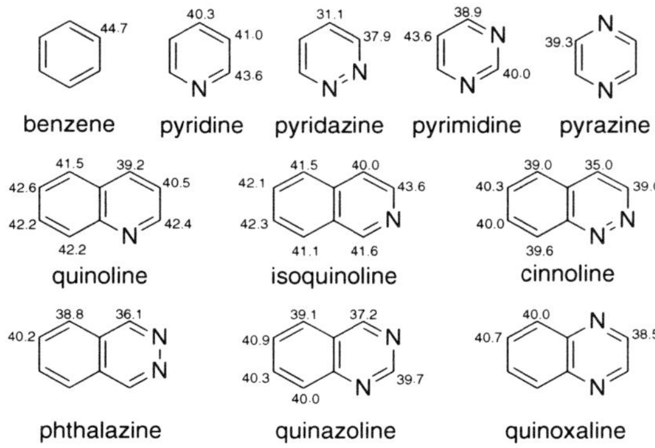

to be introduced by the use of alkylmetal compounds (mostly alkyllithium, e.g., butyllithium) or within Grignard-type reactions (when introducing, e.g., magnesium). However, this metal–halogen exchange is difficult to achieve within the diazine series. For a couple of reasons, competitive reactions often occur, namely a nucleophilic attack of the alkyllithium moiety to the diazine ring [27] or a nucleophilic substitution of the halogen (Scheme 2).

These unwanted side reactions are much less prominent or even negligible when sterically hindered or harder bases, e.g. lithium diisopropylamide (LDA) or lithium 2,2,6,6-tetramethylpiperidide (LTMP), are used instead of alkyllithium compounds. Moving from benzene via the azines to the diazines, a decrease of aromaticity can be observed, this being attributed to the weaker overlap of the p orbitals in the rings. In consequence, the acidity of hydrogens is increased in the same order. A selection of calculated pK_a values of nitrogen-containing heterocycles of interest is given in Scheme 3 [28].

Among these structures, pyridazine was identified as the most acidic compound. In contrast to five-membered heterocycles, where the most acidic proton is situated adjacent to the hetero atom, the situation is inverted when switching to six-membered heterocycles, this means that the weakest acidic proton is bound to the carbon next to the hetero atom (for annealed systems the situation gets even more complex). This fact is attributed to the increased interference of the lone pair of the nitrogen atom and the negative charge of the carbanion due to a smaller interaction angle for the electron clouds in six-membered heterocycles.

2 Lithiation of Pyridazine, Cinnoline, and Phthalazine

2.1 Unsubstituted Diazines

The depronative metalation of bare pyridazine was presented by Plé et al. in 1995 [29]. Lithiation was performed either under "in situ" reaction conditions or for a short lithiation time (6 min) and the consecutive addition of the electrophile. In general, good selectivity for a metalation in position 3 could be observed (Scheme 4, Table 1). However, when TBDMS-Cl was used as an electrophile, the 4-substituted derivative was obtained, which was completely unexpected.

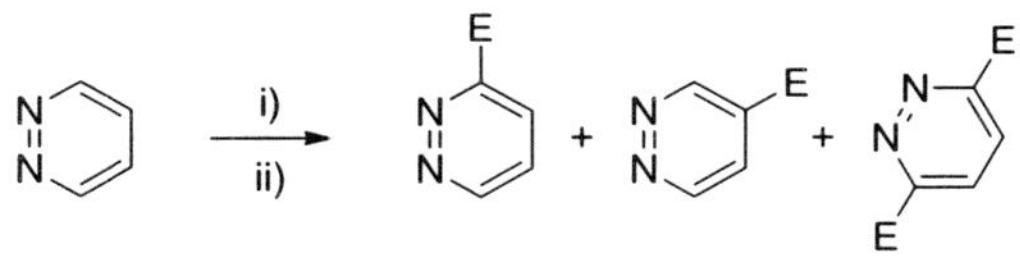

Scheme 4 Lithiation of pyridazine: (i) *n* equiv LTMP @ −75°C, THF, *t*(i) min; (ii) electrophile

2.2 Substituted Diazines

In contrast to the bare diazines, lithiation of substituted pyridazines, cinnolines, and phthalazines was exhaustively studied, as well as the use of the hereby formed lithium species in synthesis. Pioneering work within this area was already conducted in the mid-1990s mainly in the group of Quéguiner [31, 32]. In the following section, the diazines are grouped according to the nature of the DMG.

2.2.1 Halogenated and Alkoxy Substituted Derivatives

First metalations of halogen pyridazine derivatives were described by Rosseels [33]. In this study, 3,6-dichloropyridazine was metalated via halogen–metal exchange using *n*-butyllithium in benzene and subsequently carbonated to give the corresponding carboxylic acid. As already mentioned, pyridazine derivatives are generally very sensitive against nucleophilic attacks, this being an explanation for the rare use of alkyllithium compounds in the field of pyridazine chemistry. However, halogen–metal exchange reactions were studied by applying Barbier reaction conditions [34–36]. Utilizing ultrasound radiation for the activation of the lithium powder being used, iodo

Table 1 Lithiation of pyridazine (**1**)

Entry	Time *t*(i)	*n*	Electrophile	Product (yield%)
1	0	1.3	PhCHO	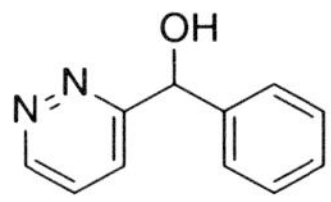 **2a** (33)
2	0	1.3	Ph_2CO	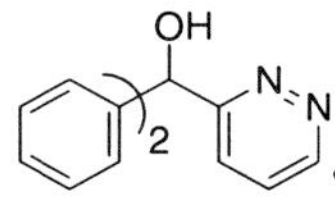 **2b** (47) 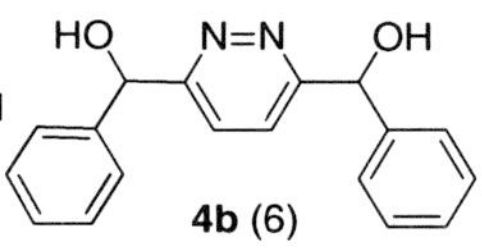 **4b** (6)
3	0	1.3	PhSSPh	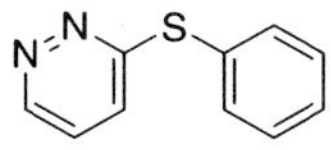 **2c** (7)
4	0	1.3	PhNCS[a]	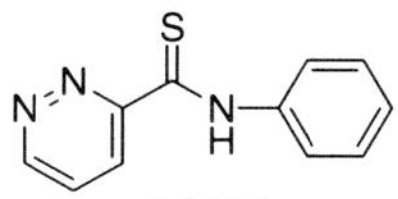 **2d** (20)
5	6	4.0	PhCHO	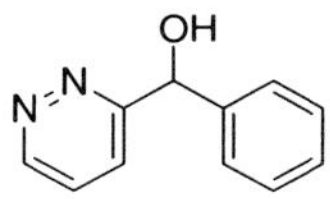 **2a** (31)
6	6	4.0	DCl	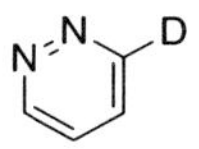 **2e** (32)
7	6	4.0	CH_3CHO	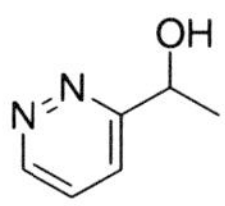 **2f** (26)
8	6	4.0	I_2	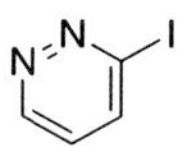 **2g** (16)
9	6	4.0	TBDMS-Cl	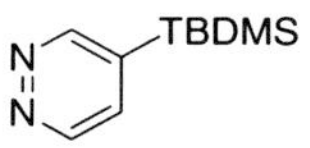 **3h** (10)

[a]Example given in [30]

pyridazines and -cinnolines were metalated at room temperature in THF [37]. Benzaldehyde, hexanal and diphenyldisulfide were used as electrophiles to quench the reaction, giving the desired products in moderate to good yields (35–72%).

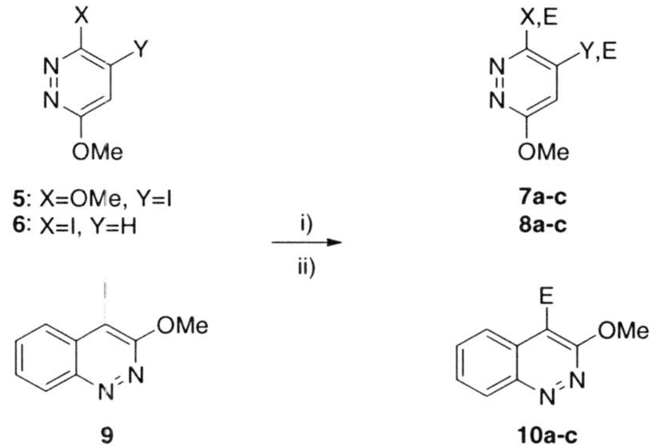

Scheme 5 Ultrasound-assisted Barbier-type reaction. (i) 2.2 equiv Li, 1.1 equiv electrophile, THF @ rt, •))), 30 min; (ii) EtOH

Interestingly, hexanal did not react with the lithiated species of 3-methoxy cinnoline, while in contrast the pyridazine derivatives reacted smoothly (Scheme 5, Table 2).

However, the common approach for the lithiation of diazines is the use of strong bases, the most prominent being LDA and LTMP. Approximately two decades ago, first investigations of deprotonative metalations of pyridazine derivatives were performed. While Turck et al. [38] studied the behavior of 3,6-dichloropyridazine **11** (Scheme 6), Mattson [39] reported the results of quenching metalated 3,6-dimethoxypyridazine **13** with various electrophiles (Scheme 7).

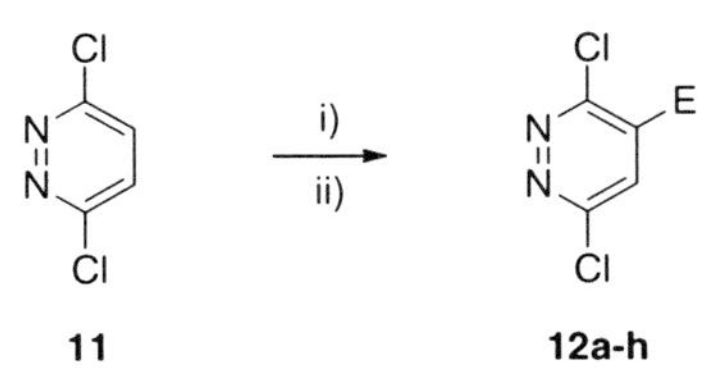

Scheme 6 Lithiation of 3,6-dichloropyridazine; (i) LTMP @ −70°C, THF, 15 min; (ii) electrophile: MeCHO, PhCHO, 4-MeOPhCHO, 2-MeOPhCHO and Ph_2CO (45–65%); $PhCONMe_2$ (28%); I_2 and TMS-Cl (32% and 20%); PhNCS (35%) [30]

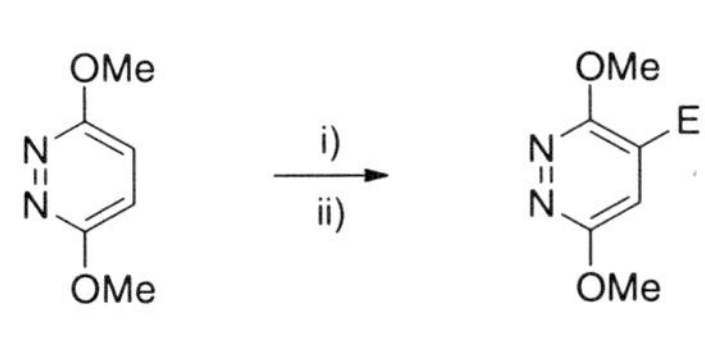

Scheme 7 Lithiation of 3,6-dimethoxypyridazine; (i) LTMP @ −70°C, THF, 90 min; (ii) electrophile: PhCHO, MeI and TMS-Cl (82–86%); PhNCS and MeNCS (85% and 24%) [30]

Table 2 Ultrasound-assisted Barbier-type reaction iodopyridazine and -cinnoline

Entry	Substrate	Electrophile	Product (yield%)
1	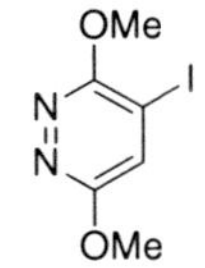 **5**	PhCHO	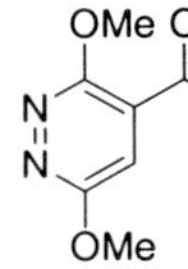 **7a** (72)
2	**5**	$C_5H_{11}CHO$	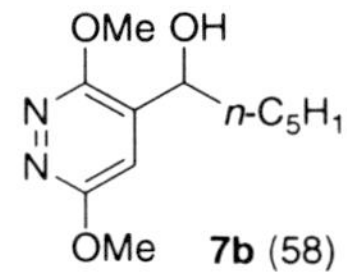 **7b** (58)
3	**5**	PhSSPh	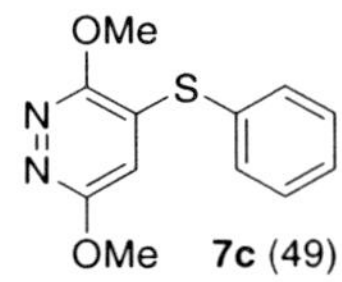 **7c** (49)
4	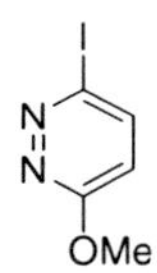 **6**	PhCHO	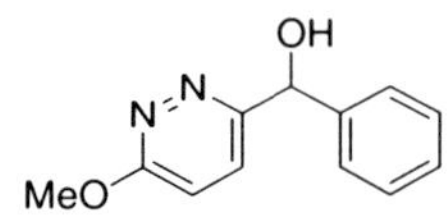 **8a** (65)
5	**6**	$C_5H_{11}CHO$	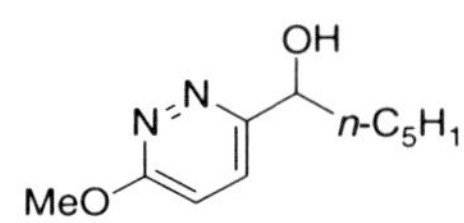 **8b** (50)
6	**6**	PhSSPh	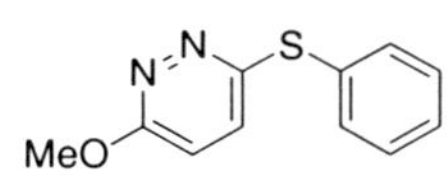 **8c** (38)
7	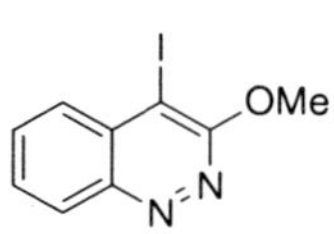 **9**	PhCHO	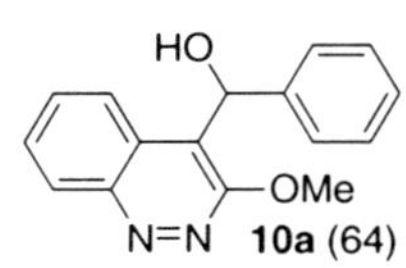 **10a** (64)
8	**9**	$C_5H_{11}CHO$	–
9	**9**	PhSSPh	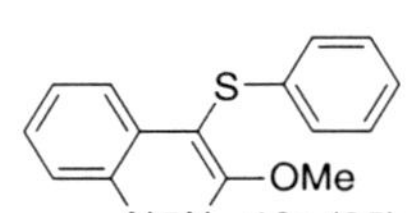 **10c** (35)

In order to investigate possibilities to influence regioselectivity of lithiation reactions using unsymmetrically substituted substrates, detailed studies have been performed. For this purpose, 3-chloro-6-methoxypyridazine was chosen and important process parameters affecting the regioselectivity were systematically altered [40]. Among the parameters of investigation (solvent; reaction temperature; lithiation time; nature of electrophile; nature, amount and concentration of lithium base), the nature of the lithium base turned out to be the key factor for tuning regioselectivity (Scheme 8, Table 3). Comparing the results using acetaldehyde as electrophile with LDA a negligible regioselectivity (**18a/19a** = 40/60) was observed, while in contrast lithiation with the sterically very hindered LB1 (*tert*-butyl-(1-isopropylpentyl) amide) [42] resulted in an excellent ratio of 3/97 with a yield of 89%.

Elongation of the methoxy substituent to methoxyethoxy leads to compounds **20** and **21** (Scheme 9) [43]. Lithiation of 3-(2methoxyethoxy)-pyridazine (**20**) with LTMP occurs selectively in position 4, but reaction with electrophiles gave only low yields of **22a** and **23a**. In contrast, high yields could be obtained with the chloro derivative **21**; however, this was accompanied by a decrease of regioselectivity. By the use of LTMP, lithiation is favored at the *o*-position to the ether functionality, whereas metalation preferably occurs in *o*-position to the chlorine when LDA is used as base (Table 4).

For comparison of the relative directing efficiency of chlorine and fluorine during lithiation reactions, 3-chloro-6-fluoro pyridazine was lithiated using LTMP. The reaction was quenched with acetaldehyde, benzaldehyde, and elemental iodine to give the *o*-fluorine derivatives exclusively, regardless of the metalating reagent being used [44].

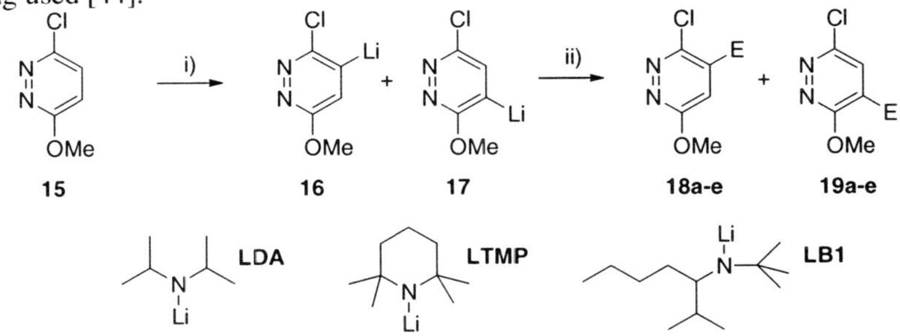

Scheme 8 Lithiation of pyridazine **15**: (i) lithium base @ −70°C, THF, *t*(i) min; (ii) electrophile

Direct competitive experiments for the determination of the regioselectivity of lithiation reactions on 3-fluoro-6-methoxy pyridazine were not performed. However, results obtained from studies incorporating 2-fluoro-6-methoxy pyrazine as substrate were used to predict these data. As hereby fluorine was determined as the most potent *o*-directing substituent within this group, the relative directing efficiency for the substituents under investigation could be given as F > OCH_3 > Cl.

Deprotonative lithiation of bromopyridazine derivative **26** was described by Decrane et al. [45]. By careful selection of the reaction conditions, good yields of **27** were obtained when metalated **26** was treated with 4-methoxybenzaldehyde.

Table 3 Lithiation of pyridazine **15** using various bases

Entry	Base	t(i)	Electrophile	Products	Yield%	Ratio **18/19**
1	LDA	30	CH_3CHO	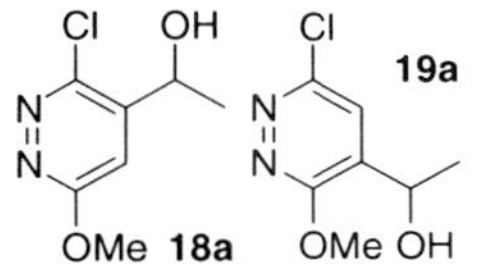	84	40/60
2	LTMP	30	CH_3CHO	**18a**, **19a**	84	15/85
3	LB1	30	CH_3CHO	**18a**, **19a**	89	3/97
4	LTMP	120	PhCHO	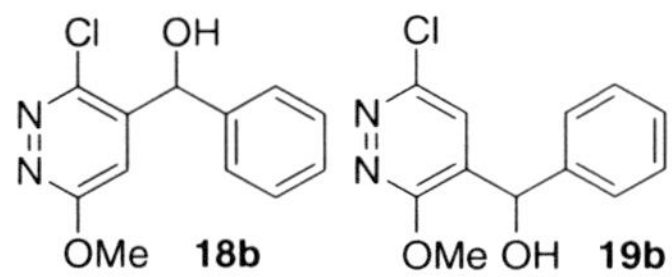	94	15/85
5	LB1	30	PhCHO	**18b, 19b**	80	8/92
6	LTMP	120	CH_3I	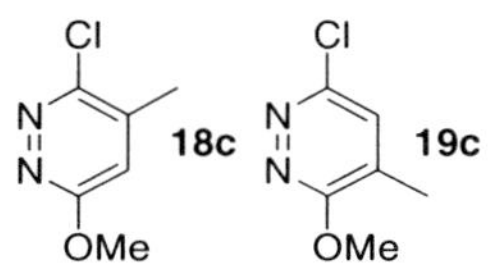	74	20/80
7	LB1	30	CH_3I	**18c, 19c**	63	1/99
8	LTMP	120	I_2	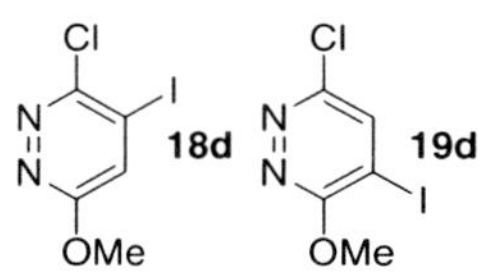	41	20/80
9	LB1	30	I_2	**18d, 19d**	68	9/91
10	LB1		$TosN_3$	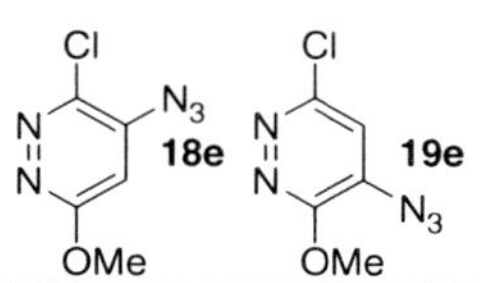	Quant[a]	0/100

[a]Data from [41]

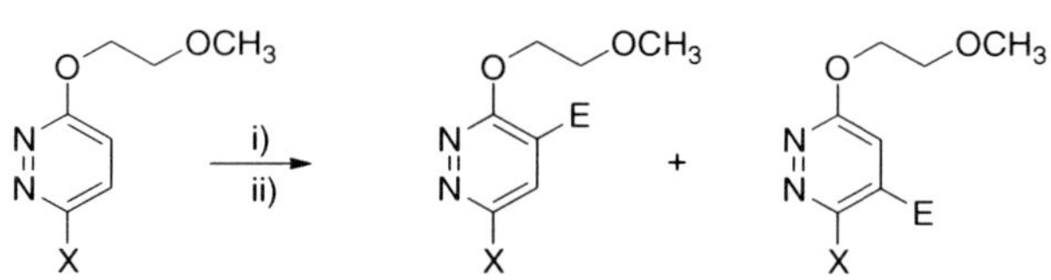

20: X=H
21: X=Cl

22a, 23a
24a, 25a

22b, 23b
24b, 25b

Scheme 9 Lithiation of (methoxyethoxy)pyridazines **20** and **21**: (i) 2.2–4.0 equiv LDA/LTMP @ −70°C, THF, 90 min; (ii) electrophile @ −70°C, THF, 90 min

Table 4 Lithiation of chloropyridazines **20** and **21** using various bases

Entry	Substrate	Base (equiv)	Electrophile	Products (yield%)
1	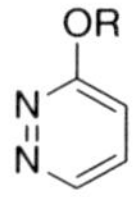 **20**	LTMP (2.2)	CH_3CHO	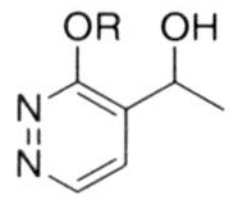 **22a** (12)
2	**20**	LTMP (2.2)	PhCHO	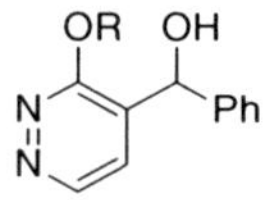 **23a** (15)
3	**20**	LTMP (4.0)	PhCHO	**23a** (26), **23b** (–)
4	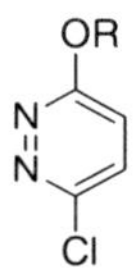 **21**	LTMP (1.2)	CH_3CHO	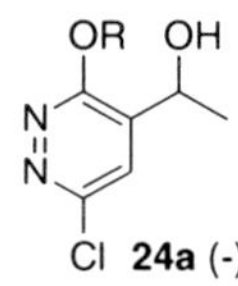 **24a** (-) 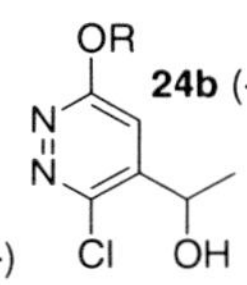 **24b** (-)
5	**21**	LTMP (2.2)	CH_3CHO	**24a** (53), **24b** (36)
6	**21**	LTMP (2.2)[a]	CH_3CHO	**24a** (31), **24b** (31)
7	**21**	LDA (2.2)	CH_3CHO	**24a** (33), **24b** (49)
8	**21**	LTMP (2.2)	PhCHO	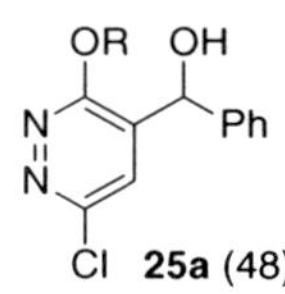 **25a** (48) 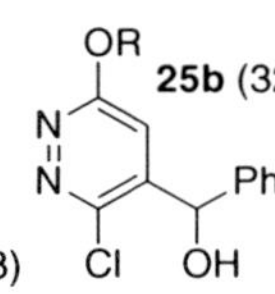 **25b** (32)
9	**21**	LDA (2.2)	PhCHO	**25a** (28), **25b** (55)

[a]Metalation time 150 min

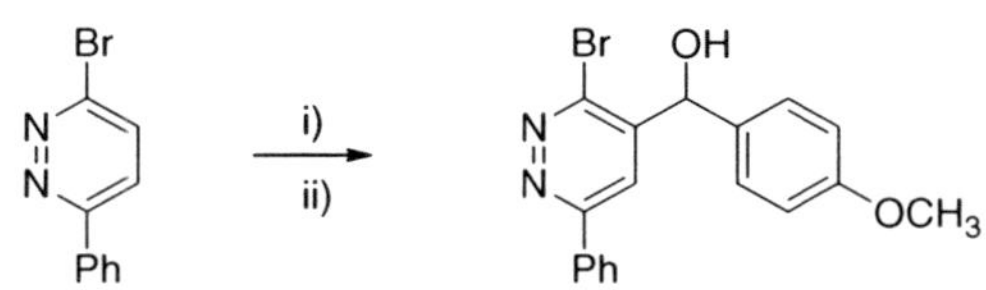

Scheme 10 Metalation of 3-bromo-phenylpyridazine: (i) 2.2 equiv LDA @ −100°C, THF, 30 min; (ii) 4-MeO-PhCHO; yield 84%

Interestingly LDA showed best performance in this particular case, especially when working at −100°C instead of −78°C (Scheme 10).

The metalation of various chloro- and methoxycinnolines was studied in detail by Turck et al. [46]. For both 4-chlorocinnoline and 4-methoxycinnoline (Scheme 11), best results were obtained when LDA was used as metalating reagent. However, in order to achieve satisfactory results when choosing aromatic aldehydes as electrophile, prolonged reaction times and/or elevated temperatures were necessary (Tables 5 and 6).

Scheme 11 Metalation of cinnolines **28** and **29**: (i) 1.1 equiv (for **28**) or 2.2 equiv (for **29**) LDA @ −75°C, THF, 30 min; (ii) electrophile

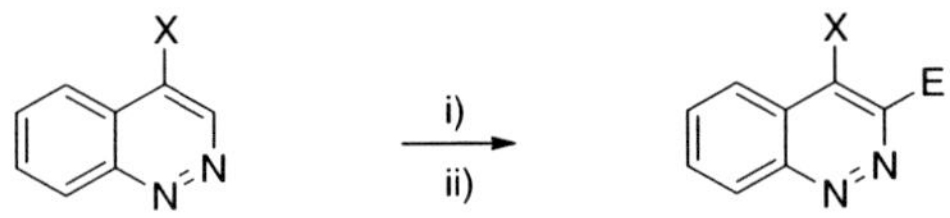

28: X=Cl
29: X=OCH_3

30a-f
31a-e

Table 5 Metalation of cinnoline **28**

Entry	Electrophile (equiv)	Conditions (ii)	Product (yield%)	Recovered s.m. %
1	CH_3CHO (20)	−75°C/2 h	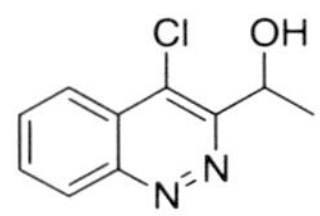 **30a** (89)	–
2	CH_3I (3.6)	−75°C/2 h	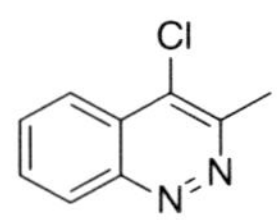 **30b** (86)	–
3	I_2 (1.1)	−75°C/2 h	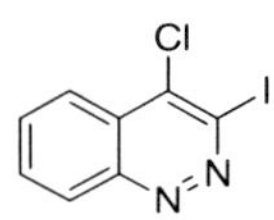 **30c** (70)	10
4	CO_2 (20)	−75°C/2 h	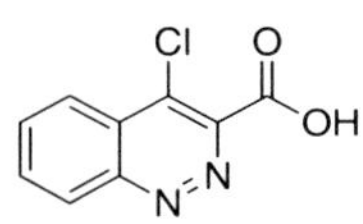 **30d** (57)	–
5	PhCHO (1.1)	−75°C/2 h	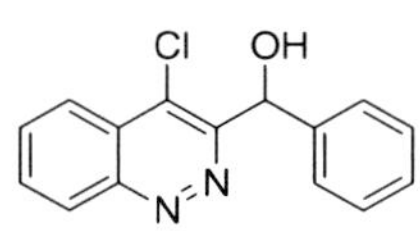 **30e** (35)	57
6	PhCHO (1.1)	−60°C/2 h	**30e** (51)	29
7	PhCHO (5.0)	−75°C/2 h	**30e** (54)	35
8	PhCHO (5.0)	−75°C/4 h	**30e** (88)	3
9	*p*-OCH_3PhCHO (1.1)	−75°C/2 h	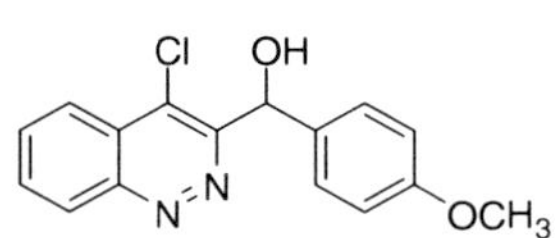 **30f** (47)	39
10	*p*-OCH_3PhCHO (1.1)	−50°C/2 h	**30f** (70)	8

Table 6 Metalation of cinnoline **29**

Entry	Electrophile (equiv)	Conditions (ii)	Product (yield%)	Recovered s.m. %
1	CH_3CHO (20)	−75°C/2 h	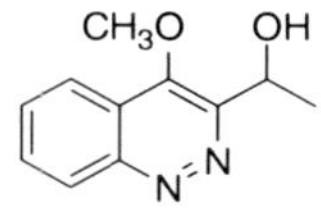 **31a** (81)	–
2	CH_3I (2.2)	−75°C/2 h	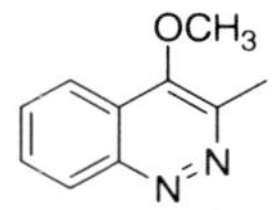 **31b** (34)[a]	18
3	I_2 (2.2)	−75°C/2 h	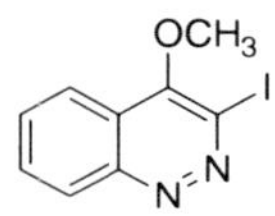 **31c** (77)	9
4	HCOOEt (2.2)	−75°C/2 h	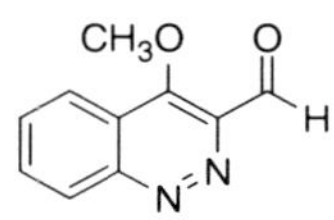 **31d** (57)	29
5	PhCHO (2.2)	−75°C/2 h	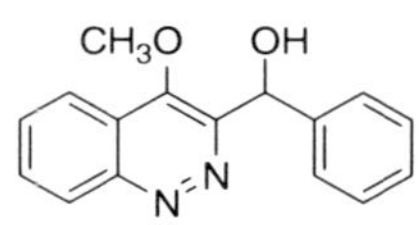 **31e** (81)	–
6	PhCHO (2.2)	−50°C/2 h	**31e** (85)	–

[a]By-product 3-ethyl-4methoxycinnoline was isolated in 24% yield, which was formed via side chain lithiation of **31b**

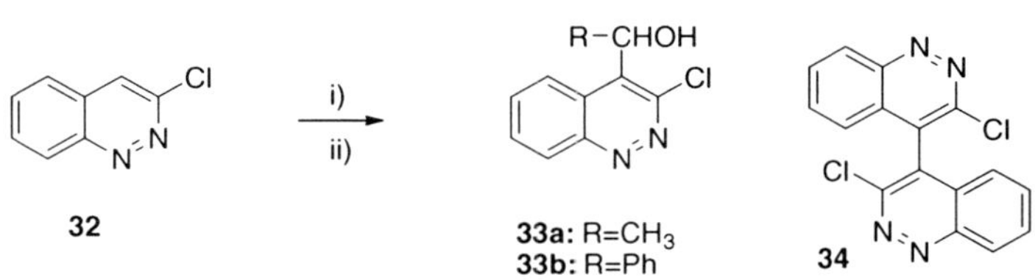

32 **33a:** R=CH3 **33b:** R=Ph **34**

Scheme 12 Metalation of cinnoline **32**: (i) 2.2 equiv LTMP @ −75°C, THF, *t*(i) min; (ii) R-CHO

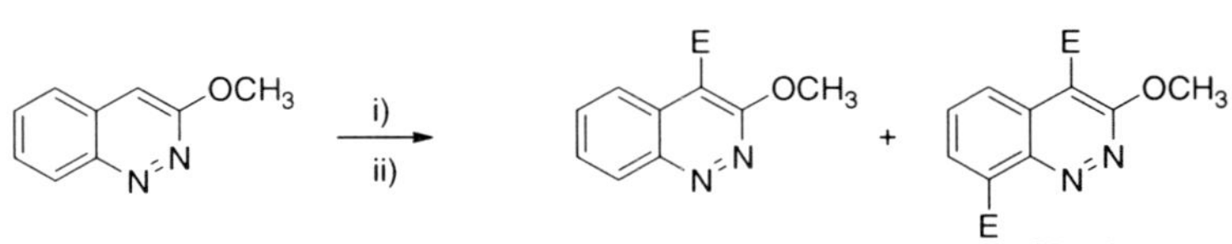

35 **36a-d** **37a-d**

Scheme 13 Metalation of cinnoline **35**: (i) 2.2 equiv LTMP @ −75°C, THF, 30 min; (ii) electrophile

Table 7 Metalation of 3-chlorocinnoline

Entry	t(i)	Electrophile (equiv)	Product (yield%)	Recovered s.m. %
1	30	CH_3CHO (20)	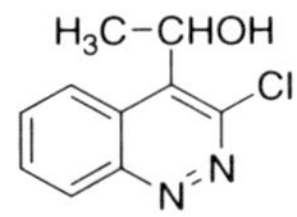 **33a** (60) 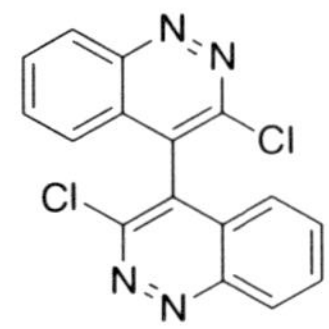 **34** (6)	14
2	60	CH_3CHO (20)	**33a** (81)	9
3	120	CH_3CHO (20)	**33a** (83)	6
4	120	PhCHO (2.2)	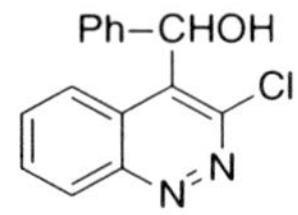 **33b** (88)	–

Regarding the metalation of 3-chloro- and 3-methoxycinnoline (Schemes 12 and 13), LTMP turned out to be the best choice for the base. In general, good yields could be obtained after optimization of reaction conditions. However, still not all side reactions (dimerization of the lithiated cinnolines and metalation of the benzene core of 3-methoxycinnolone when using trimethylsilylchloride as electrophile) could be avoided. Furthermore it is worth mentioning that for 3-chlorocinnoline an unusually high excess of LTMP (2.2 equiv) was necessary to achieve satisfactory results (Table 7). During the workup of lithiated 3-methoxycinnoline quenched with I_2 it was crucial to use water/ethanol instead of hydrochloric acid for the hydrolysis, since otherwise also chlorinated products were obtained (Table 8).

When positions 3 and 4 of cinnoline are blocked by substituents, as it is the case for compound **38**, lithiation occurs at the benzene ring. Subsequent reaction with I_2 led to 8-iodocinnoline **39** in 83% yield (Scheme 14).

In contrast to pyridazine and cinnoline, only few studies have been published for the metalation of phthalazine. In fact, to the best of our knowledge, no synthetic useful metalation at positions 1 and 4 were reported by now. However, when phthalazine was treated with various organolithium compounds, only addition products **42** (8% for **42**a, 91% for **42**b) were obtained via the air-sensitive dihydro derivatives **41** (18% for **41b**, 55% for **41c**; **41a** not isolated, Scheme 15) [27]. The vinyl derivative **42c** could not be obtained due to polymerization processes during the oxidation of **41c**.

In order to achieve at least metalation on the benzene core of phthalazine, 1,4-dimethoxyphthalazine and 1-methoxy-4-phenylphthalazine were treated with various alkyllithium reagents and LTMP (Scheme 16). For the dimethoxy derivative, either starting material was recovered or degradation was observed, whereas substitution/addition occurred for the methoxyphenyl derivative. Attaching a DMG, namely chlorine, to 1,4-dimethoxyphthalazine at position 6 resulted in smooth

Table 8 Metalation of 3-methoxycinnoline

Entry	Electrophile (equiv)	Product (yield%)
1	CH_3CHO (20)	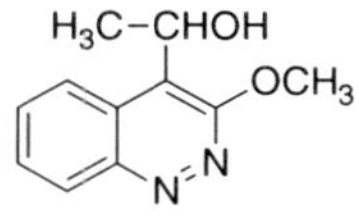 **36a** (86)
2	PhCHO (2.2)	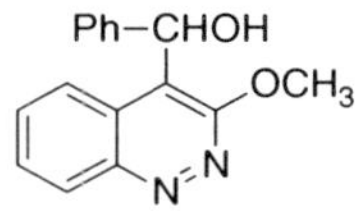 **36b** (91)
3	TMS-Cl (2.2)	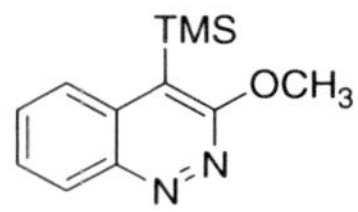 **36c** (74) 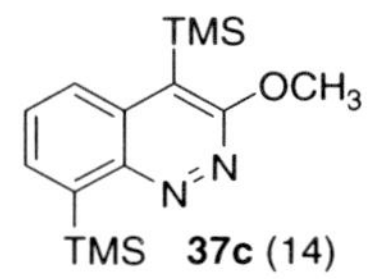 **37c** (14)
4[a]	TMS-Cl (2.2)	**36c** (63), **37c** (18)
5	I_2 (2.2)	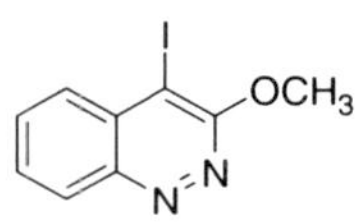 **36d** (73) 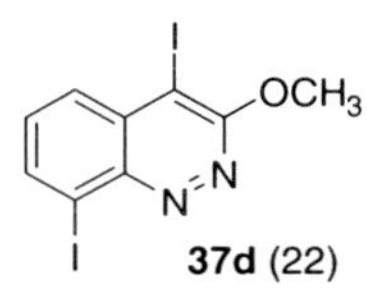 **37d** (22)

[a]In situ trapping technique used; substrate and electrophile are simultaneously added to the metalation agent

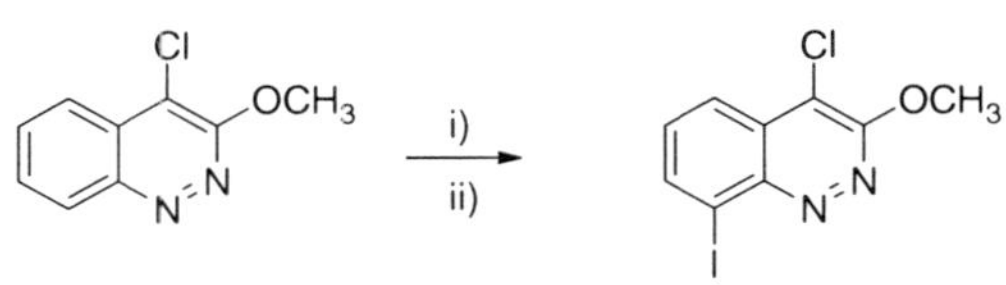

38 **39**

Scheme 14 Metalation of 4-chloro-3-methoxycinnoline: (i) 2.5 equiv TMP @ −75°C, THF, 30 min; (ii) 2.5 equiv I_2 @ −75°C, 2 h; hydrolysis with EtOH/H_2O

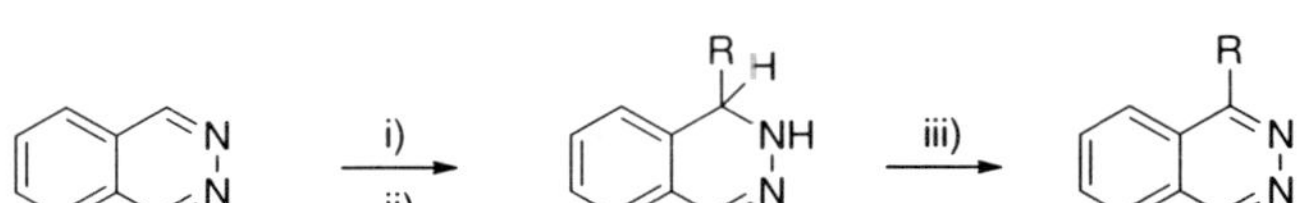

40

41a: R=CH_3
41b: R=Ph
41c: R=vinyl

42a: R=CH_3
42b: R=Ph

Scheme 15 Reaction of phthalazine with various lithium organyls: (i) lithium organyl @ 0°C, Et_2O, 60 min; (ii) H_2O; (iii) solution in ethanol (**41a**) or methanol (**41b**), stirring on air for 3–4 h

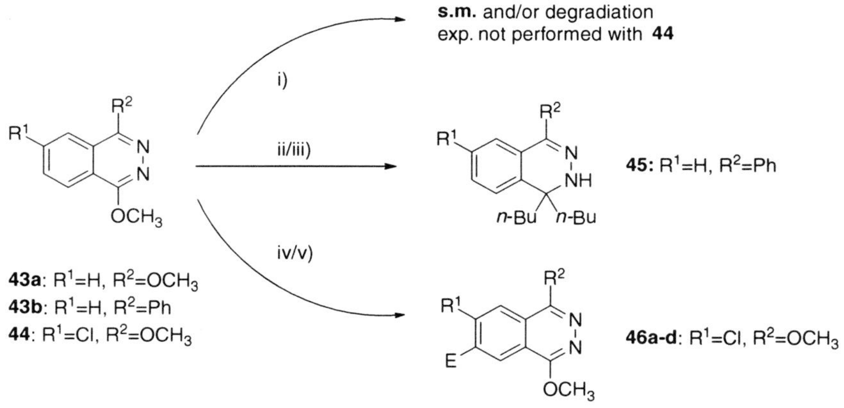

43a: R^1=H, R^2=OCH_3
43b: R^1=H, R^2=Ph
44: R^1=Cl, R^2=OCH_3

Scheme 16 Reaction of phthalazines **43a, b** and **44** with various lithium organyls: (i) LTMP, various conditions; (ii) 1.2 equiv *n*-BuLi @ −78°C, THF, 15 min; (iii) CH_3CHO; (iv) 2.2 equiv *n*-BuLi @ −78°C, THF, 30 min; (v) CH_3CHO (89%), PhCHO (78%), CH_3I (53%, s.m. recovered), I_2 (83%)

lithiation in position 7 using *n*-butyllithium as metalating reagent. Reaction with various electrophiles gave the corresponding 7-substituted phthalazine derivatives in good yields [47].

2.2.2 Carboxamides and Thiocarboxamides

Being readily available, both pyridazine derivatives are interesting starting materials for metalation reactions, as (at least) carboxamides are known to have strong *o*-directing properties. Thus, using a large excess of LTMP, lithiation of pyridazine-3-*tert*-butylcarboxamide (**47**) was achieved solely in position 4 (Scheme 17) [30]. After reaction with an electrophile, the 3,4-disubstituted products **48** were isolated in the case of acetaldehyde and benzaldehyde. However, when iodine was used as electrophile, the 5-iodo-3-carboxamide was isolated. This unexpected result could be explained by a halogen dance reaction [48–50] after the initial metalation, which occurs *ortho* to the carboxamide.

Scheme 17 Metalation of carboxamide **47**: (i) 4.0 equiv LTMP @ −75°C, THF, 60 min; (ii) electrophile @ −75°C; CH_3CHO (96%), PhCHO (89%), I_2 (65%, 20% s.m. recovered)

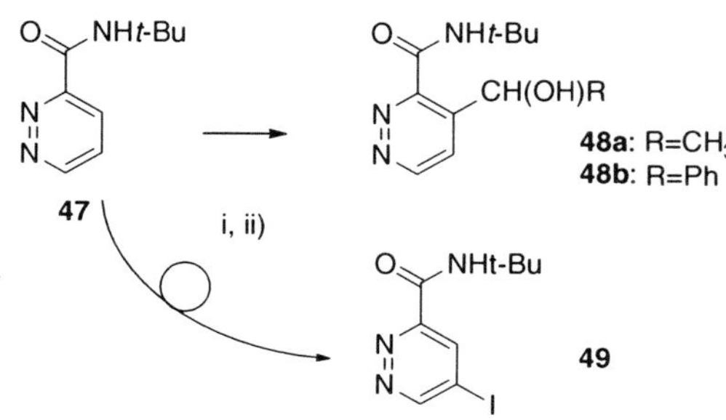

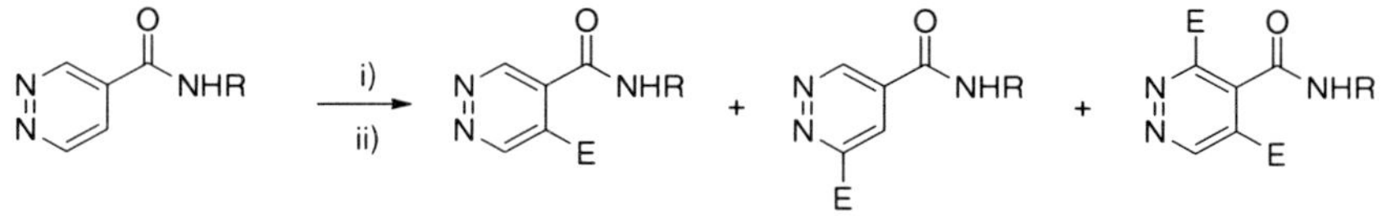

50: R=Bn **51**: R=*t*-Bu → **52a,b** / **55a-c** + **53a,b** / **56a-c** + **54a,b** / **57a-c**

Scheme 18 Metalation of 4-pyridazinecarboxamides **50** and **51**: (i) 4.0 equiv LTMP @ −70°C, THF, *t*(i) min; (ii) electrophile @ −70°C, *t*(ii) min

Table 9 Metalation of 4-pyridazinecarboxamides **50** and **51**

Substrate	*t*(i)	Electrophile	*t*(ii)	Product (yield%)
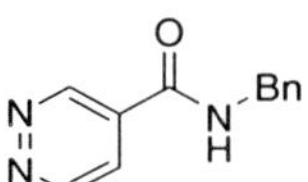 **50**	15	CH_3CHO	90	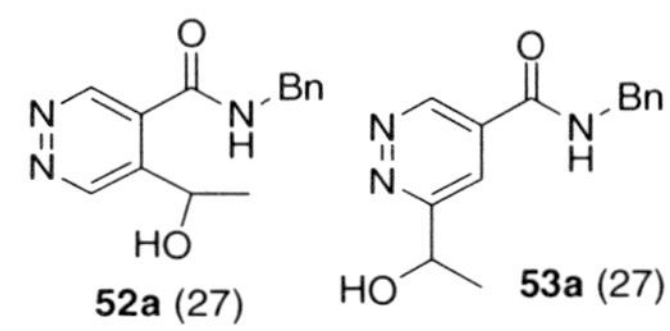 **52a** (27) **53a** (27)
50	15	PhCHO	90	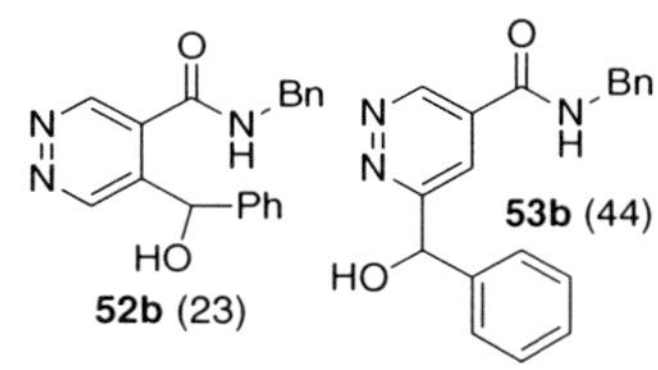 **52b** (23) **53b** (44)
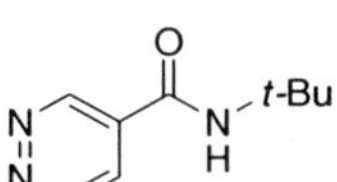 **51**	120	CH_3CHO	90	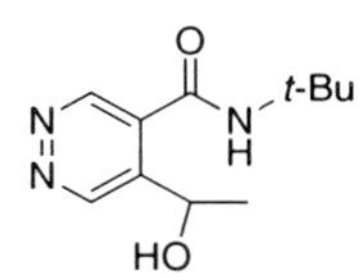 **55a** (71)
51	30	PhCHO	90	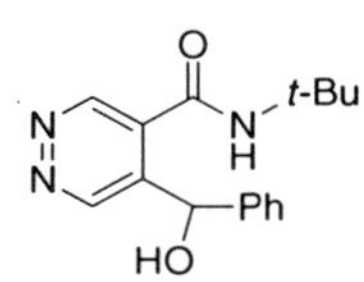 **55b** (79)
51	120	TMS-Cl	120	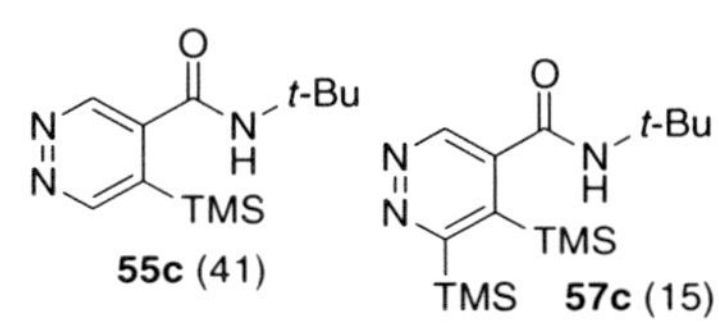 **55c** (41) **57c** (15)

Lithiation of 4-pyridazinecarboxamides was also investigated [51]; however, regioselectivity strongly depended both on the nature of the carboxamide and on the electrophile (Scheme 18). Excellent results were obtained for the *tert*-butyl

Scheme 19 Metalation of thiocarboxamide **58**: (i) 3.1 equiv LTMP @ −75°C, THF, 60 min; (ii) electrophile @ −75°C, 45 min

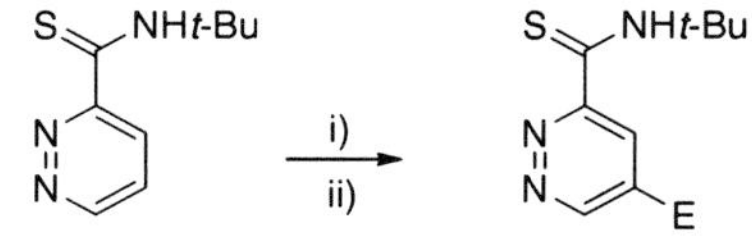

58 → **59a-g**

Table 10 Metalation of thiocarboxamide **58**

Entry	Electrophile	Product (yield)	Recovered s.m. %
1	CH_3CHO	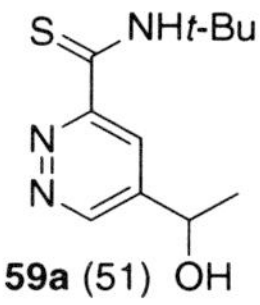 **59a** (51)	4
2	PhCHO	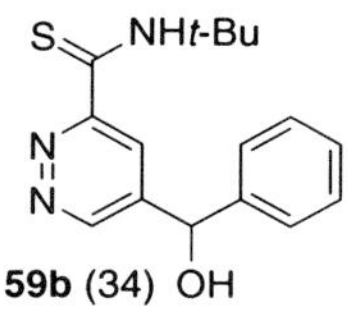 **59b** (34)	–
3	Ph_2CO	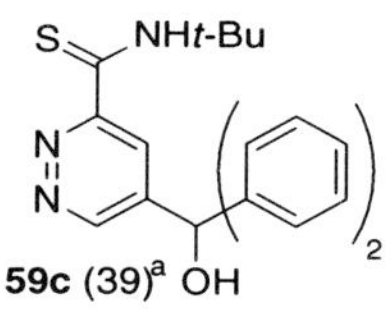 **59c** (39)[a]	–
4	n-Bu_3SnCl	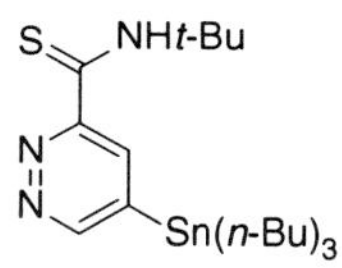 **59d** (63)	–
5	CH_3I	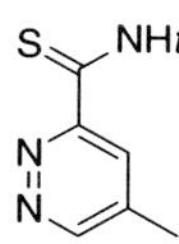 **59e** (41)	–
6	I_2	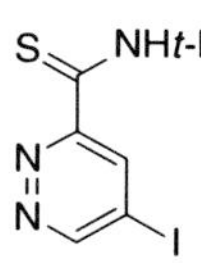 **59f** (14)	23
7	C_2Cl_6	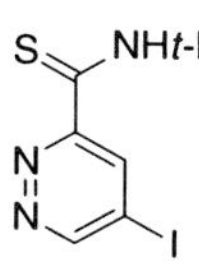 **59g** (17)[b]	–

[a]Also 4-substituted product was isolated (23%)
[b]Also 4,5-disubstituted product was isolated (20%)

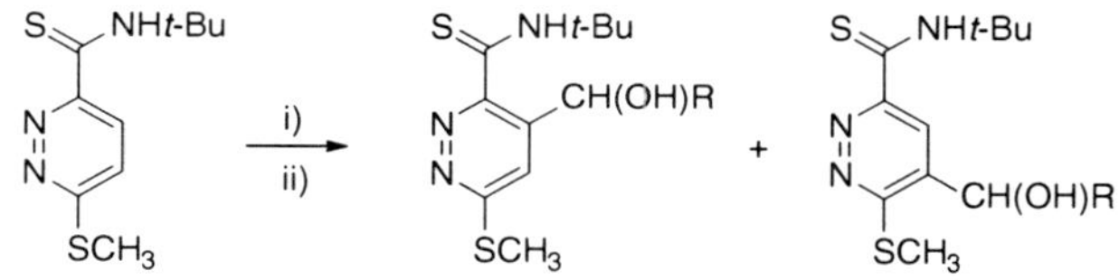

60 **61a**: R=CH_3 **61b**: R=Ph **62a**: R=CH_3 **62b**: R=Ph

Scheme 20 Metalation of thiocarboxamide **60**: (i) base @ −75°C, THF, *t*(i) min; (ii) R-CHO @ −75°C, 45 min

Table 11 Metalation of thiocarboxamide **54**

Entry	Base (equiv)	*t*(i)	Aldehyde	Product (yield%)	Recovered s.m. %
1	LTMP (2.2)	30	CH_3CHO	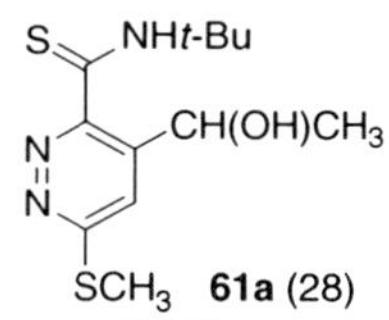 **61a** (28) 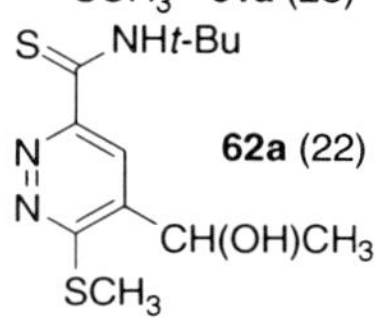 **62a** (22)	28
2	LTMP (3.1)	30	CH_3CHO	**61a** (35), **62a** (25)	–
3	LDA (3.1)	30	CH_3CHO	**61a** (28), **62a** (54)	13
4	LDA (4.1)	30	CH_3CHO	**61a** (7), **62a** (59)	17
5	LDA (3.1)	5	CH_3CHO	**62a** (38)	38
6	LDA (3.1)	90	CH_3CHO	**62a** (43)	22
7	LDA (3.1)	30	PhCHO	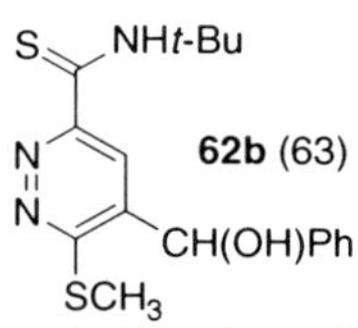 **62b** (63)	10

derivative **51** when acetaldehyde or benzaldehyde was used as electrophile. In contrast, regioselectivity dramatically dropped when either *N*-benzyl derivative was metalated or TMS-Cl was used as electrophile (Table 9).

For the metalation of pyridazine thiocarboxamides, somehow unexpected results were obtained [30]. Instead of the supposed *ortho* lithiation of substrate **58**, mainly meta substitution at position 5 occurred (Scheme 19). Only when benzophenone was used as electrophile, considerable amounts of the 4-substituted derivative were obtained in addition to the main product **59c**. In the case of hexachloroethane, nearly equal amounts of the 5-substituted product and the 5,6-dichloro derivative were isolated (Table 10).

Scheme 21 Metalation of pyridazine **63**: (i) 2.2 equiv LTMP @ −75°C, THF, 60 min; (ii) CH_3CHO @ −75°C, 60 min

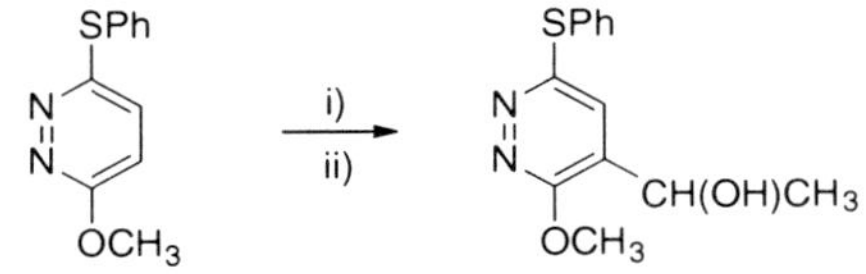

63 **64**

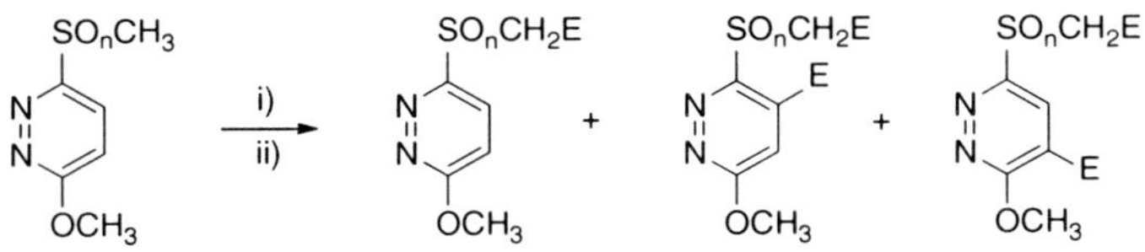

65: n=1
66: n=2

67a-c **70a,b** **68a-c** **71a,b** **69a-c** **72a,b**

Scheme 22 Metalation of pyridazines **65** and **66**: (i) base @ −75°C, THF, 30 min; (ii) electrophile @ −75°C

In order to stabilize the lithiation product of **58**, a chelating methylsulfanyl substituent was attached in position 6 of the pyridazine core (Scheme 20). This indeed increased overall yields; however, results strongly depended on the applied reaction conditions. The usage of LDA as base allowed regioselective deprotonation at position 5, but under these conditions always considerable amounts of starting material were isolated (Table 11).

2.2.3 Sulfur-Based Derivatives

Various sulfur-based derivatives of pyridazine were studied regarding their *ortho*-directing capabilities. Beside few exceptions, most examples are given for the metalation of sulfinyl and sulfonyl derivatives. However, the metalation of phenylsulfanyl pyridazine **63** occurred smoothly in *ortho* position to the methoxy group giving pyridazine **64** in excellent yield (Scheme 21) [52].

When methylsulfinyl or methylsulfonyl pyridazines (**65** and **66**) were treated with LDA or LTMP, lithiation was observed at the methyl group as well as on the pyridazine core (Scheme 22). Selectivity for a lithiation *ortho* to the sulfur-based substituent was good; however when methylsulfone derivatives were metalated, also the 5-substituted derivatives were obtained after reaction with the electrophile [52]. For the methylsulfinyl derivatives **67a** and **68a** diastereomeric mixtures were isolated (Table 12).

In order to circumvent the problem of side chain metalation, *tert*-butyl (Scheme 23, [53]) as well as phenyl (Scheme 24, [52]) sulfoxides and sulfones were lithiated using both LDA and LTMP as base. Regardless of the substituent (*tert*-butyl or phenyl), sulfoxides gave the corresponding *ortho*-metalated products in good to excellent regioselectivity, depending on the electrophile and the base being used (Tables 13 and 14).

Table 12 Metalation of pyridazines **65** and **66**

Substrate	Base (equiv)	t(i)	Electrophile	Product (yield)
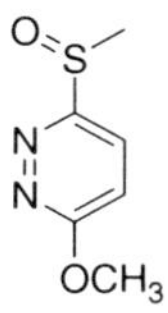 **65**	LDA (1.2)	30	CH_3CHO	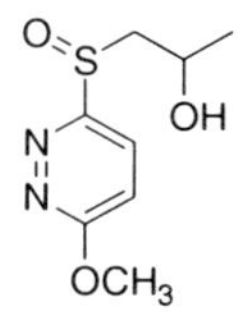 **67a** (42)[a]
65	LDA (2.2)	30	CH_3CHO	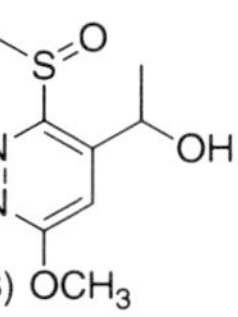 **68a** (43)
65	LDA (2.2)	30	C_2H_5OD, DCl	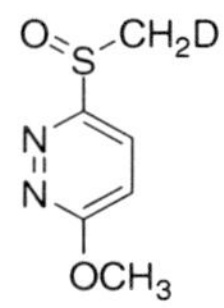 **67b** (29) 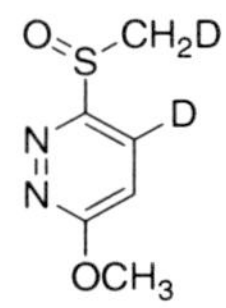 **68b** (67)
65	LDA (2.2)	30	CH_3I	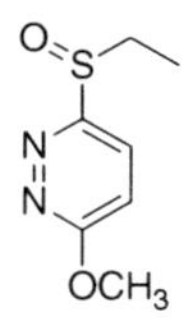 **67c** (5) 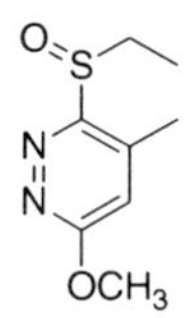 **68c** (18)
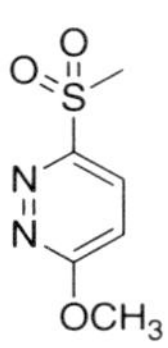 **66**	LDA (2.2)	30	CH_3CHO	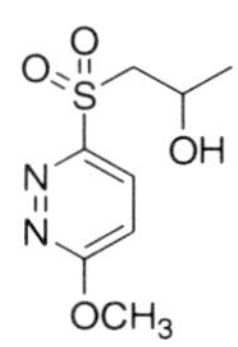 **70a** (60)
66	LTMP (2.2)	30	CH_3CHO	**70a** (43), 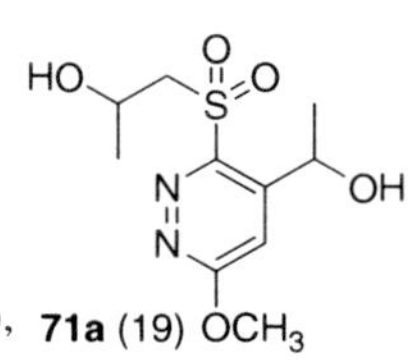 **71a** (19)
66	LTMP (2.2)	30	C_2H_5OD, DCl	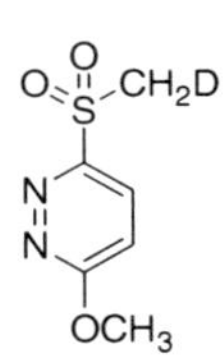 **70b** (23) 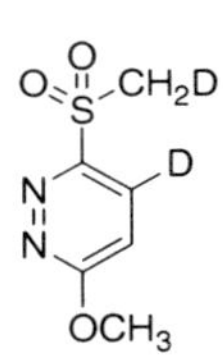 **71b** (21) 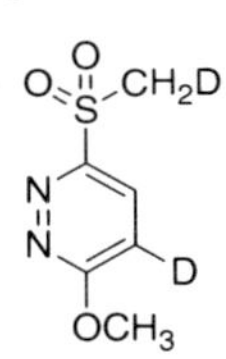 **72b** (13)

[a]Starting material was recovered (18%)

Scheme 23 Metalation of pyridazines **73** and **74**: (i) 2.2 equiv base @ −75°C, THF, 30 min; (ii) electrophile @ −75°C

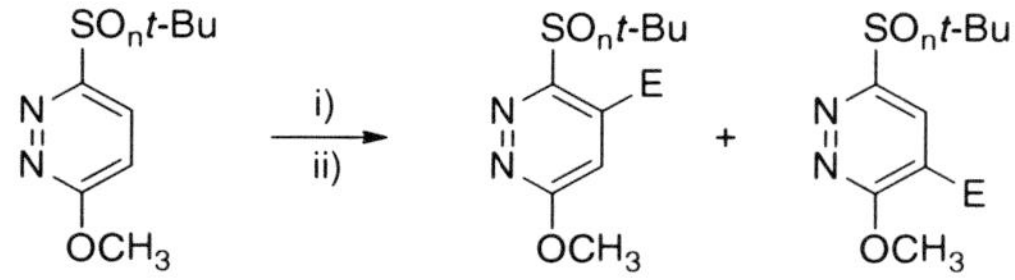

73: n=1 **75a-d** **76a-d**
74: n=2 **77a-c** **78a-c**

Scheme 24 Metalation of pyridazines **79** and **80**: (i) 2.2 equiv base @ −75°C, THF, 35 min; (ii) electrophile @ −75°C

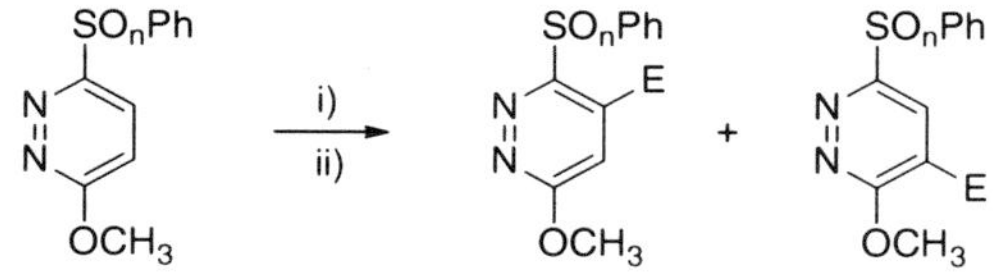

79: n=1 **81a-c** **82a-c**
80: n=2 **83a-c** **84a-c**

Table 13 Metalation of pyridazines **73** and **74**

Substrate	Base	Electrophile	Product (yield%)
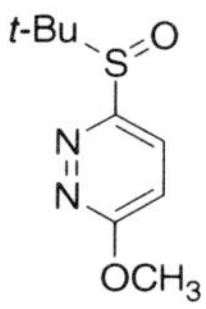 **73**	LDA	CH_3CHO	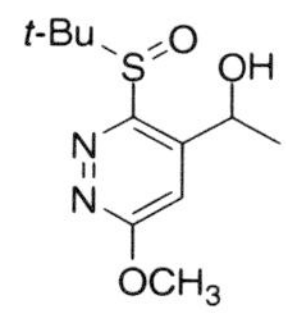 **75a** (89)
73	LTMP	CH_3CHO	**75a** (93)
73	LDA	PhCHO	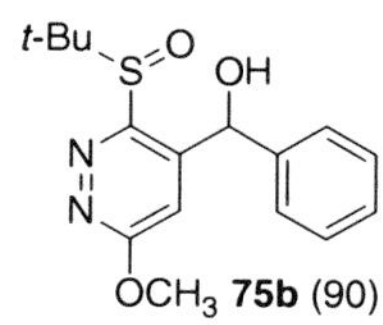 **75b** (90)
73	LTMP	PhCHO	**75b** (99)
73	LDA	CH_3I	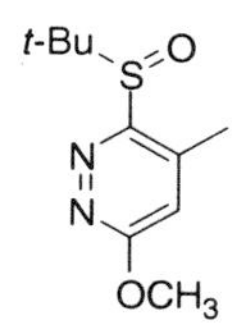 **75c** (53) 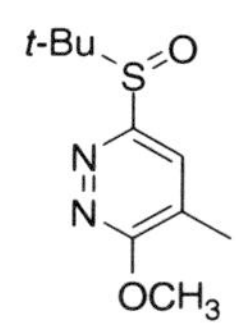 **76c** (4)
73	LTMP	CH_3I	**75c** (76), **76c** (5)
73	LDA	I_2	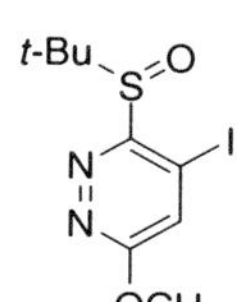 **75d** (58) 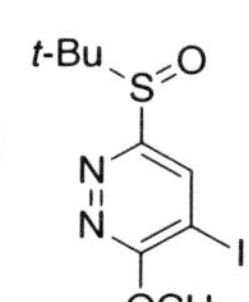 **76d** (12)

(continued)

Table 13 (continued)

Substrate	Base	Electrophile	Product (yield%)
73	LTMP	I_2	**75d** (50), **76d** (13)
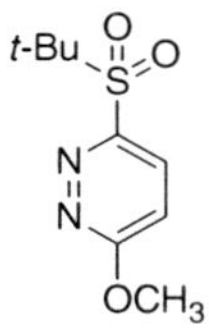 **74**	LDA	CH_3CHO	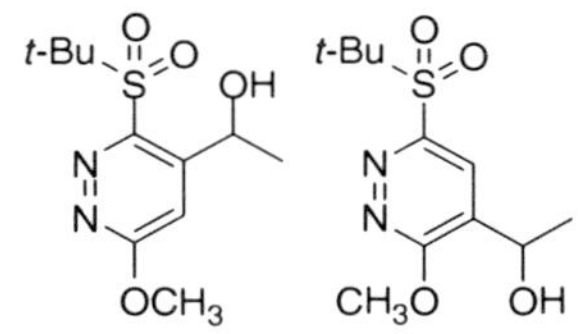 **77a** (18) **78a** (10)
74	LTMP	CH_3CHO	**77a** (59), **78a** (35)
74	LDA	PhCHO	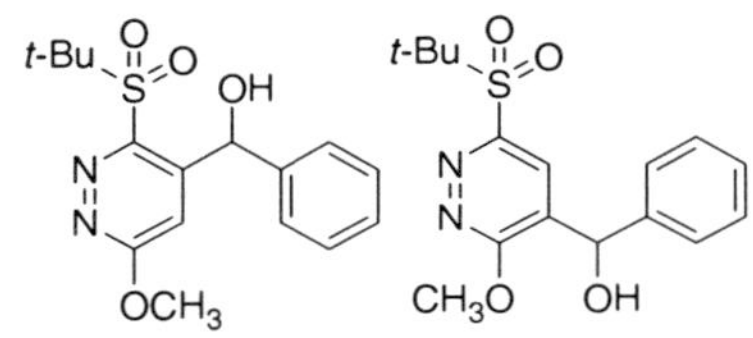 **77b** (36) **78b** (19)
74	LTMP	PhCHO	**77b** (51), **78b** (24)
74	LDA	CH_3I[a]	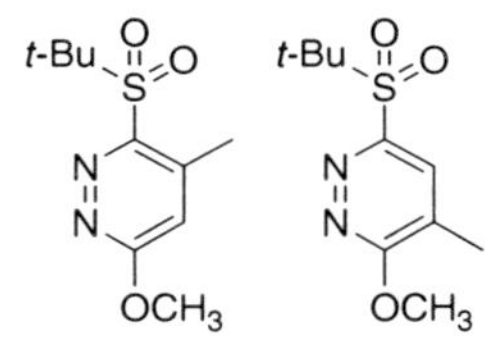 **77c** (53) **78c** (36)
74	LTMP	CH_3I	**77c** (49), **78c** (32)

For substances **75a** and **75b** diastereomeres were obtained
[a]7% of the 4,5-dimethylderivative was obtained

Table 14 Metalation of pyridazines **79** and **80**

Substrate	Base	Electrophile	Product (yield%)
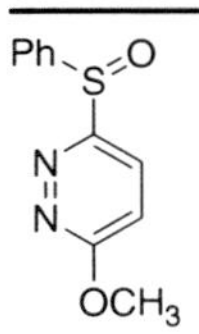 **79**	LDA	CH_3CHO	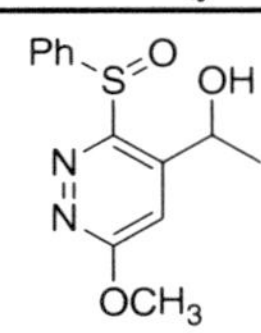 **81a** (78)
79	LTMP	CH_3CHO	**81a** (78)
79	LDA	PhCHO	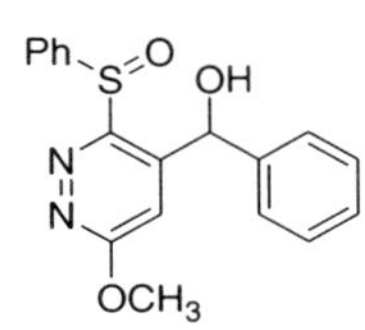 **81b** (78)

(continued)

Table 14 (continued)

Substrate	Base	Electrophile	Product (yield%)
79	LTMP	PhCHO	**81b** (65)
79	LDA	CH_3I[a]	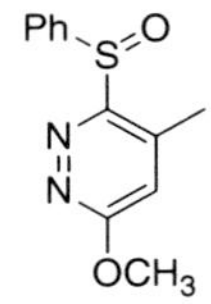 **81c** (64)
79	LTMP	CH_3I	**81c** (55)
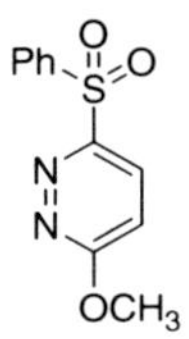 **80**	LDA	CH_3CHO	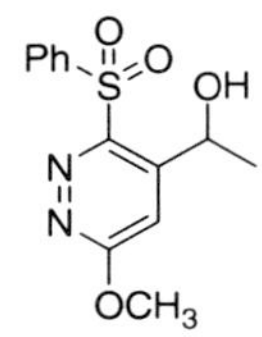 **83a** (41) 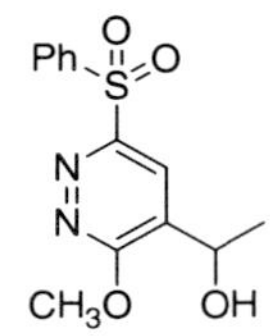 **84a** (17)
80	LTMP	CH_3CHO	**83a** (43), **84a** (13)
80	LDA	PhCHO	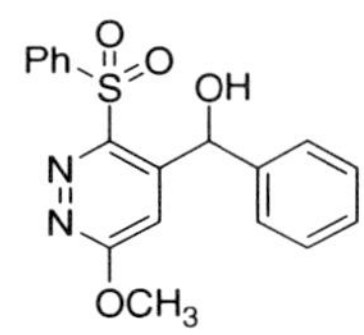 **83b** (60) 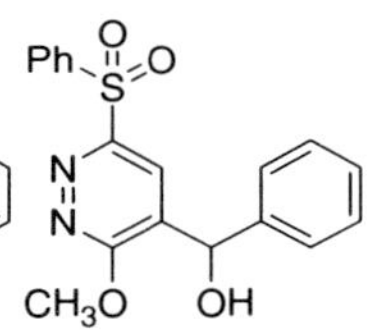 **84b** (26)
80	LTMP	PhCHO	**83b** (53), **84b** (15)
80	LDA	CH_3I[b]	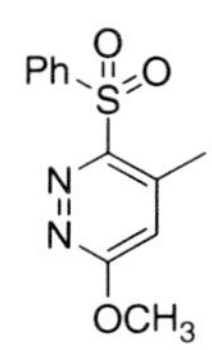 **83c** (33) 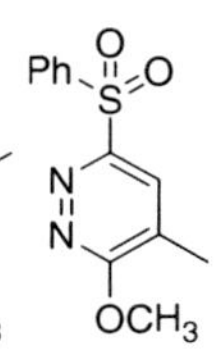 **84c** (13)
80	LTMP	CH_3I[c]	**83c** (34), **84c** (9)

For substances **81a** and **81b** diastereomeres were obtained

[a,b,c]14%, 20% and 18% of the 4,5-dimethylderivative were obtained

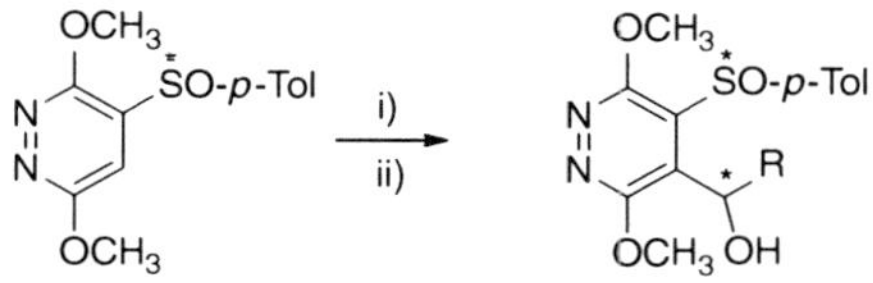

85a: (S)
85b: (R)

86a (SR), 86b (RS): R=CH3)
87a (SR), 87b (RS): R=C2H5
88a (SR), 88b (RS): R=Ph

Scheme 25 Metalation of pyridazines **85a** and **85b**: (i) 3.2 equiv LDA/LTMP @ −75°C, THF, 60 min; (ii) electrophile @ −75°C; **86**: yields 53–76%, de = 96–99%, **87**: yields 35–40%, de = 94–97%, **88**: yields 30–68%, de = 93–98%

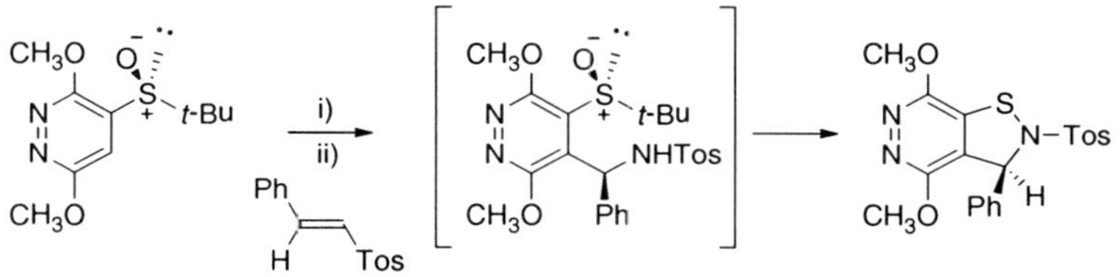

89 90 91 (S), 55%, 76% ee

Scheme 26 Metalation of pyridazine **89**: (i) 3.2 equiv base @ −75°C, THF, 60 min; (ii) tosylimine **90** @ −75°C, 90 min

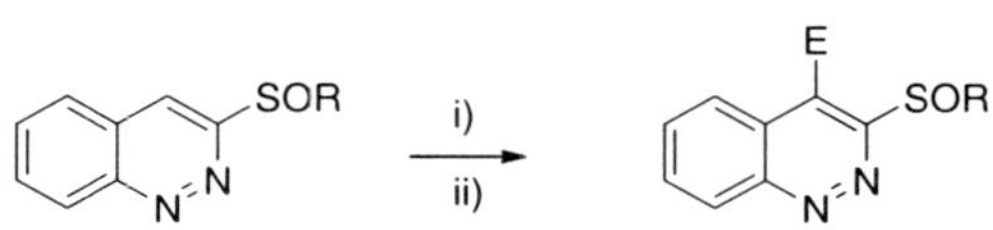

92: R=t-Bu
93: R=p-Tol

94a-g
95a,b

Scheme 27 Metalation of 3-sulfinylcinnolines **92** and **93**: (i) 2.1 equiv base @ −78°C, THF, 30 min for LDA, 60 min for LTMP; (ii) electrophile, 5 min for DCl, 60 min for all other electrophiles

Lithiation of enantiomerically pure sulfoxides was also performed. While metalation of 4-(*p*-tolylsulfinyl)-3,6-dimethoxypyridazines **85a,b** and subsequent reaction with various aliphatic aldehydes and benzaldehyde led to the expected alcohols in high yields and diastereomeric excess (Scheme 25) [54], lithiation of the 4-(*tert*-butylsulfinyl) derivative **89** and reaction with *N*-tosylimine **90** gave sulfenamide **91** via an elimination–cyclization mechanism (Scheme 26) [55].

Attempts were made to metalate sulfinyl derivatives of cinnoline [56]. Both 3-sulfinyl and 4-sulfinyl derivatives were under investigation. Due to the affinity of the sulfinyl group in position 4 to undergo nucleophilic substitutions, no or only unexpected metalation products were obtained when 4-*tert*-butylsulfinyl cinnoline was lithiated applying various reaction conditions. As cinnolines having leaving groups attached in position 3 are less prone to undergo substitution reactions, compounds **92** and **93** were tested and metalated successfully (Scheme 27). The

Table 15 Metalation of 3-sulfinylcinnolines **92** and **93**

Substrate	Base	t(i)	Electrophile	Product (yield%)	Recovered s.m %
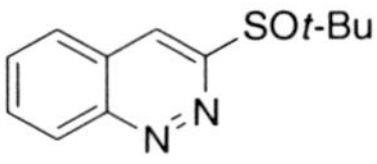 **92**	LDA	30	DCl	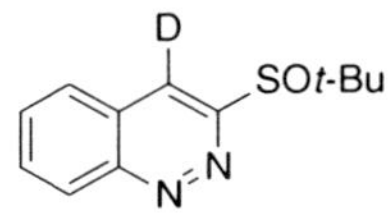 **94a** (77)	12
92	LTMP	60	CH_3CHO	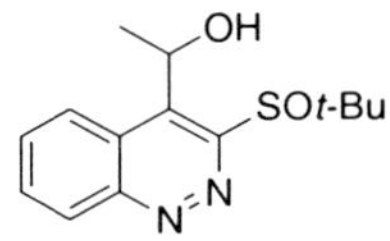 **94b** (90)	–
92	LTMP	60	PhCHO	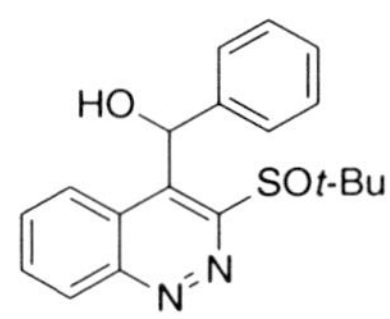 **94c** (94)	–
92	LTMP	60	*p*-MeOPhCHO	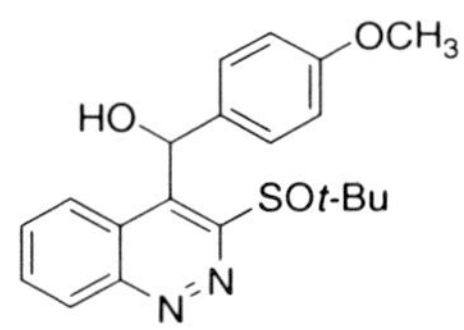 **94d** (90)	–
92	LTMP	60	$(CH_3)_3CCHO$	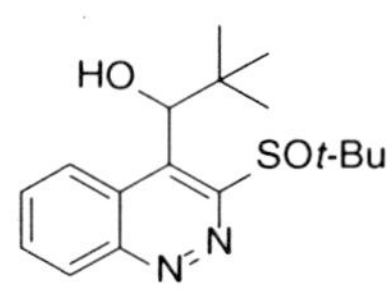 **94e** (70)	16
92	LTMP	60	I_2	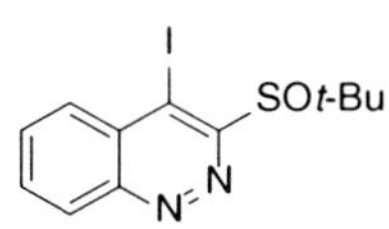 **94f** (85)	–
92	LTMP		*n*-Bu_3SnCl	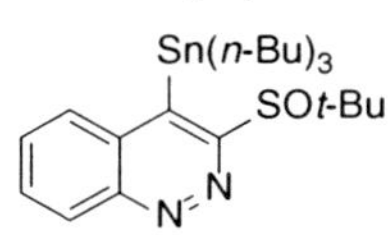 **94g** (67)	19
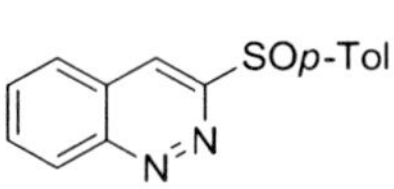 **93**	LTMP	30	DCl	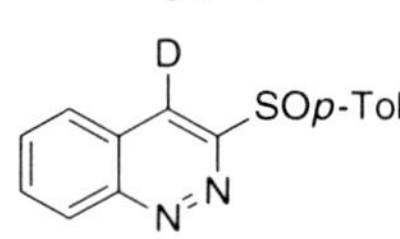 **95a** (56)	–

(continued)

Table 15 (continued)

Substrate	Base	*t*(i)	Electrophile	Product (yield%)	Recovered s.m %
93	LDA	30	DCl	**95a** (81)	–
93	LDA	30	*p*-MeOPhCHO	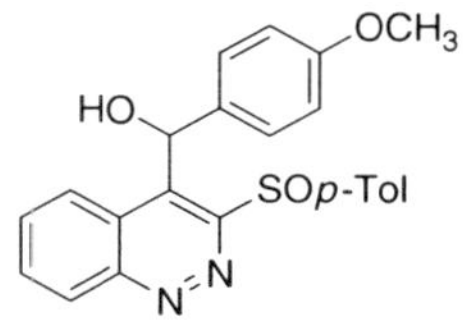 **95b** (82)	–

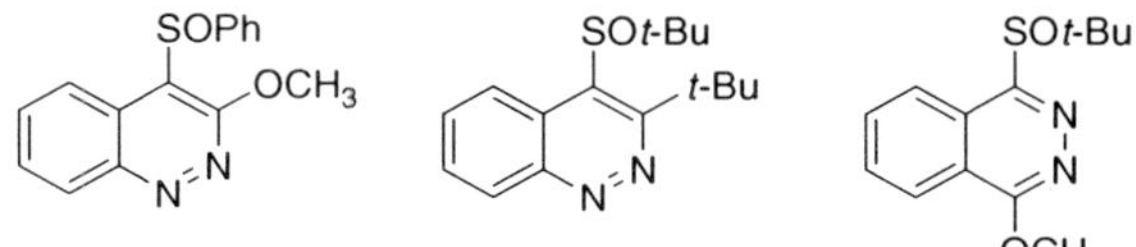

96 **97** **98**

Scheme 28 Substrates for unsuccessful *peri* metalation

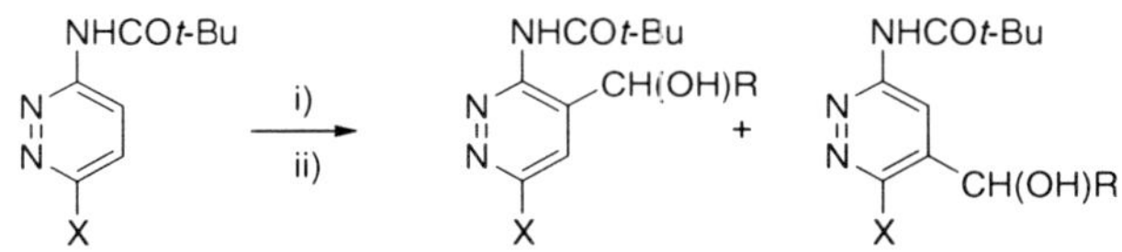

99: X=Cl
100: X=H

101a,b
103a,b

102a,b
104a,b

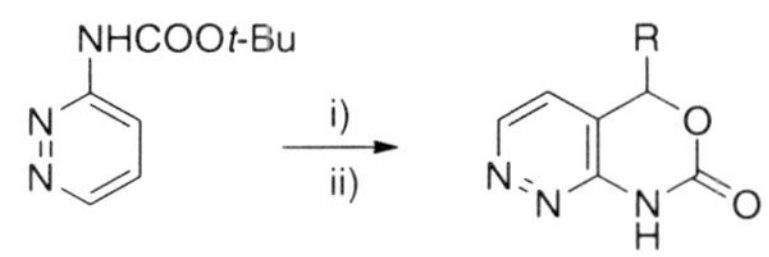

105 **106a,b**

Scheme 29 Lithiation of carbamides **99**, **100** and carbamate **105**: (i) 4.0 equiv base @ −70°C, THF, 90 min; (ii) electrophile @ −70°C, 90 min

bulkiness of the sulfinyl residue as well as the aldehyde had influence on the obtained de. However, a remarkable de of >98% was observed in case of isobutyraldehyde (Table 15).

Various attempts to achieve *peri* metalation of proper substituted cinnolines and phthalazine **96–98** failed (Scheme 28), regardless of the reaction conditions under investigation. However, in some cases starting material was recovered.

2.2.4 *N*-Acyl Derivatives

For both 3- and 4-aminopyridazines, *N*-(*tert*-butylcarbonyl) and *N*-(*tert*-butoxycarbonyl) derivatives were prepared and investigated regarding their

Table 16 Lithiation of carbamides **99**, **100** and carbamate **105**

Substrate	Base	Electrophile	Product (yield)
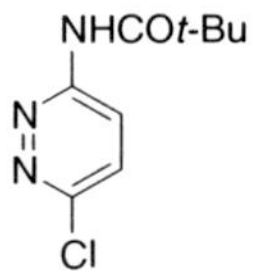 **99**	LTMP	CH_3CHO	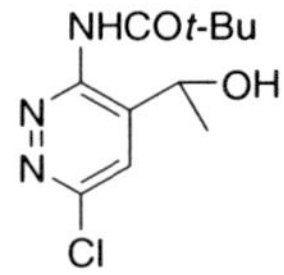 **101a** (35) 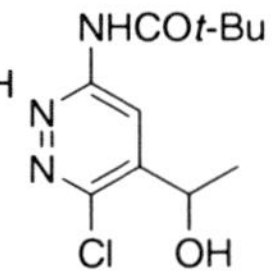 **102a** (65)
99	LTMP	PhCHO	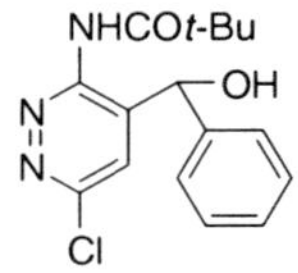 **101b** (28) 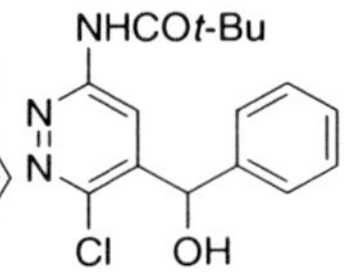 **102b** (55)
99	LDA	CH_3CHO	**102a** (82)
99	LDA[a]	PhCHO	**102b** (68)
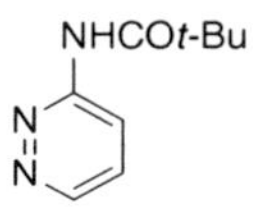 **100**	LTMP	CH_3CHO	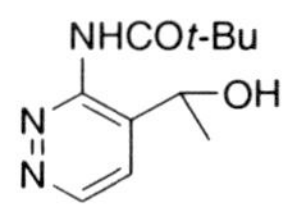 **103a** (52) 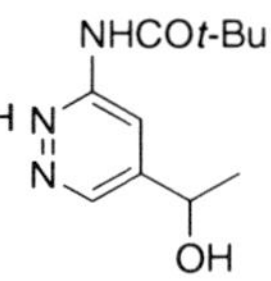 **104a** (4)
100	LTMP	PhCHO	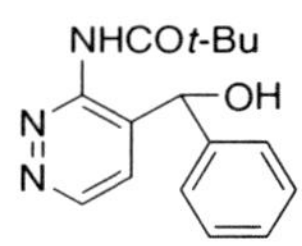 **103b** (62)
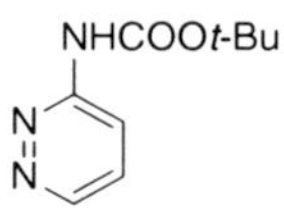 **105**	LTMP	CH_3CHO	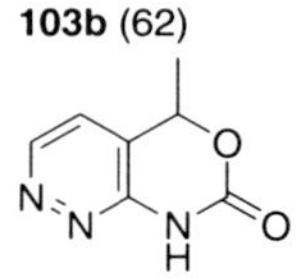 **106a** (37)
105	LTMP	PhCHO	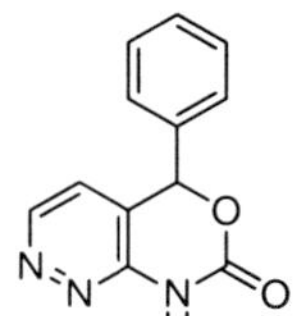 **106b** (36)

[a]180 min reaction for both lithiation and reaction with electrophile

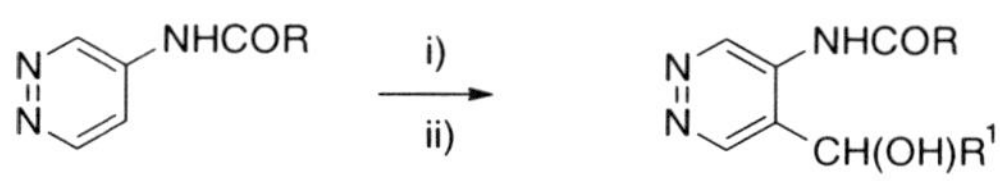

107: R=*t*-Bu
108: R=O*t*-Bu

109a,b
110a,b

Scheme 30 Lithiation of carbamide **107** and carbamate **108**: (i) 4.0 equiv LTMP @ −70°C, THF, *t*(i) min; (ii) R^1-CHO @ −70°C, 120 min

Table 17 Lithiation of carbamide **107** and carbamate **108**

Substrate	*t*(i)	Electrophile	Product (yield%)	Recovered s.m. %
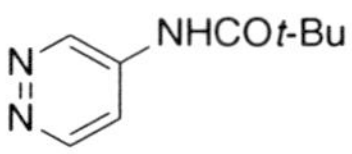 **107**	120	CH_3CHO	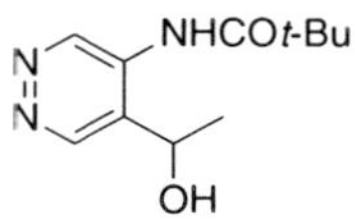 **109a** (43)	11
107	120	PhCHO	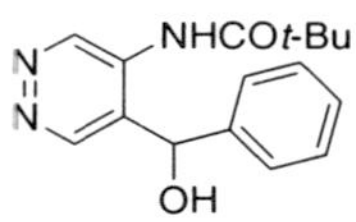 **109b** (63)	–
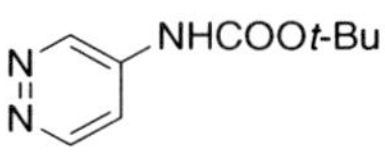 **108**	180	CH_3CHO	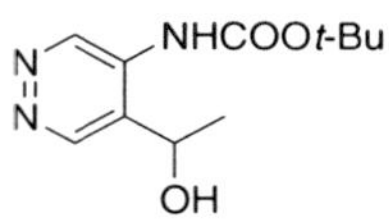 **110a** (20)	13
108	120	PhCHO	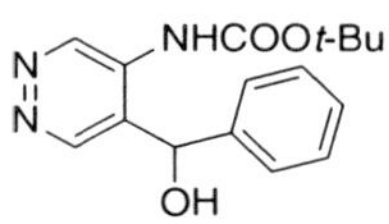 **110b** (25)	5

behavior under metalation conditions [43]. It is worth mentioning that for all lithiations a fourfold excess of metalation reagent was necessary to obtain satisfactory yields. By variation of reaction conditions (nature of base, reaction times) high regioselectivity of lithiation could be observed for 3-aminopyridazines **99** and **100** (Scheme 29). For carbamate **105**, only the annelated bicycles **106a** and **106b** were obtained due to intramolecular transesterification during workup (Table 16).

Metalation of 4-aminopyridazine derivatives (Scheme 30) results in only poor yields in the case of carbamate **108**, whereas amide **107** gave reasonable yields of alcohols **109a** and **109b**, respectively [51]. However, for both pyridazines **107** and **108** lithiation occurred regioselectively at position 5 and no further cyclication during workup was observed for products **110a** and **110b** (Table 17).

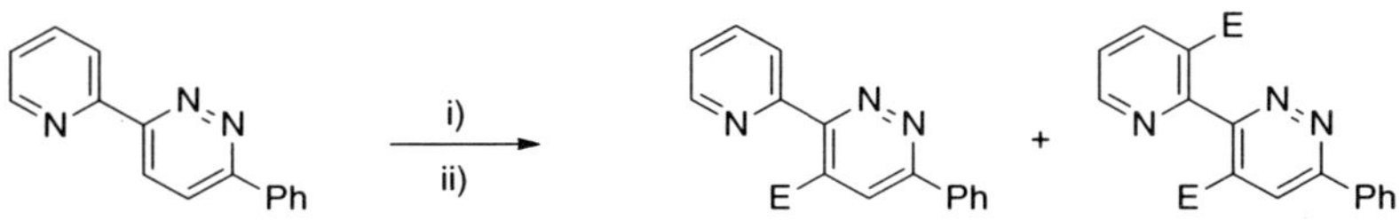

111

E = SPh — **112a**: 54% — **113a**: -
E = *n*-Bu$_3$Sn — **112b**: 44% — **113b**: 31%
E = p-OMePhCH(OH) — **112c**: 66% — **113c**: -

Scheme 31 Lithiation of pyridazine **111**: (i) 3.0 equiv LTMP @ −78°C, THF, 15 min; (ii) electrophile @ −100°C, 60–180 min

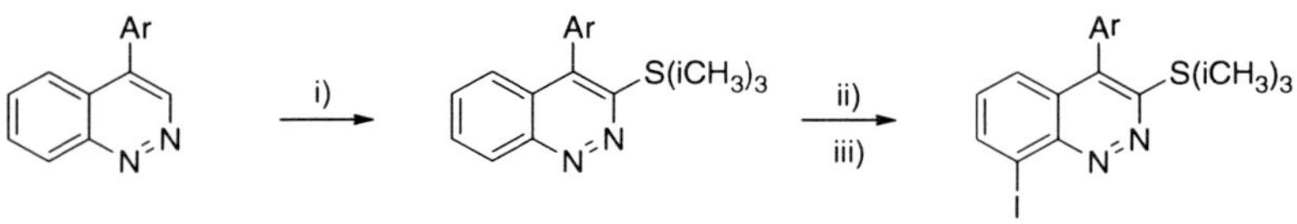

114: Ar=*p*-CF$_3$Ph — **117**: 96% — **120**: 88%
115: Ar=*m*-NO$_2$Ph — **118**: 3% — **121**: n.d.
116: Ar=*p*-CH$_3$OPh — **119**: 67% — **122**: 72%

Scheme 32 Metalation of 4-arylcinnolines: (i) in situ trapping method, 1.1 equiv TMS-Cl, 1.1 equiv LDA @ −78°C, THF, 120 min; (ii) 4.0 equiv LTMP @ −78°C, THF, 45 min; (iii) 2 equiv I_2 @ −78°C, THF, 60 min

2.2.5 Lithiation of Various Derivatives

Pyridin-2-ylpyridazine derivatives (e.g., **111**) were lithiated in order to investigate the capability of the pyridine core to act as an *ortho*-directing substituent. Indeed, a remarkable regioselectivity could be obtained (Scheme 31) [57].

On properly substituted cinnoline derivatives (**114–116**) *peri* lithiation was achieved (Scheme 32). After reaction with I_2 as electrophile, the obtained 8-iodocinnolines were used as substrates for Suzuki–Miyaura and Stille cross-coupling reactions [58].

3 Magnesation of Pyridazines

Grignard reagents are in general important and versatile intermediates in organic synthesis. Interestingly, only few examples for the magnesation of diazines and in particular pyridazine derivatives are known. To estimate appropriate metalation reagents and reaction conditions to achieve metal halogen exchange, 4-iodo-3,6-dimethoxypyridazine (**5**) was treated with different Grignard reagents under various conditions (Scheme 33) [59]. Reaction of the hereby formed magnesium pyridazine derivative with electrophiles gave the 4-substituted pyridazines in moderate to good yields (Table 18). Experiments using 4,5-dihalo-3,6-dimethoxypyridazines **125** and

Scheme 33 Metalation of pyridazine **5**: (i) 1.0 equiv *i*-PrMgCl @ 20°C, THF, 30 min; (ii) electrophile @ 20°C, 60–720 min

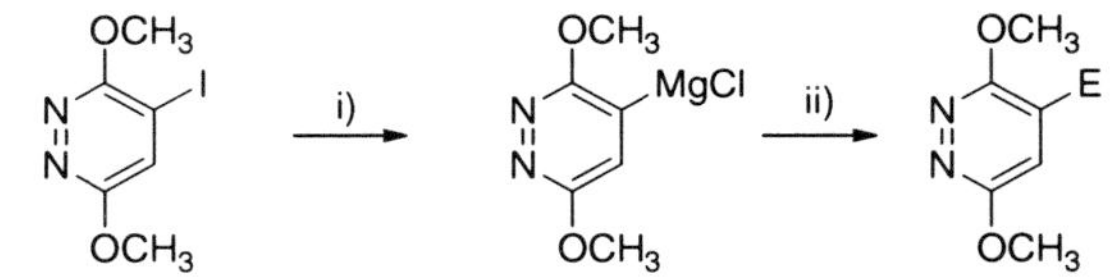

5 **123** **7a-c, 14b** **124a-c**

Table 18 Metalation of pyridazine **5**

Entry	Electrophile	Product (yield%)
1	PhCHO	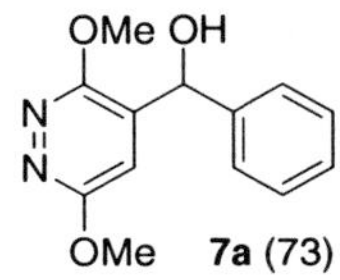 **7a** (73)
2	$C_5H_{11}CHO$	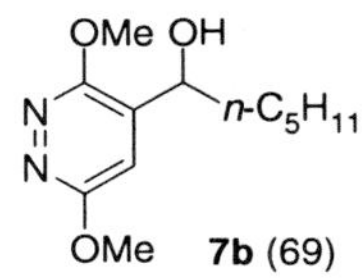 **7b** (69)
3	PhSSPh	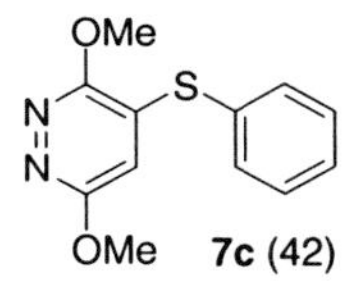 **7c** (42)
4	TMS-Cl	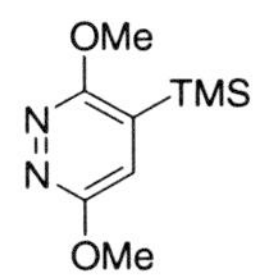 **14b** (41)[a]
5	CH_3CHO	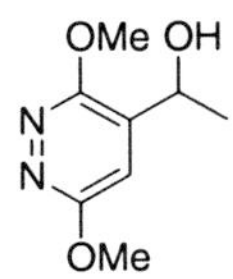 **124a** (35)[b]
6	NC-COOEt	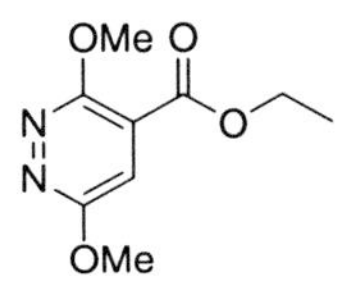 **124b** (30)[c]

(continued)

Table 18 (continued)

Entry	Electrophile	Product (yield%)
7	DMF	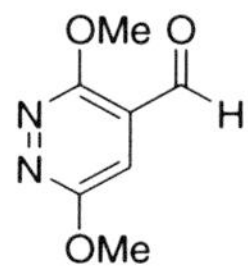 **124c** (38)

[a]Hydrolysis product (E = H) was obtained in 28%
[b]Hydrolysis product (E = H) was obtained in 49%
[c]Hydrolysis product (E = H) was obtained in 5%

Scheme 34 Metalation of pyridazines **125** and **126**: (i) 1.0 equiv *i*-PrMgCl @ 20°C, THF, 30 min; (ii) electrophile @ 20°C, 60–720 min

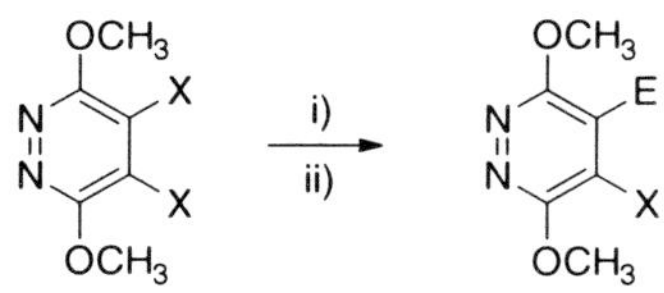

125: X=I
126: X=Br

127a-f
128a-f

Table 19 Metalation of pyridazines **125** and **126**

Substrate	Electrophile	Product (yield%)
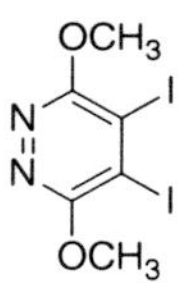 **125**	PhCHO	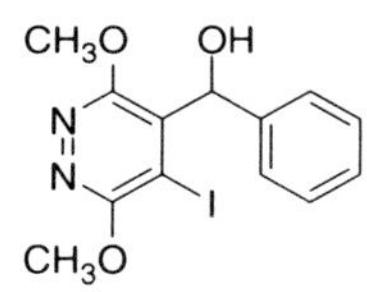 **127a** (75)
125	CH_3CHO	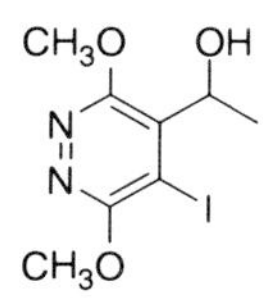 **127b** (-)[a]
125	$C_5H_{11}CHO$	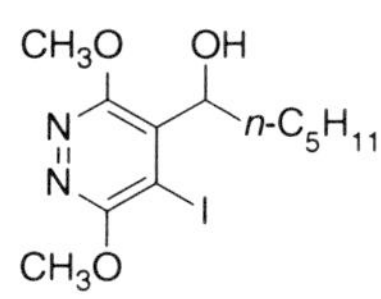 **127c** (74)

(continued)

Table 19 (continued)

Substrate	Electrophile	Product (yield%)
125	TMS-Cl	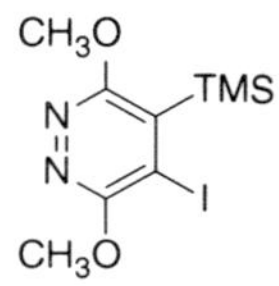 **127d** (70)
125	NC-COOEt	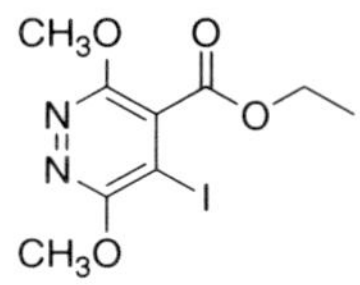 **127e** (30)[b]
125	PhSSPh	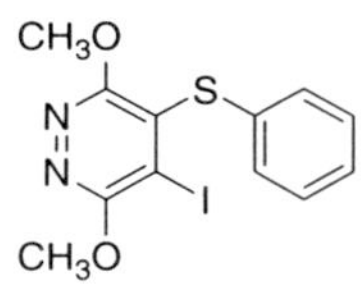 **127f** (40)
125	DMF	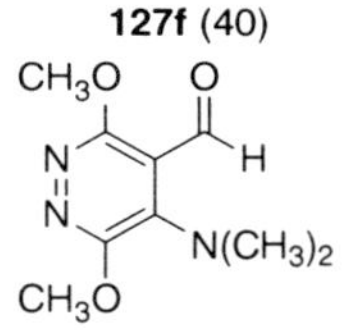 **129** (38)
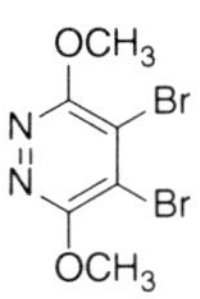 **126**	H_2O	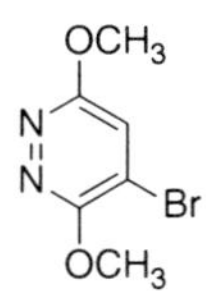 **128a** (80)
126	EtOD	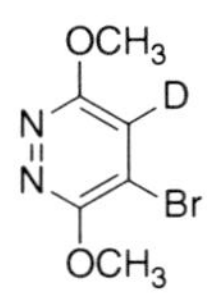 **128b** (80)
126	PhCHO	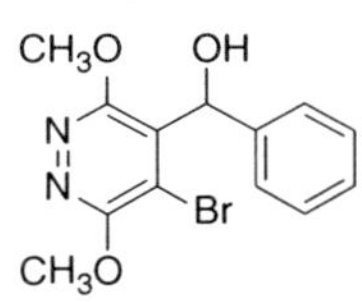 **127c** (65)

(continued)

Table 19 (continued)

Substrate	Electrophile	Product (yield%)
126	$C_5H_{11}CHO$	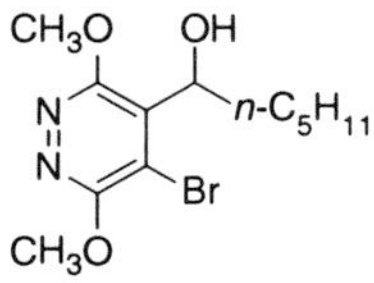 **127d** (60)
126	NC-COOEt	 **127e** (50)
126	I_2	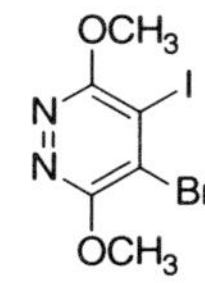 **128f** (80)
126	DMF	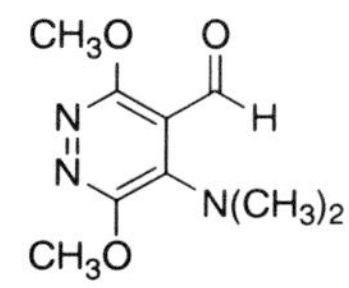 **129** (47)

[a]Hydrolysis product **5** (E = H) was obtained in 45%
[b]Hydrolysis product (E = H) was obtained in 5%

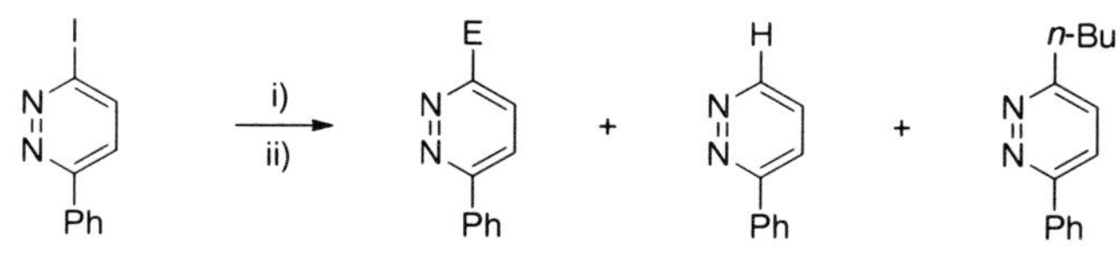

130 **131a-e** **132** **133**

Scheme 35 Magnesation of pyridazine **130**: (i) 0.35 equiv. n-Bu_3MgLi @ −10°C, THF, 150 min; (ii) electrophile @ −10°C to 35°C, 18 h

Table 20 Magnesation of pyridazine **130**

Entry	Electrophile	Product (yield%)
1	*p*-CH_3O-PhCHO	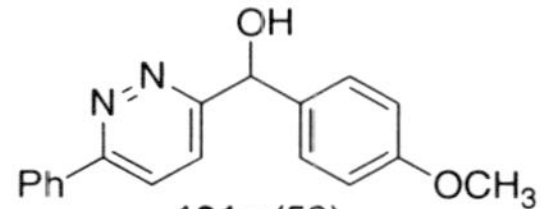 **131a** (59) 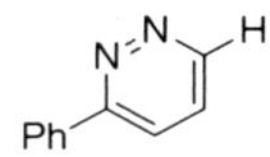 **132** (3) 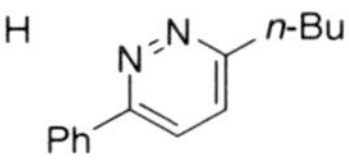 **133** (7)
2	PhSSPh	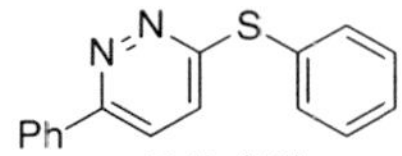 **131b** (62) **132** (2), **133** (4)
3	Ph_2CO	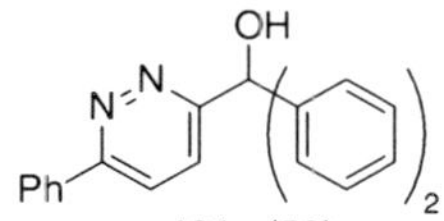 **131c** (50) **133** (5)
4	$(CH_3)_2CHCHO$	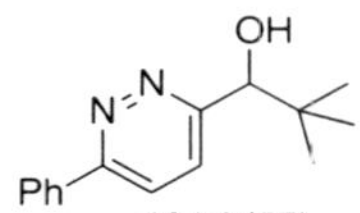 **131d** (57) **132** (2), **133** (7)
5	Cy-CHO	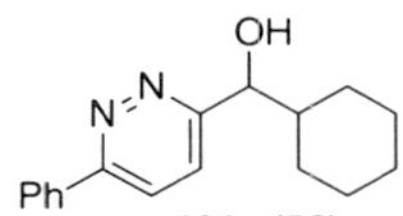 **131e** (56) **132** (1), **133** (8)

126 were also conducted, leading to comparable yields as for **5** (Scheme 34). No significant difference between the two halogens under investigation was observed. It is worth mentioning that when DMF was used as electrophile, the formed dimethylamine lead to a nucleophilic substitution of the remaining halogen atom of the primary reaction product giving 3,6-dimethoxy-5-(*N*,*N*-dimethylamino)-4-formylpyridazine (**129**) as final reaction product (Table 19).

The use of "Ate Complexes" [60], in particular lithium tri-*n*-butylmagnesate was also investigated for the magnesium-halogen exchange on 3-iodo-6-phenylpyridazine (Scheme 35) [61]. It was found out that the order of reagent addition is crucial to obtain high yields. Tri-*n*-butyllithiummagnesate solution had to be slowly added to a solution of pyridazine **130**; otherwise after addition of *p*-methoxybenzaldehyde as electrophile and workup, the dehalogenation product 3-phenylpyridazine was isolated as major product in 60% yield (Table 20).

4 Zincation

Despite their extensive applications, the use of organo-lithium and Grignard reagents (which are in general highly reactive intermediates) is limited by several drawbacks, the most prominent being the need for cryogenic temperatures, limited functional group tolerance, and sometimes restricted regioselectivity. The use of

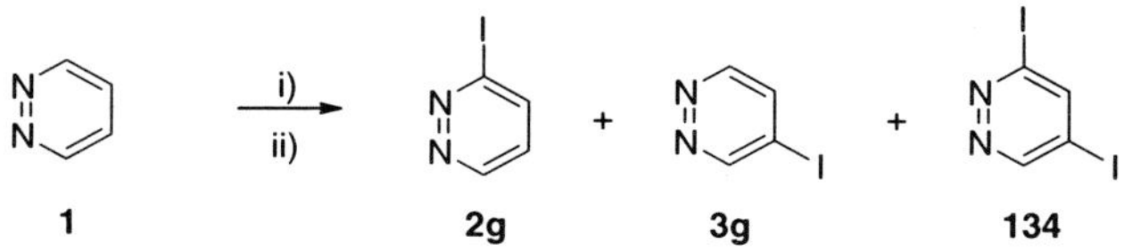

Scheme 36 Zincation of pyridazine: (i) 0.5 $ZnCl_2$·TMEDA/1.5 equiv LTMP @ rt, 120 min. THF/ 5 equiv TMEDA. (ii) I_2; yield 66%; ratio **1**/**2g**/**3g**/134 = 11/74/8/7

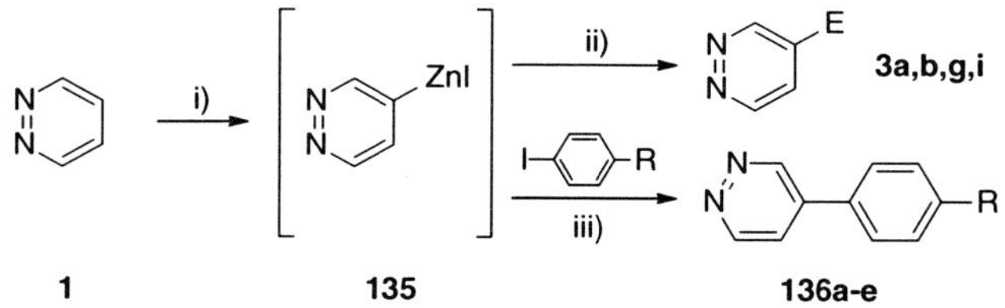

Scheme 37 Zincation of pyridazine: (i) *t*-Bu-P4/ZnI_2 @ −78°C to rt, THF or toluene; (ii) electrophile; (iii) $Pd(PPh_3)_4$ @ rt, 24 h; **136a** (R = H): 76%, **136b** (R = COOEt): 91%, **136c** (R = NO_2): 89%, **136d** (R = OCH_3): 92%, **136e** (R = CN): 71%

Table 21 Zincation of pyridazine using *t*-Bu-P4/ZnI_2

Entry	Equiv *t*-Bu-P4	Equiv ZnI_2	Solvent	Electrophile	Product (yield%)
1	1.0	1.0	THF	I_2	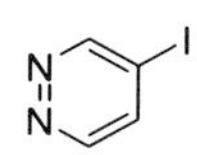 **3g** (trace)
2	1.0	1.6	THF	I_2	**3g** (trace)
3	1.0	2.0	THF	I_2	**3g** (14)
4	1.5	3.0	THF	I_2	**3g** (46)
5	3.0	6.0	THF	I_2	**3g** (83)
8	1.5	1.0	Toluene	PhCHO	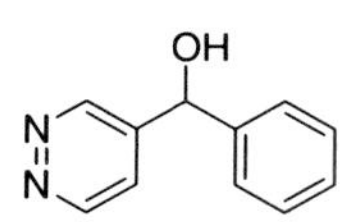 **3a** (-)[a]
6	1.0	0.0	Toluene	Ph_2CO	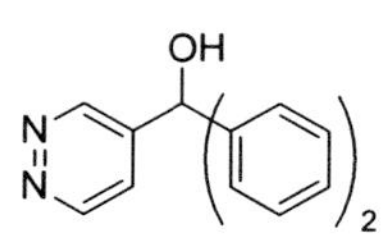 **3b** (4)
7	1.5	1.0	Toluene	Ph_2CO	**3b** (91)
9	1.5	1.0	Toluene	*t*-BuCHO	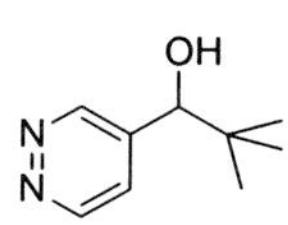 **3i** (73)

[a]4-Benzoylpyridazine was obtained in 36% yield

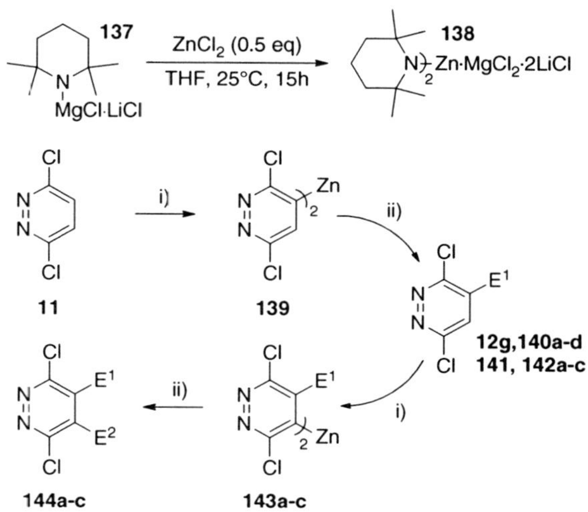

Scheme 38 Metalation of pyridazines **11**, **12g** and **140a–c** with $(tmp)_2Zn$ (**138**): (i) 0.55 equiv $(tmp)_2Zn$ @ −78°C, THF, 120–180 min; (ii) electrophile/Negishi cross-coupling

organo-zincates can circumvent these problems in many cases. Furthermore, (hetero)aryl zincates are valuable substrates for the palladium-catalyzed Negishi cross-coupling reaction, which makes various diazine derivatives easily accessible. Hence, zincates of the diazines under discussion are subject of recent research. In particular, the use of mixed-metal complex bases, e.g. LTMP/$ZnCl_2$·TMEDA formulations has proven to be a very successful approach for the synthesis of (hetero)aryl zincates [62].

Zincation of bare pyridazine could be accomplished by an in situ prepared mixture of $ZnCl_2$·TMEDA (0.5 equiv) and LTMP (1.5 equiv). Using THF as solvent and TMEDA as cosolvent, moderate yields and acceptable regioselectivity could be obtained for the synthesis of 3-iodopyridazine (Scheme 36) [62].

In contrast to the "regular" regioselectivity for deprotonation in position adjacent to the nitrogen atom(s) by using metallic bases, switching to the extremely strong nonmetallic *t*-Bu-P4 base [63] in combination with ZnI_2 allowed zincation at position 4 of pyridazine with remarkable regioselectivity [64, 65]. By variation of the *t*-Bu-P4/ZnI_2 ratio, in general good yields of 4 substituted pyridazines were obtained after reaction with electrophiles or Negishi coupling with various iodobenzenes (Scheme 37, Table 21).

The zincation of 3,6-dichloropyridazine was also reported [66]. In this study the readily available mixed metal complex base $(tmp)_2Zn·2MgCl_2·2LiCl$ (**138**) was used, this being known to be particularly suitable for heterocycles sensitive against nucleophilic attack or ring-opening reactions. Hence, zincation performed smoothly and the obtained pyridazinyl-zinc species was reacted with various nucleophiles, used as substrate for Negishi cross-coupling reactions or dimerized by the addition of chloranil (Scheme 38) [67]. The 4-substituted 3,6-dichloropyridazines were obtained in good yields. By applying a second zincation step, the last vacant position, namely carbon C5 could be functionalized after adding appropriate electrophiles (Table 22).

Table 22 Metalation of pyridazine **11**, **12 g** and **140a–c** with $(tmp)_2Zn$ (**138**)

Entry	Substrate	Electrophile	Product (yield%)
1	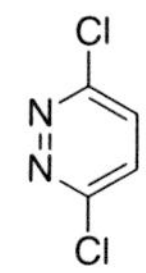 **11**	I_2	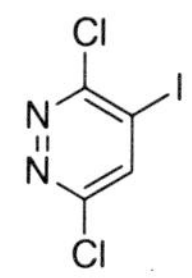 **12g** (82)
2	**11**	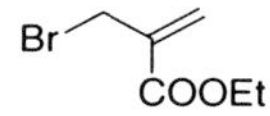	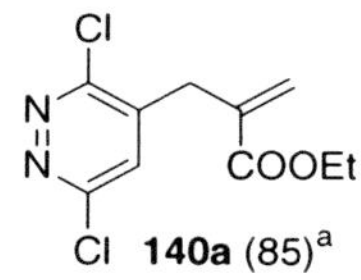 **140a** (85)[a]
3	**11**	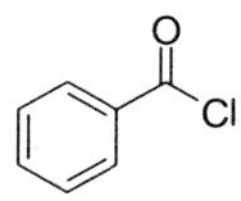	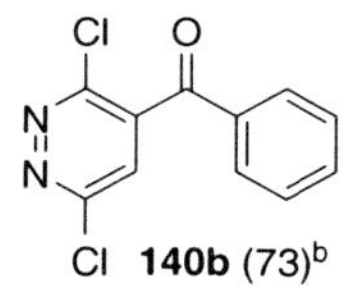 **140b** (73)[b]
4	**11**	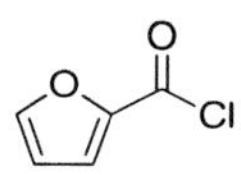	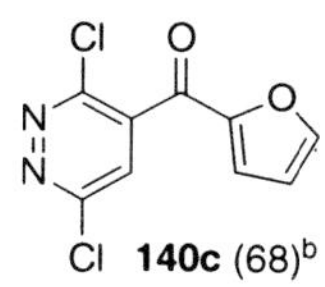 **140c** (68)[b]
5	**11**	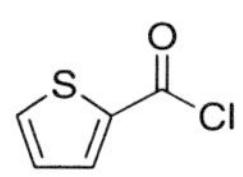	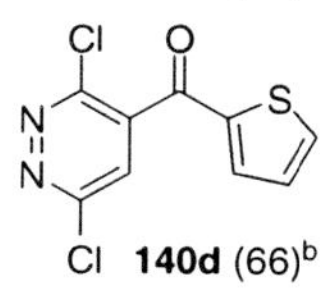 **140d** (66)[b]
6	**11**	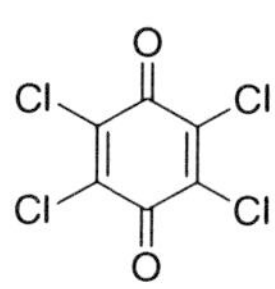	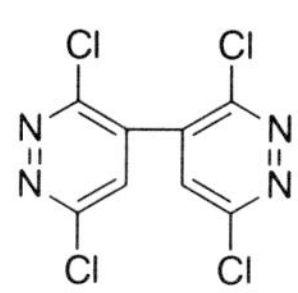 **141** (88)[c]
7	**11**	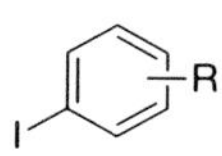	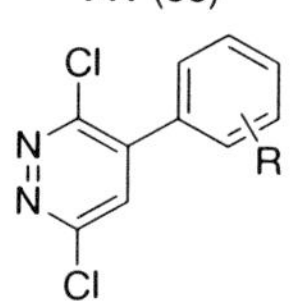 R=4-COOEt: **142a** (81)[d] R=4-OCH$_3$: **142b** (76)[d] R=3-NO$_2$: **142c** (77)[d]
8	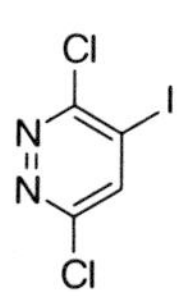 **12g**	I_2	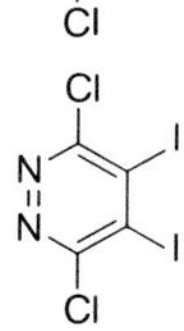 **144a** (56)

(continued)

Table 22 (continued)

Entry	Substrate	Electrophile	Product (yield%)
9	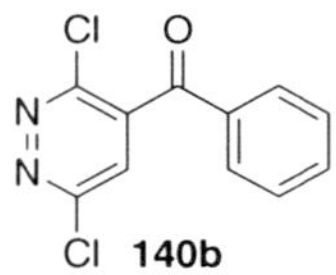 **140b**	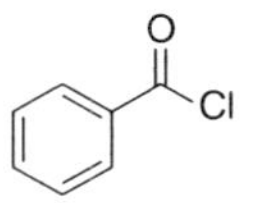	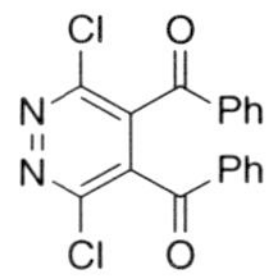 **144b** (77)[b]
10	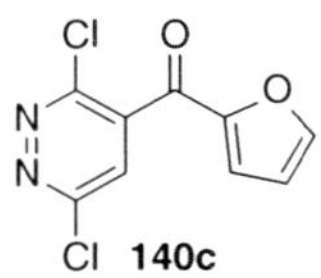 **140c**	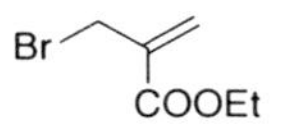	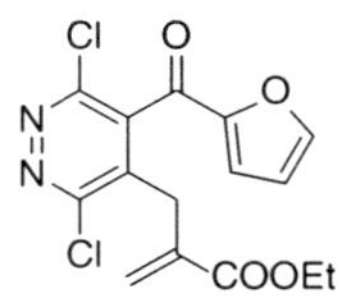 **144c** (75)[a]

[a]Addition of 0.25 equiv CuCN·2LiCl
[b]Addition of 1.1 equiv CuCN·2LiCl
[c]0.6 equiv of Chloranil were used
[d]Palladium-catalyzed cross-coupling: 5 mol% $Pd(dba)_2$, 10 mol% $P(o\text{-furyl})_3$

Scheme 39 Metalation of phthalazines **146** and **148**: (i) 1.1 equiv **145** @ 25°C, THF, 1 min; (ii) 4.0 equiv morpholine @ 85°C, toluene, 12 h; (iii) 1.1 equiv **145** @ 60°C (mw), THF, 45 min; (iv) electrophile/palladium catalyzed cross-couplings

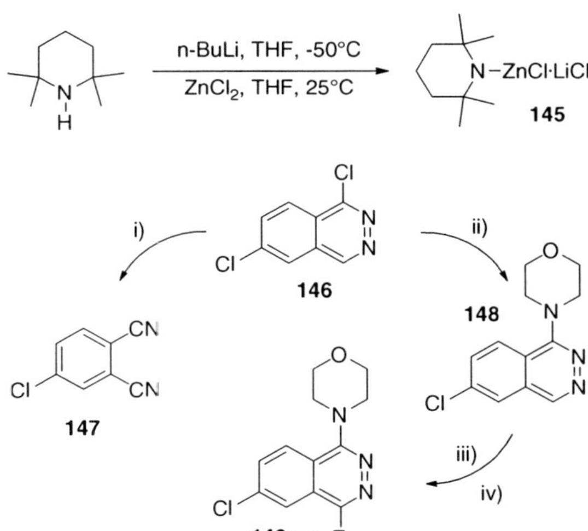

Phthalazine derivatives were also investigated regarding their reactivity against zinc-amide-type bases [68]. When dichlorophthalazine **146** was subjected to metalation conditions, only the ring-fragmentation product **147** was obtained. However, when a morpholino group was introduced by nucleophilic substitution of chlorine in position 1, the more stable phthalazine derivative **148** was obtained (Scheme 39). Zincation could be achieved successfully under microwave conditions, as reaction under normal reaction conditions (25°C) required 48 h and led to the zincated product in only low yields. The formed zinc organyl was treated with various

Table 23 Metalation of phthalazine **148**

Entry	Electrophile	Product (yield%)	
1	I_2	**149a** (71) 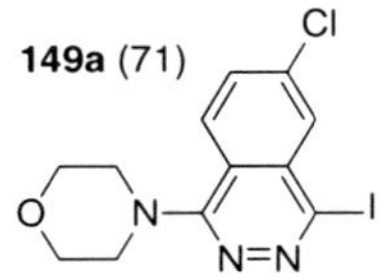	
2	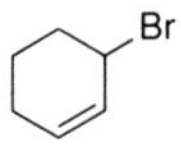	**149b** (60)[a] 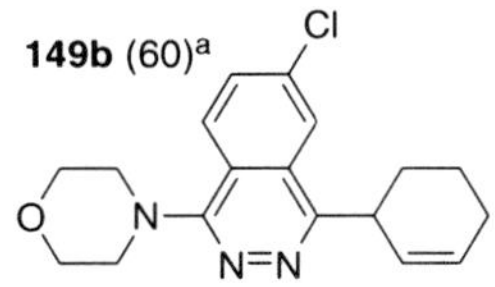	
3	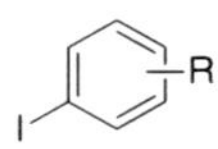	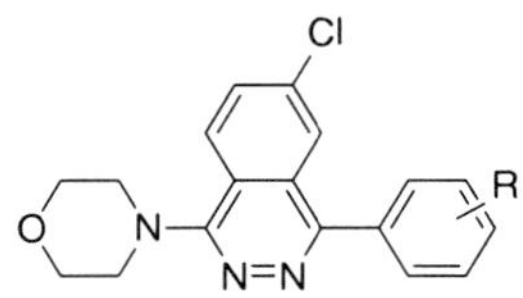	R=4-COOEt: **149c** (83)[b] R=4-CN: **149d** (74)[b] R=3-CF_3: **149e** (84)[b] R=4-(CO)cy-hex: **149f** (72)[b]
4	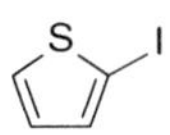	**149g** (86)[b] 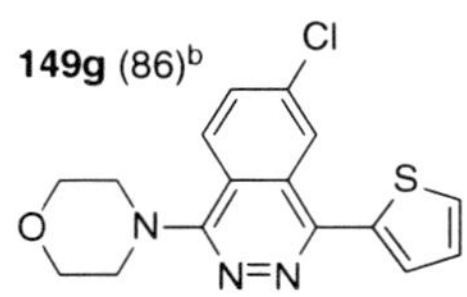	
5	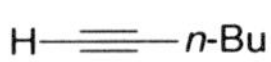	**149h** (89)[c] 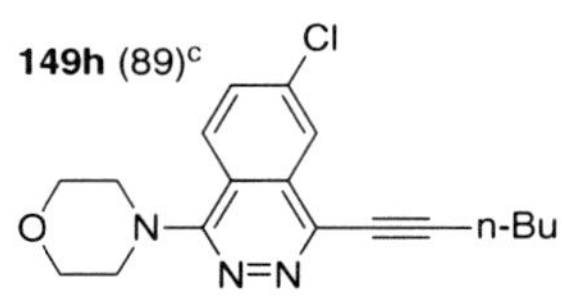	

[a]Addition of 0.05 equiv CuCN·2LiCl
[b]Palladium-catalyzed cross-coupling: 2 mol% $Pd(dba)_2$, 4 mol% $P(o\text{-furyl})_3$
[c]Sonogashira reaction with in situ generated compound **149a**

electrophiles and proved to be a potent reagent for Negishi cross-coupling reactions (Table 23).

5 Borylation and Stannylation

Both organoboron and organotin compounds are versatile building blocks in modern organic synthesis. By applying palladium-catalyzed cross-coupling reactions, they allow easy and efficient C–C bond formation and are therefore widely used in all fields of synthetic chemistry. Interestingly, only few examples are given for the synthesis of boron or tin pyridazine derivatives. In fact, to the best of our knowledge, no direct metalation procedure is described in literature by now, only transmetalating or alternative synthetic procedures are reported.

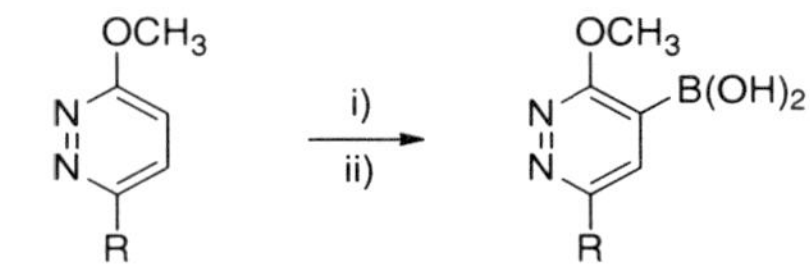

150a-d, 13 → **151a-e**

Scheme 40 Synthesis of pyridazine boronic acids **151a–e**: (i) 2.0 equiv LDA @ −78°C, THF or Et_2O, 30–60 min; (ii) 3.0 equiv triisopropylborate @ −78°C, 90 min, acidic workup

Table 24 Synthesis of pyridazine boronic acids **151a–e**

Entry	Substrate	Product (yield%)
1	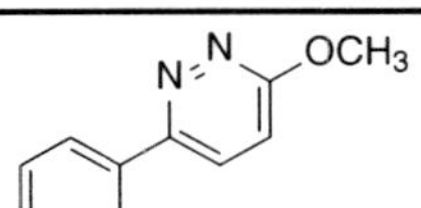 **150a**	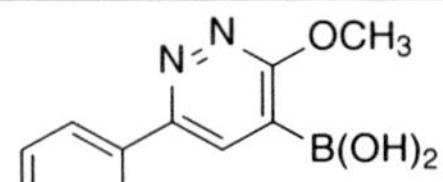 **151a** (96)
2	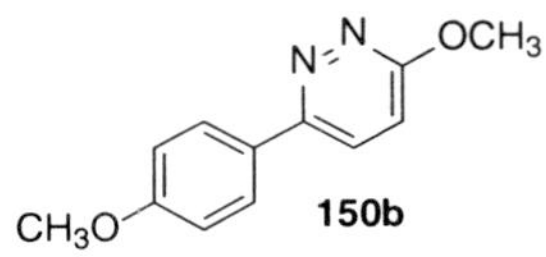 **150b**	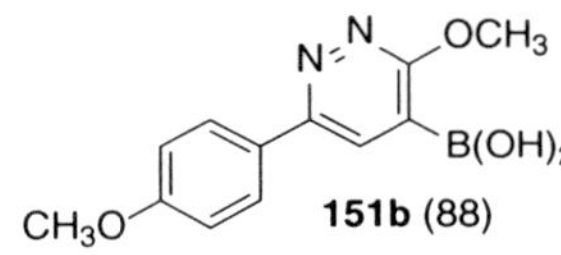 **151b** (88)
3	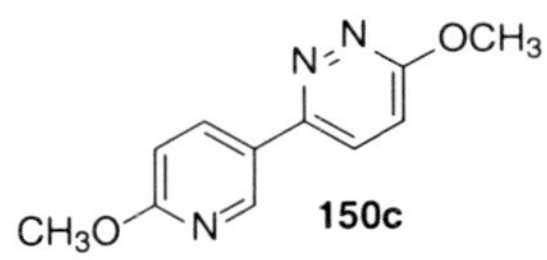 **150c**	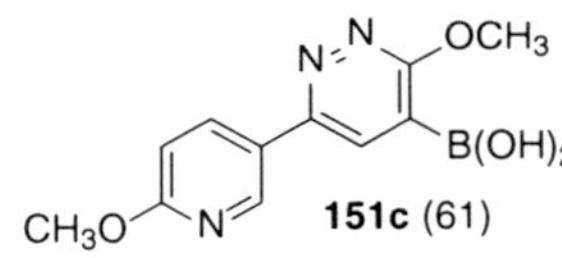 **151c** (61)
4	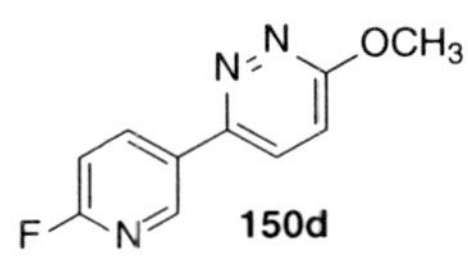 **150d**	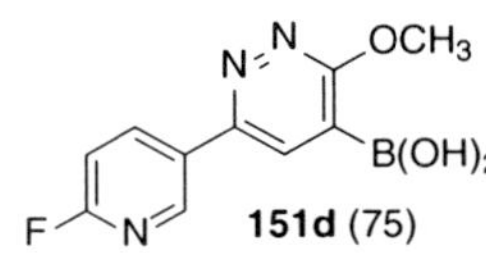 **151d** (75)
5	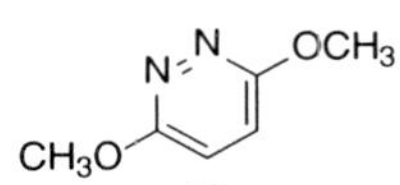 **13**	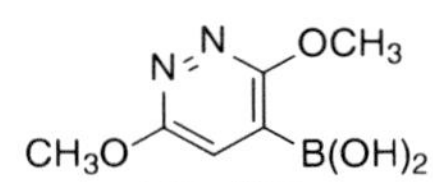 **151e** (88)

Lithiation with LDA of various 3-methoxypyridazine derivatives and subsequent reaction of the formed organo-lithium intermediate with triisopropylborate led to pyridazinyl boronic acids **151a–e** in good yields (Scheme 40, Table 24). Suzuki cross-coupling of the boronic acids with a wide variety of (hetero)aryl bromides required microwave conditions in some cases; however, polyfunctionalized pyridazines were obtained in moderate to good yields [69].

Examples for the synthesis of tri-*n*-butylstannylpyridazines (compounds **59d**, **112b**, and **113b**) and cinnolines (**94g**) have already been mentioned. In order to achieve stannylation, the lithiated diazine derivatives were reacted with tri-*n*-butyltinchloride. Nucleophilic attack of the tri-*n*-butylstannyl anion (prepared from tri-*n*-butylstannane

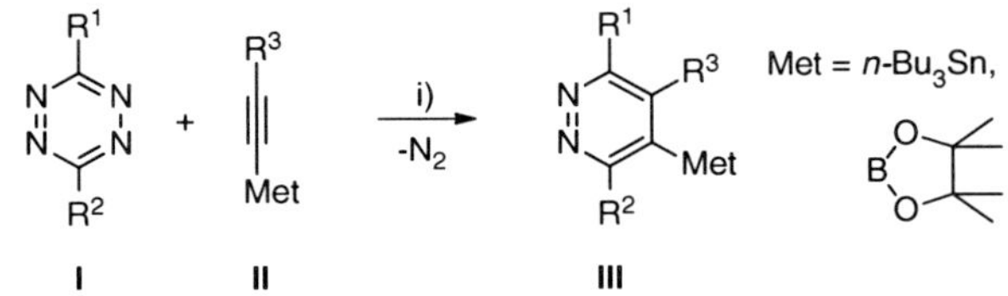

Scheme 41 Synthesis of boronic esters and tri-*n*-butylstannyl derivatives of pyridazine. (i) Various solvents, ΔT

and LDA) on 3,6-dichloropyridazine (**11**) and 3-chloro-6-methoxypyridazine (**15**) was also investigated [70]; however, this approach led only to low product yields or failed completely.

Considering the above-mentioned importance of boron and tin derivatives, another approach for these target compounds shall be briefly discussed, although no metalation step is incorporated. Based on an inverse-electron-demand Diels–Alder reaction of 3,6-disubstituted 1,2,4,5-tetrazines [71], an effective method for the synthesis of highly substituted pyridazines bearing boronic esters [72, 73] and the tri-*n*-butylstannyl group [74, 75] was developed (Scheme 41).

Both type **I** and **II** of starting materials were readily available. The yields were (with few exceptions) approximately 60% or even better. In case of unsymmetrically substituted compounds **I** or **II**, high regioselectivity for one product **III** was observed. Substances **III** were used in palladium-catalyzed cross-coupling reactions for the synthesis of highly functionalized pyridazine derivatives.

6 Conclusion

In this chapter the broad applicability of metalation reactions to pyridazine, cinnoline, and phthalazine derivatives was presented. While for the bare diazines under discussion only few examples are published, the metalation of substituted scaffolds is widely used for further functionalization. By the usage of different bases and careful selection of reaction conditions high regioselectivity can be achieved, thus leading to valuable synthetic building blocks. However, only few is known upon the direct borylation and stannylation of the discussed diazines using, e.g., functionalized boranes or stannanes to obtain boronic esters or (trialkyl) stannyl derivatives. Due to the fact that those moieties can serve as coupling partners in frequently used transition metal-catalyzed cross-coupling reactions, this synthetic strategy might be an interesting alternative to classical base-mediated metallation reactions.

References

1. Wermuth CG (2011) Are pyridazines privileged structures? Med Chem Commun 2:935
2. Grote R, Chen Y, Zeeck A, Chen ZX, Zähner H, Mischnick-Lübbecke P, König WA (1988) Metabolic products of microorganisms. 243. Pyridazomycin, a new antifungal antibiotic produced by *Streptomyces violaceoniger*. J Antibiot 41:595
3. Nakamura Y, Ito A, Shin C-G (1994) Dehydrooligopeptides. XVI. Convenient syntheses of two kinds of antrimycins Av and Dv containing dehydrovaline residues. Bull Chem Soc Jpn 67:2151

4. Hughes P, Clardy J (1989) Total synthesis of 3(S)-carboxy-4(S)-hydroxy-2,3,4,5-tetrahydropyridazine, an unusual amino acid constituent of luzopeptin A. J Org Chem 54:3260
5. Shimada N, Morimoto K, Naganawa H, Takita T, Hamada M, Maeda K, Takeuchi T, Umezawa H (1981) Antrimycin, a new peptide antibiotic. J Antibiot 34:1613
6. Morimoto K, Shimada N, Naganawa H, Takita T, Umezawa H (1981) The structure of antrimycin. J Antibiot 34:1615
7. Shiroza T, Ebisawa N, Furihata K, Endo T, Seto H, Otake N (1982) The structures of cirratiomycin A and B, the new peptide antibiotics. Agric Biol Chem 46:1891
8. Shiroza T, Ebisawa N, Kojima A, Furihata K, Shimazu A, Endo T, Seto H, Otake N (1982) Taxonomy of producing organism, and production, isolation, physico-chemical properties and biological activities of cirratiomycin A and B. Agric Biol Chem 46:1885
9. Wermuth CG (1998) Search for new lead compounds: the example of the chemical and pharmacological dissection of aminopyridazines. J Heterocycl Chem 35:1091
10. Rognan D, Boulanger T, Hoffmann R, Vercauteren DP, Andre JM, Durant F, Wermuth CG (1992) Structure and molecular modeling of GABAA receptor antagonists. J Med Chem 35:1969
11. Akahane A, Katayama H, Mitsunaga T, Kato T, Kinoshita T, Kita Y, Kusunoki T, Terai T, Yoshida K, Shiokawa Y (1999) Discovery of 6-oxo-3-(2-phenylpyrazolo[1,5-a]pyridin-3-yl)-1(6H)-pyridazinebutanoic acid (FK 838): a novel non-xanthine adenosine A1 receptor antagonist with potent diuretic activity. J Med Chem 42:779
12. Pal M, Batchu VR, Khanna S, Yeleswarapu KR (2002) Regioselective aluminium chloride induced heteroarylation of pyrrolo[1,2-b]pyridazines: its scope and application. Tetrahedron 58:9933
13. Okushima H, Narimatsu A, Kobayashi M, Furuya R, Tsuda K, Kitada Y (1987) A novel class of cardiotonics. Synthesis and pharmacological properties of [4-(substituted-amino)phenyl] pyridazinones and related derivatives. J Med Chem 30:1157
14. Herter R, Engler H, Moersdorf P, Schickaneder H, Pfahlert V, Ahrens KH (1989) Preparation and testing of 6-phenyldihydro-3(2H)-pyridazinones as cardiovascular agents. Appl EP, EP 327800
15. Pinna GA, Curzu MM, Loriga M, Cignarella G, Barlocco D, Malandrino S (1985) Rigid congeners of arylpyridazinones. IV. Synthesis and activity of derivatives of the new heterocyclic system 9H-indeno[2,1-c]pyridazine. Farmaco Sci 40:979
16. Cignarella G, Loriga M, Pinna GA, Pirisi MA, Schiatti P, Selva D (1982) Unexpected anti-inflammatory activity of rigid structures derived from antihypertensive 6-arylpyridazinones. III. Synthesis and activity of 7-fluoro- and 5-keto-5H-indeno(1,2-c)pyridozines. Farmaco Sci 37:133
17. Sircar I, Duell BL, Cain MH, Burke SE, Bristol JA (1986) Cardiotonic agents. 4. Synthesis and biological evaluation of N-substituted 2,4,4a,5-tetrahydro-3H-indeno[1,2-c]pyridazin-3-ones: rigid structures derived from CI-930 and analogs. J Med Chem 29:2142
18. Murineddu G, Cignarella G, Chelucci G, Loriga G, Pinna G (2002) Synthesis and cytotoxic activities of pyrrole[2,3-d]pyridazin-4-one derivatives. Chem Pharm Bull 50:754
19. Barlocco D, Cignarella G, Piaz VD, Giovannoni MP, De Benedetti PG, Fanelli F, Montesano F, Poggesi E, Leonardi A (2001) Phenylpiperazinylalkylamino substituted pyridazinones as potent α1 adrenoceptor antagonists. J Med Chem 44:2403
20. Pitarch L, Coronas R, Mallol J (1974) Chemical and pharmacological study of a series of substituted pyridazones. Eur J Med Chem 9:644
21. Schacht E, Kurmeier H-A, Gante J, Lissner R, Melzer G, Orth D (1980) 6-Arylpyridazine-3-ones and pharmaceutical compositions containing them. Appl EP, EP 10156
22. Black LA, Basha A, Kolasa T, Kort ME, Liu H, McCarty CM, Patel MV, Rohde JJ, Coghlan MJ, Stewart AO (2000) Preparation of arylpyridazinones as prostaglandin endoperoxide H synthase biosynthesis inhibitors. Appl. PI, WO 2000024719
23. Sauter F, Stanetty P, Blaschke A, Vyplel H (1981) 1-Pyridyl- (and -pyrimidinyl-) pyridazin-6(1H)-ones as potential herbicides. J Chem Res Synop 4:103

24. Krämer W, Schirmer U, Jeschke P, Witschel M (eds) (2011) Modern crop protection compounds. Wiley, London. ISBN 978-3-527-32965-6
25. Eicher T, Hauptmann S, Speicher A (2012) The chemistry of heterocycles – structure, reactions, synthesis, and applications. Wiley-VCH, Weinheim. ISBN 978-3-527-32868-0
26. Joule JA, Mills K (2010) Heterocyclic chemistry. Wiley-Blackwell, United Kingdom. ISBN 978-1-4051-9365-8
27. Hisrch A, Orphanos DG (1966) The action of organolithium compounds on phthalazine and I (2H)-phthalazinone. J Heterocycl Chem 3:38
28. Shen K, Fu Y, Li J-N, Liu L, Guo Q-X (2007) What are the pKa values of C–H bonds in aromatic heterocyclic compounds in DMSO? Tetrahedron 63:1568
29. Plé N, Turck A, Couture K, Queguiner G (1995) Diazines 13: metalation without ortho-directing group – functionalization of diazines via direct metalation. J Org Chem 60:3781
30. Fruit C, Turck A, Plé N, Mojovic L, Quéguiner G (2002) Syntheses and metalation of pyridazinecarboxamides and thiocarboxamides. Diazines. Part 32. Tetrahedron 58:2743
31. Quéguiner G (2000) Efficient synthesis of organometallics of pyridines, quinolines and diazines: new synthetic methodologies for azaaromatic biomolecules. J Heterocycl Chem 37:615
32. Turck A, Plé N, Mongin F, Quéguiner G (2001) Advances in the directed metallation of azines and diazines (pyridines, pyrimidines, pyrazines, pyridazines, quinolines, benzodiazines and carbolines). Part 2: metallation of pyrimidines, pyrazines, pyridazines and benzodiazines. Tetrahedron 57:4489
33. Rosseels G (1966) Acids of the pyridazine series. IV. New synthetic alkaloid, 2-azanicotine. Bull Soc Chim Belg 75:5
34. Barbier P (1899) Synthesis of dimethylheptenol. C R Acad Sci Paris 128:110
35. Blomberg C, Hartog FA (1977) The Barbier reaction – a one-step alternative for syntheses via organomagnesium compounds. Synthesis 1977:18
36. Cintas P, Palmisano G, Cravotto G (2011) Power ultrasound in metal-assisted synthesis: from classical Barbier-like reactions to click chemistry. Ultrason Sonochem 18:836
37. Leprêtre A, Turck A, Plé N, Quéguiner G (2000) Syntheses in the nitrogen π-deficient heterocycles series using a Barbier type reaction under sonication. Diazines. Part 29. Tetrahedron 56:3709
38. Turck A, Plé N, Mojovic L, Quéguiner G (1990) Metallation of diazines. II. First metallation of pyridazine, metallation of 2,4-dichloropyrimidine. J Heterocycl Chem 27:1377
39. Mattson RJ, Sloan CP (1990) Ortho-directed lithiation in π-deficient diazinyl heterocycles. J Org Chem 55:3410
40. Mojovic L, Turck A, Plé N, Dorsy M, Ndzi B, Quéguiner G (1996) Metalation of diazines. XVII. Very hindered bases as new metalating agents, improvement of regioselectivity for the metalation of 3-chloro-6-methoxypyridazine. Tetrahedron 52:10417
41. Plé N, Turck A, Couture K, Quéguiner G (1996) A new route to aminodiazines via metalation reaction. Synthesis of an aza analogue of nevirapine: diazines XV. Synthesis 1996:838
42. Fraser RR, Mansour TS (1984) Acidity measurements with lithiated amines: steric reduction and electronic enhancement of acidity. J Org Chem 49:3442
43. Turck A, Plé N, Ndzi B, Quéguiner G, Haider N, Schuller H, Heinisch G (1993) On the metalation of 3-substituted and 3,6-disubstituted pyridazines. Tetrahedron 49:599
44. Toudic F, Turck A, Plé N, Quéguiner G, Darabantu M, Lequeux T, Pommelet JC (2003) Relative ortho-directing power of fluorine, chlorine and methoxy group for the metalation reaction in the diazine series. Diazines XXXV. J Heterocycl Chem 40:855
45. Decrane L, Plé N, Turck A (2005) Metalation of bromodiazines. Diazines XL. J Heterocycl Chem 42:509
46. Turck A, Pié N, Tallon V, Quéguiner G (1995) Metallation of diazines. XIV. First O-directed metallation of cinnolines. Metallation of 3-, 4-chloro and 3-, 4-methoxycinnolines. Tetrahedron 51:13045

47. Chapoulaud VG, Salliot I, Plé N, Turck A, Quéguiner G (1999) Functionalization by metalation of the benzene moiety of benzodiazines. Determination of structures by long-range 1H-15N correlation at natural abundance. Diazines XXV. Tetrahedron 55:5389
48. Schnürch M, Spina M, Khan AF, Mihovilovic MD, Stanetty P (2007) Halogen dance reactions – a review. Chem Soc Rev 36:1046
49. Schnürch M (2011) In: Iskra J (ed) Halogenated heterocycles, vol 27, Topics in heterocyclic chemistry. Springer, Berlin, p 185
50. Fröhlich J (1994) Substituted heterocyclic compounds by selective control of halogen-dance reactions. Prog Heterocycl Chem 6:1
51. Turck A, Plé N, Mojovic L, Ndzi B, Quéguiner G, Haider N, Schuller H, Heinisch G (1995) On the metalation of 4-substituted pyridazines. J Heterocycl Chem 32:841
52. Turck A, Plé N, Pollet P, Mojovic L, Duflos J, Quéguiner G (1997) Use of sulfur derivatives as an ortho directing group for the metalation of diazines. Metalation of diazines. XVIII. J Heterocycl Chem 34:621
53. Turck A, Plé N, Pollet P, Quéguiner G (1998) Metalation of t-butyl sulfoxides, sulfones and sulfonamides of pyridazine and pyrazine. Metalation of diazines. XX. J Heterocycl Chem 35:429
54. Pollet P, Turck A, Plé N, Quéguiner G (1999) Synthesis of chiral diazine and pyridine sulfoxides. asymmetric induction by chiral sulfoxides in an "aromatic ortho-directed metalation–reaction with electrophiles sequence". Diazines. 24. J Org Chem 64:4512
55. Le Fur N, Mojovic L, Plé N, Turck A, Reboul V, Metzner P (2006) Ortho-metalation of enantiopure aromatic sulfoxides and stereocontrolled addition to imines. J Org Chem 71:2609
56. Fur NL, Mojovic L, Plé N, Turck A, Marsais F (2005) Metalation of sulfoxides in the benzodiazine series. Diazines. Part 44. Tetrahedron 61:8924
57. Berghian C, Darabantu M, Turck A, Plé N (2005) Metallation of pyridin-2-yldiazines. Use of pyridine ring as ortho-directing group. Diazines. Part 45. Tetrahedron 61:9637
58. Chapoulaud VG, Plé N, Turck A, Quéguiner G (2000) Synthesis of 4,8-diarylcinnolines and quinazolines with potential applications in nonlinear optics. Diazines. Part 28. Tetrahedron 56:5499
59. Leprêtre A, Turck A, Plé N, Knochel P, Quéguiner G (2000) First study of syntheses and reactivity of Grignard compounds in the diazine series. Diazines. Part 27. Tetrahedron 56:265
60. Mulvey RE, Mongin F, Uchiyama M, Kondo Y (2007) Deprotonative metalation using ate compounds: synergy, synthesis, and structure building. Angew Chem Int Ed Engl 46:3802
61. Buron F, Plé N, Turck A, Marsais F (2006) Metal-halogen exchange using tri-n-butyl lithium magnesate in the diazine series: diazine 49. Synlett 2006:1586
62. Seggio A, Chevallier F, Vaultier M, Mongin F (2007) Lithium-mediated zincation of pyrazine, pyridazine, pyrimidine, and quinoxaline. J Org Chem 72:6602
63. Schwesinger R, Schlemper H (1987) Peralkylated polyaminophosphazenes – extremely strong, neutral nitrogen bases. Angew Chem Int Ed 26:1167
64. Imahori T, Kondo Y (2003) A new strategy for deprotonative functionalization of aromatics: transformations with excellent chemoselectivity and unique regioselectivities using t-Bu-P4 base. J Am Chem Soc 125:8082
65. Imahori T, Suzawa K, Kondo Y (2008) Deprotonative zincation of heteroaromatics using ZnI_2 and tert-Bu-P4 base. Heterocycles 76:1057
66. Wunderlich S, Knochel P (2008) Efficient mono- and bis-functionalization of 3,6-dichloropyridazine using (tmp)2Zn·2MgCl2·2LiCl. Chem Commun (Camb) 6387
67. Krasovskiy A, Tishkov A, Del Amo V, Mayr H, Knochel P (2006) Transition-metal-free homocoupling of organomagnesium compounds. Angew Chem Int Ed 45:5010
68. Crestey F, Knochel P (2010) Regioselective functionalization of chlorophthalazine derivatives. Synthesis 2010:1097
69. Clapham KM, Batsanov AS, Greenwood RDR, Bryce MR, Smith AE, Tarbit B (2008) Functionalized heteroarylpyridazines and pyridazin-3(2H)-one derivatives via palladium-catalyzed cross-coupling methodology. J Org Chem 73:2176

70. Darabantu M, Boully L, Turck A, Plé N (2005) Synthesis of new polyaza heterocycles. Part 42: diazines. Tetrahedron 61:2897
71. Carboni RA, Lindsey RV (1959) Reactions of tetrazines with unsaturated compounds. a new synthesis of pyridazines. J Am Chem Soc 81:4342
72. Helm MD, Moore JE, Plant A, Harrity JP (2005) Synthesis of highly substituted pyridazines through alkynyl boronic ester cycloaddition reactions. Angew Chem Int Ed Engl 44:3889
73. Helm MD, Plant A, Harrity JP (2006) A novel approach to functionalised pyridazinone arrays. Org Biomol Chem 4:4278
74. Heldmann DK, Sauer J (1997) Synthesis of metallated (metal = Si, Ge, Sn) pyridazines by cycloaddition of metal substituted alkynes to 1,2,4,5-tetrazine. Tetrahedron Lett 38:5791
75. Sauer J, Heldmann DK (1998) Synthesis of 3,5-disubstituted pyridazines by regioselective [4+2] cycloadditions with ethynyltributyltin and subsequent replacement of the organotin substituent. Tetrahedron 54:4297

Index

CPSIA information can be obtained at www.ICGtesting.com
Printed in the USA
LVOW07*2122280813

350119LV00006B/58/P

9 783642 350214